Springer Series in Statistics

Springer Series in Statistics

Alho/Spencer: Statistical Demography and Forecasting
Andersen/Borgan/Gill/Keiding: Statistical Models Based on Counting Processes
Atkinson/Riani: Robust Diagnostic Regression Analysis
Atkinson/Riani/Ceriloi: Exploring Multivariate Data with the Forward Search
Berger: Statistical Decision Theory and Bayesian Analysis, 2nd edition
Borg/Groenen: Modern Multidimensional Scaling: Theory and Applications, 2nd edition
Brockwell/Davis: Time Series: Theory and Methods, 2nd edition
Bucklew: Introduction to Rare Event Simulation
Cappé/Moulines/Rydén: Inference in Hidden Markov Models
Chan/Tong: Chaos: A Statistical Perspective
Chen/Shao/Ibrahim: Monte Carlo Methods in Bayesian Computation
Coles: An Introduction to Statistical Modeling of Extreme Values
Devroye/Lugosi: Combinatorial Methods in Density Estimation
Diggle/Ribeiro: Model-based Geostatistics
Dudoit/Van der Laan: Multiple Testing Procedures with Applications to Genomics
Efromovich: Nonparametric Curve Estimation: Methods, Theory, and Applications
Eggermont/LaRiccia: Maximum Penalized Likelihood Estimation, Volume I: Density Estimation
Fahrmeir/Tutz: Multivariate Statistical Modeling Based on Generalized Linear Models, 2nd edition
Fan/Yao: Nonlinear Time Series: Nonparametric and Parametric Methods
Ferraty/Vieu: Nonparametric Functional Data Analysis: Theory and Practice
Ferreira/Lee: Multiscale Modeling: A Bayesian Perspective
Fienberg/Hoaglin: Selected Papers of Frederick Mosteller
Frühwirth-Schnatter: Finite Mixture and Markov Switching Models
Ghosh/Ramamoorthi: Bayesian Nonparametrics
Glaz/Naus/Wallenstein: Scan Statistics
Good: Permutation Tests: Parametric and Bootstrap Tests of Hypotheses, 3rd edition
Gouriéroux: ARCH Models and Financial Applications
Gu: Smoothing Spline ANOVA Models
Györfi/Kohler/Krzyźak/Walk: A Distribution-Free Theory of Nonparametric Regression
Haberman: Advanced Statistics, Volume I: Description of Populations
Hall: The Bootstrap and Edgeworth Expansion
Härdle: Smoothing Techniques: With Implementation in S
Harrell: Regression Modeling Strategies: With Applications to Linear Models, Logistic Regression, and Survival Analysis
Hart: Nonparametric Smoothing and Lack-of-Fit Tests
Hastie/Tibshirani/Friedman: The Elements of Statistical Learning: Data Mining, Inference, and Prediction
Hedayat/Sloane/Stufken: Orthogonal Arrays: Theory and Applications
Heyde: Quasi-Likelihood and Its Application: A General Approach to Optimal Parameter Estimation
Huet/Bouvier/Poursat/Jolivet: Statistical Tools for Nonlinear Regression: A Practical Guide with S-PLUS and R Examples, 2nd edition
Iacus: Simulation and Inference for Stochastic Differential Equations

(continued after index)

Stefano M. Iacus

Simulation and Inference for Stochastic Differential Equations

With R Examples

Springer

Stefano M. Iacus
Dept. Economics, Business and Statistics
University of Milan
Via Conservatorio, 7
20122 Milano
Italy
stefano.iacus@unimi.it

ISBN: 978-0-387-75838-1 e-ISBN: 978-0-387-75839-8
DOI: 10.1007/978-0-387-75839-8

Library of Congress Control Number: 2008924170

Printed on acid-free paper

9 8 7 6 5 4 3 2 1

springer.com

To Ilia, Lucy, and Ludo

Preface

Stochastic differential equations model stochastic evolution as time evolves. These models have a variety of applications in many disciplines and emerge naturally in the study of many phenomena. Examples of these applications are physics (see, e.g., [176] for a review), astronomy [202], mechanics [147], economics [26], mathematical finance [115], geology [69], genetic analysis (see, e.g., [110], [132], and [155]), ecology [111], cognitive psychology (see, e.g., [102], and [221]), neurology [109], biology [194], biomedical sciences [20], epidemiology [17], political analysis and social processes [55], and many other fields of science and engineering. Although stochastic differential equations are quite popular models in the above-mentioned disciplines, there is a lot of mathematics behind them that is usually not trivial and for which details are not known to practitioners or experts of other fields. In order to make this book useful to a wider audience, we decided to keep the mathematical level of the book sufficiently low and often rely on heuristic arguments to stress the underlying ideas of the concepts introduced rather than insist on technical details. Mathematically oriented readers may find this approach inconvenient, but detailed references are always given in the text.

As the title of the book mentions, the aim of the book is twofold. The first is to recall the theory and implement methods for the simulation of paths of stochastic processes $\{X_t, t \geq 0\}$ solutions to stochastic differential equations (SDEs). In this respect, the title of the book is too ambitious in the sense that only SDEs with Gaussian noise are considered (i.e., processes for which the writing $\mathrm{d}X_t = S(X_t)\mathrm{d}t + \sigma(X_t)\mathrm{d}W_t$ has a meaning in the Itô sense). This part of the book contains a review of well-established results and their implementations in the R language, but also some fairly recent results on simulation.

The second part of the book is dedicated to the review of some methods of estimation for these classes of stochastic processes. While there is a well-established theory on estimation for continuous-time observations from these processes [149], the literature about discrete-time observations is dispersed (though vaste) in several journals. Of course, real data (e.g., from finance [47],

[88]) always lead to dealing with discrete-time observations $\{X_{t_i}, i = 1, \dots, n\}$, and many of the results from the continuous-time case do not hold or cannot be applied (for example, the likelihood of the observations is almost always unavailable in explicit form). It should be noted that only the observations are discrete whilst the underlying model is continuous; hence most of the standard theory on discrete-time Markov processes does not hold as well.

Different schemes of observations can be considered depending on the nature of the data, and the estimation part of the problem is not necessarily the same for the different schemes. One case, which is considered "natural," is the fixed-Δ scheme, in which the time step between two subsequent observations X_{t_i} and $X_{t_i+\Delta_n}$ is fixed; i.e., $\Delta_n = \Delta$ (or is bounded away from zero) and independent from n. In this case, the process is observed on the time interval $[0, T = n\Delta]$ and the asymptotics considered as $n \to \infty$ (*large-sample asymptotics*). The underlying model might be ergodic or stationary and possibly homogeneous. For such a scheme, the time step Δ might have some influence on estimators because, for example, the transition density of the process is usually not known in explicit form and has to be approximated via simulations. This is the most difficult case to handle.

Another scheme is the "high frequency" scheme, in which the observational step size Δ_n decreases with n and two cases are possible: the time interval is fixed, say $[0, T = n\Delta_n]$, or $n\Delta_n$ increases as well. In the first case, neither homogeneity nor erogidicy are needed, but consistent estimators are not always available. On the contrary, in the "*rapidly increasing experimental design*," when $\Delta_n \to 0$ and $n\Delta_n \to \infty$ but $n\Delta_n^2 \to 0$, consistent estimators can be obtained along with some distributional results.

Other interesting schemes of partially observed processes, missing at random [75], thresholded processes (see, e.g., [116], [118]), observations with errors (quantized or interval data, see, e.g., [66], [67], [97]), or large sample and "small diffusion" asymptotics have also recently appeared in the literature (see, e.g., [222], [217]). This book covers essentially the parametric estimation under the large-sample asymptotics scheme ($n\Delta_n \to \infty$) with either fixed $\Delta_n = \Delta$ or $\Delta_n \to 0$ with $n\Delta_n^k \to 0$ for some $k \geq 2$. The final chapter contains a miscellaneous selection of results, including nonparametric estimation, model selection, and change-point problems.

This book is intended for practitioners and is not a theoretical book, so this second part just recalls briefly the main results and the ideas behind the methods and implements several of them in the R language. A selection of the results has necessarily been made. This part of the book also shows the difference between the theory of estimation for discrete-time observations and the actual performance of such estimators once implemented. Further, the effect of approximation schemes on estimators is investigated throughout the text. Theoretical results are recalled as "Facts" and regularity conditions as "Assumptions" and numbered by chapter in the text.

So what is this book about?

This book is about ready to be used, R-efficient code for simulation schemes of stochastic differential equations and some related estimation methods based on discrete sampled observations from such models. We hope that the code presented here and the updated survey on the subject might be of help for practitioners, postgraduate and PhD students, and researchers in the field who might want to implement new methods and ideas using R as a statistical environment.

What this book is not about

This book is not intended to be a theoretical book or an exhaustive collection of all the statistical methods available for discretely observed diffusion processes. This book might be thought of as a companion book to some advanced theoretical publication (already available or forthcoming) on the subject. Although this book is not even a textbook, some previous drafts of it have been used with success in mathematical finance classes for the numerical simulation and empirical analysis of financial time series.

What comes with the book

All the algorithms presented in the book are written in pure R code but, because of the speed needed in real-world applications, we have rewritten some of the R code in the `C` language and assembled everything in a package called `sde` freely available on CRAN, the Comprehensive R Archive Network. R and `C` functions have the same end-user interface; hence all the code of the examples in the book will run smoothly regardless of the underlying coding language. A minimal knowledge of the R environment at the introductory level is assumed, although brief recalls to the main R concepts, limited to what is relevant to this text, are given at the end of the book. Some crucial aspects of implementation are discussed in the main body of the book to make them more effective.

What is missing?

This book essentially covers one-dimensional diffusion processes driven by the Wiener process. Today's literature is vast and wider than this choice. In particular, it focuses also on multidimensional diffusion processes and stochastic differential equations driven by Lévy processes. To keep the book self-contained and at an introductory level and to preserve some homogeneity within the text, we decided to restrict the field. This also allows simple and easy-to-understand R code to be written for each of the techniques presented.

Acknowledgments

I am grateful to Alessandro De Gregorio and Ilia Negri for careful reading of the manuscript, as well as to the anonymous referees for their constructive comments and remarks on several drafts of the book. All the remaining errors contained are of course my responsibility. I also wish to thank John Kimmel for his great support and patience with me during the last three years.

Milan, November 2007 *Stefano M. Iacus*

Contents

Notation

Var : the variance operator

$\mathbb{E}$: the expected value operator

$N(\mu, \sigma^2)$: the Gaussian law with mean μ and variance σ^2

χ^2_d : chi-squared distribution with d degrees of freedom

t_d : Student t distribution with d degrees of freedom

$I_\nu(x)$: modified Bessel function of the first kind of order ν

$\mathbb{R}$: the real line

$\mathbb{N}$: the set of natural numbers $1, 2, \ldots$

$\xrightarrow{d}$: convergence in distribution

$\xrightarrow{p}$: convergence in probability

$\xrightarrow{a.s.}$: almost sure convergence

$\mathcal{B}(\mathbb{R})$: Borel σ-algebra on $\mathbb{R}$

$\chi_A, \mathbf{1}_A$: indicator function of the set A

$x \wedge y$: $\min(x, y)$

$x \vee y$: $\max(x, y)$

$(f(x))_+$: $\max(f(x), 0)$

$\Phi(z)$: cumulative distribution function of standard Gaussian law

$[x]$: integer part of x

$<X, X>_t, [X, X]_t$: quadratic variation process associated to X_t

$\mathrm{V}_t(X)$: simple variation of process X

$\propto$: proportional to

$f_{v_i}(v_1, v_2, \ldots, v_n)$: $\frac{\partial}{\partial v_i} f(v_1, v_2, \ldots, v_n)$

$f_{v_i,v_j}(v_1, v_2, \ldots, v_n)$: $\frac{\partial^2}{\partial v_i v_j} f(v_1, v_2, \ldots, v_n)$, etc.

$\partial_\theta f(v_1, v_2, \ldots, v_n; \theta)$: $\frac{\partial}{\partial \theta} f(v_1, v_2, \ldots, v_n; \theta)$

$\partial_\theta^k f(v_1, v_2, \ldots, v_n; \theta)$: $\frac{\partial^k}{\partial \theta^k} f(v_1, v_2, \ldots, v_n; \theta)$

$\Pi_n(A)$: partition of the interval $A = [a, b]$ in n subintervals of $[a = x_0, x_1)$, $[x_1, x_2)$, ..., $[x_{n-1}, x_n = b]$

$||\Pi_n||$: $\max_j |x_{j+1} - x_j|$

$C_0^2(\mathbb{R})$: space of functions with compact support and continuous derivatives up to order 2

$L^2([0, T])$: space of functions from $[0, T] \to \mathbb{R}$ endowed by the L^2 norm

$||f||_2$: the L^2 norm of f

W_t : Brownian motion or Wiener process

i.i.d. : independent and identically distributed

AIC : Akaike information criterion

CIR : Cox-Ingersoll-Ross

CRAN : the Comprehensive R Archive Network

CKLS : Chan-Karolyi-Longstaff-Sanders

EA : exact algorithm

GMM : generalized method of moments

MCMC : Markov chain Monte Carlo

MISE : mean integrated square error

1

Stochastic Processes and Stochastic Differential Equations

This chapter reviews basic material on stochastic processes and statistics as well as stochastic calculus, mostly borrowed from [170], [130], and [193]. It also covers basic notions on simulation of commonly used stochastic processes such as random walks and Brownian motion and also recalls some Monte Carlo concepts. Even if the reader is assumed to be familiar with these basic notions, we will present them here in order to introduce the notation we will use throughout the text. We will limit our attention mainly to one-dimensional, real random variables and stochastic processes. We also restrict our attention to parametric models with multidimensional parameters.

1.1 Elements of probability and random variables

A probability space is a triple $(\Omega, \mathcal{A}, P)$ where Ω is the *sample space* of possible outcomes of a random experiment; $\mathcal{A}$ is a σ*-algebra*: i.e., $\mathcal{A}$ is a collection of sets such that *i)* the empty set $\emptyset$ is in $\mathcal{A}$; *ii)* if $A \in \mathcal{A}$, then the complementary set $\bar{A} \in \mathcal{A}$; *iii)* if $A_1, A_2, \ldots \in \mathcal{A}$, then

$$\bigcup_{i=1}^{\infty} A_i \in \mathcal{A}\,.$$

P is a probability measure on $(\Omega, \mathcal{A})$. In practice, $\mathcal{A}$ forms the collection of events for which a probability can be assigned. Given a probability space $(\Omega, \mathcal{A}, P)$, a *random variable* X is defined as a *measurable* function from Ω to $\mathbb{R}$,

$$X : \Omega \mapsto \mathbb{R}\,.$$

In the above, the term measurable intuitively means that it is always possible to calculate probabilities related to the random variable X. More precisely, denote by $\mathcal{B}(\mathbb{R})$ the Borel σ-algebra on $\mathbb{R}$ (i.e., the σ-algebra generated by the open sets of $\mathbb{R}$) and let X^{-1} be the inverse function of X. Then, X is *measurable* if

S.M. Iacus, *Simulation and Inference for Stochastic Differential Equations*,
doi: 10.1007/978-0-387-75839-8_1, © Springer Science+Business Media, LLC 2008

$$\forall A \in \mathcal{B}(\mathbb{R}), \quad \exists B \in \mathcal{A} : X^{-1}(A) = B;$$

i.e., such that it is always possible to measure the set of values assumed by X using the probability measure P on the original space Ω,

$$P(X \in A) = P(\{\omega \in \Omega : X(\omega) \in A\}) = P(\{\omega \in \Omega : \omega \in X^{-1}(A)\}) = P(B),$$

for $A \in \mathcal{B}(\mathbb{R})$ and $B \in \mathcal{A}$.

Distribution and density function

The function $F(x) = P(X \leq x) = P(X(\omega) \in (-\infty, x])$ is called the cumulative *distribution function*: it is a nondecreasing function such that $\lim_{x\to-\infty} F(x) = 0$, $\lim_{x\to+\infty} F(x) = 1$, and F is right continuous. If F is absolutely continuous, its derivative $f(x)$ is called a *density function*, which is a Lebesgue integrable nonnegative function whose integral over the real line is equal to one. Loosely speaking, if $F(x)$ is the probability that the random variable X takes values less than or equal to x, the quantity $f(x)\mathrm{d}x$ can be thought of as the probability that the random variable takes values in the infinitesimal interval $[x, x + \mathrm{d}x)$. If the random variable takes only a countable set of values, then it is said to be discrete and its density at point x is defined as $P(X = x)$. In the continuous case, $P(X = x) = 0$ always.

Independence

Two random variables X and Y are *independent* if

$$P(X \in A, Y \in B) = P(X \in A)P(Y \in B)$$

for any two sets A and B in $\mathbb{R}$.

1.1.1 Mean, variance, and moments

The mean (or expected value) of a continuous random variable X with distribution function F is defined as

$$\mathbb{E}X = \int_\Omega X(\omega)\mathrm{d}P(\omega) = \int_\mathbb{R} x\mathrm{d}F(x)$$

provided that the integral is finite. If X has a density, then $\mathbb{E}X = \int_\mathbb{R} xf(x)\mathrm{d}x$ and the integral is the standard Riemann integral; otherwise integrals in $\mathrm{d}P$ or $\mathrm{d}F$ should be thought of as integrals in the abstract sense. If Ω is countable, the expected value is defined as

$$\mathbb{E}X = \sum_{\omega\in\Omega} X(\omega)P(\omega)$$

or, equivalently, when X is a discrete random variable, the expected value reduces to $\mathbb{E}X = \sum_{x\in I} xP(X = x)$, where I is the set of possible values of X. The variance is defined as

$$\operatorname{Var} X = \mathbb{E}\,(X - \mathbb{E}X)^2 = \int_\Omega (X(\omega) - \mathbb{E}X)^2 \mathrm{d}P(\omega)\,,$$

and the kth moment is defined as

$$\mathbb{E}X^k = \int_\Omega X^k(\omega)\mathrm{d}P(\omega)\,.$$

In general, for any measurable function $g(\cdot)$, $\mathbb{E}g(X)$ is defined as

$$\mathbb{E}g(X) = \int_\Omega g(X(\omega))\mathrm{d}P(\omega)\,,$$

provided that the integral is finite.

Types of convergence

Let $\{F_n\}_{n\in\mathbb{N}}$ be a sequence of distribution functions for the sequence of random variables $\{X_n\}_{n\in\mathbb{N}}$. Assume that

$$\lim_{n\to\infty} F_n(x) = F(x)$$

for all $x \in \mathbb{R}$ such that $F(\cdot)$ is continuous in x, where F is the distribution function of some random variable X. Then, the sequence X_n is said to *converge in distribution* to the random variable X, and this is denoted by $X_n \stackrel{d}{\to} X$. This only means that the distributions F_n of the random variables converge to another distribution F, but nothing is said about the random variables itself. So this convergence is only about the probabilistic behavior of the random variables on some intervals $(-\infty, x]$, $x \in \mathbb{R}$.

A sequence of random variables X_n is said to *converge in probability* to a random variable X if, for any $\epsilon > 0$,

$$\lim_{n\to\infty} P(|X_n - X| \geq \epsilon) = 0.$$

This is denoted by $X_n \stackrel{p}{\to} X$ and it is a pointwise convergence of the probabilities. This convergence implies the convergence in distribution. Sometimes we use the notation

$$p - \lim_{n\to\infty} |X_n - X| = 0$$

for the convergence in probability. A stronger type of convergence is defined as the probability of the limit in the sense $P(\lim_{n\to\infty} X_n = X) = 1$ or, more precisely,

$$P(\{\omega \in \Omega : \lim_{n\to\infty} X_n(\omega) = X(\omega)\}) = 1.$$

When this happens, X_n is said to converge to X *almost surely* and is denoted by $X_n \stackrel{a.s.}{\to} X$. Almost sure convergence implies convergence in probability.

A sequence of random variables X_n is said to *converge in the rth mean* to a random variable X if

$$\lim_{n\to\infty} \mathbb{E}|X_n - X|^r = 0, r \geq 1.$$

The convergence in the r-th mean implies the convergence in probability thanks to Chebyshev's inequality, and if X_n converges to X in the rth mean, then it also converges in the sth mean for all $r > s \geq 1$. *Mean square convergence* is a particular case of interest and corresponds to the case $r = 2$. This type of convergence will be used in Section 1.9 to define the Itô integral.

1.1.2 Change of measure and Radon-Nikodým derivative

In some situations, for example in mathematical finance, it is necessary to reassign the probabilities to the events in Ω, switching from a measure P to another one $\tilde{P}$. This is done with the help of a random variable, say Z, which reweights the elements in Ω. This change of measure should be done set-by-set instead of ω-by-ω (see, e.g., [209]) as

$$\tilde{P}(A) = \int_A Z(\omega)\mathrm{d}P(\omega), \tag{1.1}$$

where Z is assumed to be almost surely nonnegative and such that $\mathbb{E}Z = 1$. The new $\tilde{P}$ is then a true probability measure and, for any nonnegative random variable X, the equality

$$\tilde{\mathbb{E}}X = \mathbb{E}(XZ)$$

holds, where $\tilde{\mathbb{E}}X = \int_\Omega X(\omega)\mathrm{d}\tilde{P}(\omega)$. Two measures P and $\tilde{P}$ are said to be *equivalent* if they assign probability 0 to the same sets. The previous change of measure from P to $\tilde{P}$ trivially guarantee that the two measures are equivalent when Z is strictly positive. Another way to read the change of measure in (1.1) is to say that Z is the *Radon-Nikodým derivative* of $\tilde{P}$ with respect to P. Indeed, a formal differentiation of (1.1) allows us to write

$$Z = \frac{\mathrm{d}\tilde{P}}{\mathrm{d}P}\,. \tag{1.2}$$

Fact 1.1 (Theorem 1.6.7 [209]) *Let P and $\tilde{P}$ be two equivalent measures on $(\Omega, \mathcal{A})$. Then, there exists a random variable Z, almost surely positive, such that $\mathbb{E}Z = 1$ and*

$$\tilde{P}(A) = \int_A Z(\omega)\mathrm{d}P(\omega)$$

for every $A \in \mathcal{A}$.

The Radon-Nikodým derivative is an essential requirement in statistics because Z plays the role of the likelihood ratio in the inference for diffusion processes.

1.2 Random number generation

Every book on simulation points the attention of the reader to the quality of random number generators. This is of course one central point in simulation studies. R developers and R users are in fact quite careful in the implementations and use of random number generators. We will not go into details, but we just warn the reader about the possibilities available in R and what is used in the examples in this book.

The random number generator can be specified in R via the `RNGkind` function. The default generator of uniform pseudo random numbers is the *Mersenne-Twister* and is the one used throughout the book. Other methods available as of this writing are *Wichmann-Hill*, *Marsaglia-Multicarry*, *Super-Duper*, and two versions of *Knuth-TAOCP* random number generators. The user can implement and provide his own method as well. Specifically, for the normal random number generators, available methods are *Kinderman-Ramage*, *Buggy Kinderman-Ramage*, *Ahrens-Dieter*, *Box-Muller*, and the default *Inversion* method, as explained in [229]. For this case as well, the user can provide her own algorithm. For other than normal variates, R implements quite advanced pseudo random number generators. For each of these, the reader has to look at the manual page of the corresponding `r*` functions (e.g., `rgamma`, `rt`, `rbeta`, etc.).

For reproducibility of all the numerical results in the book we chose to use a fixed initialization seed before any listing of R code. We use everywhere the function `set.seed(123)`, and the reader should do the same if she wants to obtain the same results.

1.3 The Monte Carlo method

Suppose we are given a random variable X and are interested in the evaluation of $\mathbb{E}g(X)$ where $g(\cdot)$ is some known function. If we are able to draw n pseudo random numbers $x_1, \dots, x_n$ from the distribution of X, then we can think about approximating $\mathbb{E}g(X)$ with the sample mean of the $g(x_i)$,

$$\mathbb{E}g(X) \simeq \frac{1}{n}\sum_{i=1}^{n} g(x_i) = \bar{g}_n \,. \tag{1.3}$$

The expression (1.3) is not just symbolic but holds true in the sense of the law of large numbers whenever $\mathbb{E}|g(X)| < \infty$. Moreover, the central limit theorem guarantees that

$$\bar{g}_n \xrightarrow{d} N\left(\mathbb{E}g(X), \frac{1}{n}\mathrm{Var}(g(X))\right),$$

where $N(m, s^2)$ denotes the distribution of the Gaussian random variable with expected value m and variance s^2. In the end, the number we estimate with simulations will have a deviation from the true expected value $\mathbb{E}g(X)$ of order $1/\sqrt{n}$. Given that $P(|Z| < 1.96) \simeq 0.95$, $Z \sim N(0,1)$, one can construct an interval for the estimate $\bar{g}_n$ of the form

$$\left(\mathbb{E}g(X) - 1.96\frac{\sigma}{\sqrt{n}}, \mathbb{E}g(X) + 1.96\frac{\sigma}{\sqrt{n}}\right),$$

with $\sigma = \sqrt{\mathrm{Var} g(X)}$, which is interpreted that the Monte Carlo estimate of $\mathbb{E}g(X)$ above is included in the interval above 95% of the time. The confidence interval depends on $\mathrm{Var} g(X)$, and usually this quantity has to be estimated through the sample as well. Indeed, one can estimate it as the sample variance of Monte Carlo replications as

$$\hat{\sigma}^2 = \frac{1}{n-1}\sum_{i=1}^{n}(g(x_i) - \bar{g}_n)^2$$

and use the following 95% level Monte Carlo confidence interval[1] for $\mathbb{E}g(X)$:

$$\left(\bar{g}_n - 1.96\frac{\hat{\sigma}}{\sqrt{n}}, \bar{g}_n + 1.96\frac{\hat{\sigma}}{\sqrt{n}}\right).$$

The quantity $\hat{\sigma}/\sqrt{n}$ is called the *standard error*. The standard error is itself a random quantity and thus subject to variability; hence one should interpret this value as a "qualitative" measure of accuracy.

One more remark is that the rate of convergence $\sqrt{n}$ is not particularly fast but at least is independent of the smoothness of $g(\cdot)$. Moreover, if we need to increase the quality of our approximation, we just need to draw additional samples[2] instead of rerunning the whole simulation.

About Monte Carlo intervals length

In some cases, Monte Carlo intervals are not very informative if the variance of $Y = g(X)$ is too large. The next example, taken from [156], is one such case. Let $Y = g(X) = e^{\beta X}$ with $X \sim N(0,1)$, and assume we are interested in $\mathbb{E}g(X)$ with $\beta = 5$. The analytical value can be calculated as $e^{\beta^2/2} = 268337.3$, and the true standard deviation $\sigma = \sqrt{e^{2\beta^2} - e^{\beta^2}} = 72004899337$, quite a big number with respect to the mean of Y. Suppose we want to estimate $\mathbb{E}Y$ via the Monte Carlo method using 100000 replications and construct 95% confidence intervals using the true standard deviation σ and the estimated standard error. The following R code does the job.

[1] Again, this means that the interval covers the true value 95% of the time.

[2] A warning note: Of course one should take care of the seed of the random number generator to avoid duplicated samples. If we have already run n replications and we want to add n' new samples, we cannot simply rerun the algorithm for a length of n' with the same original seed because in this case we are just replicating the first n' samples among the n original ones, hence inducing bias without increasing accuracy.

```
> # ex1.01.R
> set.seed(123)
> n <- 1000000
> beta <-5
> x <- rnorm(n)
> y <- exp(beta*x)
>
> # true value of E(Y)
> exp(beta^2/2)
[1] 268337.3
> # MC estimation of E(Y)
> mc.mean <- mean(y)
> mc.mean
[1] 199659.2
> mc.sd <- sd(y)
> true.sd <- sqrt(exp(2*beta^2) - exp(beta^2))
>
> # MC conf. interval based on true sigma
> mc.mean - true.sd*1.96/sqrt(n)
[1] -140929943
> mc.mean + true.sd*1.96/sqrt(n)
[1] 141329262
>
> # MC conf. interval based on estimated sigma
> mc.mean - mc.sd*1.96/sqrt(n)
[1] 94515.51
> mc.mean + mc.sd*1.96/sqrt(n)
[1] 304802.9
>
> plot(1:n,cumsum(y)/(1:n),type="l",axes=F,xlab="n",
+   ylab=expression(hat(g)[n]),ylim=c(0,350000))
> axis(1,seq(0,n,length=5))
> axis(2,seq(0,350000,length=6))
> abline(h=268337.3)  # true value
> abline(h=mc.mean-mc.sd*1.96/sqrt(n),lty=3) # MC conf interval
> abline(h=mc.mean+mc.sd*1.96/sqrt(n),lty=3)
> abline(h=mc.mean,lty=2) # MC estimate
> box()
```

Running this code in R, we obtain the two intervals

$$(-140929943; 141329262) \quad \text{using } \sigma$$

and

$$(94515.51; 304802.9) \quad \text{using } \hat{\sigma}$$

with an estimated value of $\mathbb{E}g(X)$, $\hat{g}_n = 199659.2$. As one can see, the confidence interval based on σ contains the true value of $\mathbb{E}g(X)$ but is too large and hence meaningless. The confidence interval based on $\hat{\sigma}$ is smaller but still large. The first effect is due to the big variance of $g(X)$, while the second is due to the fact that the sample variance underestimates the true one ($\hat{\sigma} = 53644741$). The reason is that, in this particular case, the state of asymptotic normality after $n = 1000000$ replications is not yet reached (the reader is invited to look at this with a `plot(density(y))`) and thus the estimator $\hat{\sigma}$ is not necessarily an unbiased estimator of the true σ. Looking at Figure 1.1 one can expect that the Monte Carlo confidence interval for smaller values of n (the reader can try with $n = 100000$) does not even contain the true value.

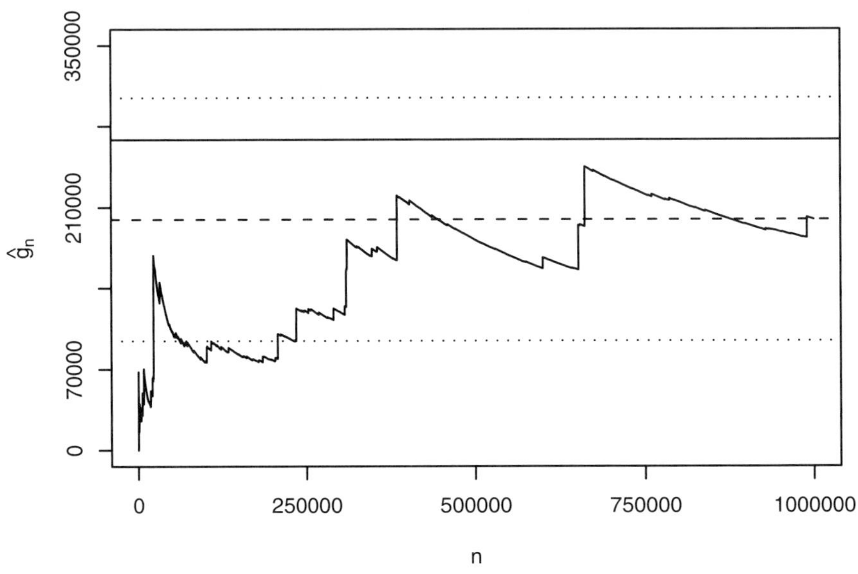

Fig. 1.1. The very slow convergence of the Monte Carlo estimate associated with a target true value with too high variability (see Section 1.3). The solid line is the target value, dotted lines are upper and lower limits of the Monte Carlo 95% confidence interval, and the dashed line is the estimated value $\hat{g}_n$.

1.4 Variance reduction techniques

The example in the last section gives evidence that in order to have less variability in Monte Carlo methods and hence use a smaller number of replications in simulations, one needs to try to reduce variability with some workaround. There are several methods of variance reduction for Monte Carlo estimators. We review here just the ones that can be applied in our context, but interesting reviews on methods for other classes of problems and processes can be found, for example, in [156] and [125]. Here we just show the basic ideas, while applications to stochastic differential equations are postponed to Section 2.15. We do not include the treatment of sequences with *low discrepancy*[3] because this is beyond the scope of this book.

[3] Discrepancy is a measure of goodness of fit for uniform random variates in high dimensions. Low-discrepancy sequences are such that numerical integration on this grid of points allows for a direct variance reduction. The reader can refer to the review paper [153].

1.4.1 Preferential sampling

The idea of this method is to express $\mathbb{E}g(X)$ in a different form in order to reduce its variance. Let $f(\cdot)$ be the density of X; thus

$$\mathbb{E}g(X) = \int_{\mathbb{R}} g(x)f(x)\mathrm{d}x.$$

Introduce now another strictly positive density $h(\cdot)$. Then,

$$\mathbb{E}g(X) = \int_{\mathbb{R}} \frac{g(x)f(x)}{h(x)} h(x)\mathrm{d}x$$

and

$$\mathbb{E}g(X) = \mathbb{E}\left(\frac{g(Y)f(Y)}{h(Y)}\right) = \mathbb{E}\tilde{g}(Y),$$

with Y a random variable with density $h(\cdot)$, and denote $\tilde{g}(\cdot) = g(\cdot)f(\cdot)/h(\cdot)$. If we are able to determine an $h(\cdot)$ such that $\mathrm{Var}\tilde{g}(Y) < \mathrm{Var}g(X)$, then we have reached our goal. But let us calculate $\mathrm{Var}\tilde{g}(Y)$,

$$\mathrm{Var}\tilde{g}(Y) = \mathbb{E}\tilde{g}(Y)^2 - (\mathbb{E}\tilde{g}(Y))^2 = \int_{\mathbb{R}} \frac{g^2(x)f^2(x)}{h(x)}\mathrm{d}x - (\mathbb{E}g(X))^2 .$$

If $g()$ is strictly positive, by choosing $h(x) = g(x)f(x)/\mathbb{E}g(X)$, we obtain $\mathrm{Var}\tilde{g}(Y) = 0$, which is nice only in theory because, of course, we don't know $\mathbb{E}g(X)$. But the expression of $h(x)$ suggests a way to obtain a useful approximation: just take $\tilde{h}(x) = |g(x)f(x)|$ (or something close to it), then normalize it by the value of its integral, and use

$$h(x) = \frac{\tilde{h}(x)}{\int_{\mathbb{R}} \tilde{h}(x)\mathrm{d}x} .$$

Of course this is simple to say and hard to solve in specific problems, as integration should be done analytically and not using the Monte Carlo technique again. Moreover, the choice of $h(\cdot)$ changes from case to case. We show an example, again taken from [156], which is quite interesting and is a standard application of the method in finance. Suppose we want to calculate $\mathbb{E}g(X)$ with $g(x) = \max(0, K - e^{\beta x}) = (K - e^{\beta x})_+$, K and β constants, and $X \sim N(0,1)$. This is the price of a *put option* in the Black and Scholes framework [36, 162], and the explicit solution, which is known, reads as

$$\mathbb{E}\left(K - e^{\beta X}\right)_+ = K\Phi\left(\frac{\log(K)}{\beta}\right) - e^{\frac{1}{2}\beta^2}\Phi\left(\frac{\log(K)}{\beta} - \beta\right),$$

where Φ is the cumulative distribution function of the standard Gaussian law; i.e., $\Phi(x) = P(Z < z)$ with $Z \sim N(0,1)$. The true value, in the case $K = \beta = 1$, is $\mathbb{E}g(X) = 0.2384217$. Let's see what happens in Monte Carlo simulations.

```
> # ex1.02.R
> set.seed(123)
> n <- 10000
> beta <-1
> K <- 1
> x <- rnorm(n)
> y <- sapply(x, function(x) max(0,K-exp(beta*x)))
>
> # the true value
> K*pnorm(log(K)/beta)-exp(beta^2/2)*pnorm(log(K)/beta-beta)
[1] 0.2384217
>
> t.test(y[1:100]) # first 100 simulations

    One Sample t-test

data:  y[1:100]
t = 7.701, df = 99, p-value = 1.043e-11
alternative hypothesis: true mean is not equal to 0
95 percent confidence interval:
 0.1526982 0.2586975
sample estimates:
mean of x
0.2056978

> t.test(y[1:1000]) # first 1000 simulations

    One Sample t-test

data:  y[1:1000]
t = 24.8772, df = 999, p-value < 2.2e-16
alternative hypothesis: true mean is not equal to 0
95 percent confidence interval:
 0.2131347 0.2496388
sample estimates:
mean of x
0.2313868

> t.test(y) # all simulation results

    One Sample t-test

data:  y
t = 80.3557, df = 9999, p-value < 2.2e-16
alternative hypothesis: true mean is not equal to 0
95 percent confidence interval:
 0.2326121 0.2442446
sample estimates:
mean of x
0.2384284
```

Results of simulations are reported in Table 1.1 (a). Note that on the algorithm we have used the function `sapply` instead of the (apparently) more natural (but wrong) line of code

```
y <- max(0, K-exp(beta*x))
```

as this will only return one value, actually the maximum value among 0 and all the $y_i = K - e^{\beta * x_i}$. This is one place where, vector-wise functions need to be used in the correct order. Note also the use of `t.test`, which actually performs both estimation and construction of the confidence intervals for $Y = \mathbb{E}g(X)$.

We now try to rewrite $\mathbb{E}g(X)$ as $\mathbb{E}g'(Y)$ (where g' is a function different from g) in order to reduce its variance. Indeed, $\mathbb{E}g(X)$ can be rewritten as

Table 1.1. Evaluation of the price of a put option with the Monte Carlo method. The true value is 0.2384217, with (b) and without (a) applying the variance reduction technique of Section 1.4.1.

n	$\hat{g}_n$	95% conf. interval
100	0.206	(0.153 ; 0.259)
1000	0.231	(0.213 ; 0.250)
10000	0.238	(0.232 ; 0.244)

(a)

n	$\hat{g}_n$	95% conf. interval
100	0.234	(0.222 ; 0.245)
1000	0.236	(0.233 ; 0.240)
10000	0.238	(0.237 ; 0.239)

(b)

$$\int_{\mathbb{R}} \frac{(1-e^{\beta x})_+}{\beta|x|}\beta|x|\frac{e^{-\frac{1}{2}x^2}}{\sqrt{2\pi}}\mathrm{d}x,$$

setting $K = 1$ and noticing that $e^x - 1 \simeq x$ for x close to 0. By the change of variable $x = \sqrt{y}$ for $x > 0$ and $x = -\sqrt{y}$ for $x < 0$, the integral above can be rewritten as

$$\int_0^\infty \frac{\left(1-e^{\beta\sqrt{y}}\right)_+ + \left(1-e^{-\beta\sqrt{y}}\right)_+}{\sqrt{2\pi}\sqrt{y}}\frac{e^{-\frac{1}{2}y}}{2}\mathrm{d}y,$$

from which we remark that $f(y) = \lambda e^{-\lambda y}$, with $\lambda = \frac{1}{2}$, is the density of the exponential distribution. Therefore,

$$\mathbb{E}g(X) = \mathbb{E}\left(\frac{\left(1-e^{\beta\sqrt{Y}}\right)_+ + \left(1-e^{-\beta\sqrt{Y}}\right)_+}{\sqrt{2\pi}\sqrt{Y}}\right)$$

can be evaluated as the expected value of a function of the exponential random variable Y. The following algorithm executes the calculation, and results are reported in Table 1.1 (b), from which the reduction in variance is quite evident.

```
> # ex1.03.R
> set.seed(123)
> n <- 10000
> beta <-1
> K <- 1
>
> x <- rexp(n,rate=0.5)
> h <- function(x) (max(0,1-exp(beta*sqrt(x))) +
+    max(0,1-exp(-beta*sqrt(x))))/sqrt(2*pi*x)
> y <- sapply(x, h)
>
> # the true value
> K*pnorm(log(K)/beta)-exp(beta^2/2)*pnorm(log(K)/beta-beta)
[1] 0.2384217
>
> t.test(y[1:100]) # first 100 simulations
> t.test(y[1:1000]) # first 1000 simulations
> t.test(y) # all simulation results
```

1.4.2 Control variables

The very simple case of variance reduction via *control variables* is as follows. Suppose that we want to calculate $\mathbb{E}g(X)$. If we can rewrite it in the form

$$\mathbb{E}g(X) = \mathbb{E}(g(X) - h(X)) + \mathbb{E}h(X),$$

where $\mathbb{E}h(X)$ can be calculated explicitly and $g(X) - h(X)$ has variance less than $g(X)$, then by estimating $\mathbb{E}(g(X) - h(X))$ via the Monte Carlo method, we obtain a reduction in variance.

Call-put parity example

Continuing with the example of the previous section, consider the price of a *call option*

$$c(X) = \mathbb{E}\left(e^{\beta X} - K\right)_+ .$$

It is easy to show that $c(X) - p(X) = e^{\frac{1}{2}\beta^2} - K$, where p is the price of the put option. Hence we can write $c(X) = p(X) + e^{\frac{1}{2}\beta^2} - K$. It is also known (see, e.g., [154]) that the variance of $p(X)$ is less than the variance of $c(X)$. Thus we obtained an estimator of $c(X)$ with reduced bias. The exact formula for $c(X)$ is also known and reads as

$$\mathbb{E}\left(e^{\beta X} - K\right)_+ = e^{\frac{1}{2}\beta^2}\Phi\left(\beta - \frac{\log(K)}{\beta}\right) - K\Phi\left(-\frac{\log(K)}{\beta}\right).$$

The following R code shows this empirically, and the results are reported in Table 1.2.

```
> # ex1.04.R
> set.seed(123)
> n <- 10000
> beta <-1
> K <- 1
>
> x <- rnorm(n)
> y <- sapply(x, function(x) max(0,exp(beta*x)-K))
>
> # the true value
> exp(beta^2/2)*pnorm(beta-log(K)/beta)-K*pnorm(-log(K)/beta)
>
> t.test(y[1:100]) # first 100 simulations
> t.test(y[1:1000]) # first 1000 simulations
> t.test(y) # all simulation results
>
> set.seed(123)
> x <- rexp(n,rate=0.5)
> h <- function(x) (max(0,1-exp(beta*sqrt(x))) +
+   max(0,1-exp(-beta*sqrt(x))))/sqrt(2*pi*x)
> y <- sapply(x, h)
>
> # variance reduction
> # CALL = PUT + e^{0.5*beta^2} - K
> z <- y +exp(0.5*beta^2) - K
>
> t.test(z[1:100]) # first 100 simulations
> t.test(z[1:1000]) # first 1000 simulations
> t.test(z) # all simulation results
```

Table 1.2. Evaluation of the price of a call option with the Monte Carlo method. The true value is 0.887143 with (b) and without (a) applying the variance reduction technique of Section 1.4.2.

n	$\hat{g}_n$	95% conf. interval
100	0.858	(0.542 ; 1.174)
1000	0.903	(0.780 ; 1.026)
10000	0.885	(0.844 ; 0.925)

(a)

n	$\hat{g}_n$	95% conf. interval
100	0.882	(0.871 ; 0.894)
1000	0.885	(0.881 ; 0.889)
10000	0.887	(0.886 ; 0.888)

(b)

Table 1.3. Evaluation of the price of a put option with the Monte Carlo method. The true value is 0.2384217 with (b) and without (a) applying the variance reduction technique of Section 1.4.3.

n	$\hat{g}_n$	95% conf. interval
100	0.206	(0.153 ; 0.259)
1000	0.231	(0.213 ; 0.250)
10000	0.238	(0.232 ; 0.244)

(a)

n	$\hat{g}_n$	95% conf. interval
100	0.226	(0.202 ; 0.250)
1000	0.235	(0.226 ; 0.242)
10000	0.238	(0.235 ; 0.240)

(b)

1.4.3 Antithetic sampling

The idea of antithetic sampling can be applied when it is possible to find transformations of X that leave its measure unchanged (for example, if X is Gaussian, then $-X$ is Gaussian as well). Suppose that we want to calculate

$$I = \int_0^1 g(x)\mathrm{d}x = \mathbb{E}g(X),$$

with $X \sim U(0,1)$. The transformation $x \mapsto 1-x$ leaves the measure unchanged (i.e., $1 - X \sim U(0,1)$), and I can be rewritten as

$$I = \frac{1}{2}\int_0^1 (g(x)+g(1-x))\mathrm{d}x = \frac{1}{2}\mathbb{E}(g(X)+g(1-X)) = \frac{1}{2}\mathbb{E}(g(X)+g(h(X))).$$

Therefore, we have a variance reduction if

$$\mathrm{Var}\left(\frac{1}{2}\left(g(X)+g(h(X))\right)\right) < \mathrm{Var}\left(\frac{1}{2}g(X)\right),$$

which is equivalent to saying that $\mathrm{Cov}(g(X), g(h(X))) < 0$. If $h(x)$ is a monotonic function of x (as in the example above), this is always the case. This way of proceeding has the effect of reducing the variance but also increasing the accuracy of the calculation of the mean.[4] Going back to the example of

[4] It does not correct higher-order moment estimation, though.

the calculation of the price of a put option, one should calculate it using X and $-X$ and then averaging as follows:

```
> # ex1.05.R
> set.seed(123)
> n <- 10000
> beta <-1
> K <- 1
> x <- rnorm(n)
> y1 <- sapply(x, function(x) max(0,K-exp(beta*x)))
> y2 <- sapply(-x, function(x) max(0,K-exp(beta*x)))
>
> y <- (y1+y2)/2
> # the true value
> K*pnorm(log(K)/beta)-exp(beta^2/2)*pnorm(log(K)/beta-beta)
>
> t.test(y[1:100]) # first 100 simulations
> t.test(y[1:1000]) # first 1000 simulations
> t.test(y) # all simulation results
```

The results are reported in Table 1.3. Notice that we have applied this method to the naive Monte Carlo estimator and not the one built on the exponential distribution Y, as in that case $-Y$ is no longer an exponential distribution.

1.5 Generalities of stochastic processes

Let $(\omega, \mathcal{A}, P)$ a probability space. A real valued *stochastic process* is a family of random variables $\{X_\gamma, \gamma \in \Gamma\}$ defined on $\Omega \times \Gamma$ taking values in $\mathbb{R}$. Thus, the random variables of the family (measurable for every $\gamma \in \Gamma$) are functions of the form

$$X(\gamma, \omega) : \Gamma \times \Omega \mapsto \mathbb{R}\,.$$

For $\Gamma = \mathbb{N}$, we have a *discrete-time* process, and for $\Gamma \subset \mathbb{R}$ we have a *continuous-time* process. We are mainly interested in continuous-time processes with $\Gamma = [0, \infty)$, and we always think of $[0, \infty)$ as the time axis. We will denote a continuous-time stochastic process as $X = \{X_t, t \geq 0\}$. Sometimes, to avoid multiple subscripts, we will also adopt the usual notation $X(t)$ to denote X_t. For a fixed value of ω, say $\bar{\omega}$, $\{X(t, \bar{\omega}), t \geq 0\}$ (respectively $\{X(n, \bar{\omega}), n \in \mathbb{N}\}$ for the discrete case) is called the *path* or *trajectory* of the process and represents one possible evolution of the process. For a fixed t, say $\bar{t}$, the set of values $\{X(\bar{t}, \omega), \omega \in \Omega\}$ (respectively $\{X(\bar{n}, \omega), \omega \in \Omega\}$) represents the set of possible states of the process at time $\bar{t}$ (respectively n).

1.5.1 Filtrations

Consider the probability space $(\Omega, \mathcal{A}, P)$. A *filtration* $\{\mathcal{F}_t, t \geq 0\}$ is an increasing family of sub-σ-algebras of $\mathcal{A}$ indexed by $t \geq 0$; i.e., for each $s, t \geq 0$ such that $s < t$, we have $\mathcal{F}_s \subset \mathcal{F}_t$ with $\mathcal{F}_0 = \{\Omega, \emptyset\}$. To each process $\{X(t), t \geq 0\}$ and for each t, we can associate a σ-algebra denoted by $\mathcal{F}_t = \sigma(X(s); 0 \leq s \leq t)$, which is the σ-algebra generated by the process X up to time t; i.e., the smallest σ-algebra of $\mathcal{A}$ that makes $X(s, \omega)$ measurable for every $0 \leq s \leq t$. This σ-algebra is the smallest set of subsets of Ω that makes it possible to assign probabilities to all the events related to the process X up to time t.

Given a stochastic process $\{X_t, t \geq 0\}$ and a filtration $\{\mathcal{F}_t, t \geq 0\}$ (not necessarily the one generated by X), the process X is said to be *adapted* to $\{\mathcal{F}_t, t \geq 0\}$ if for every $t \geq 0$, $X(t)$ is $\mathcal{F}_t$-measurable.

1.5.2 Simple and quadratic variation of a process

The notion of *total variation* or *first order* variation of a process $V(X)$ is linked to the differentiability of its paths. Let $\Pi_n = \Pi_n([0,t]) = \{0 = t_0 < t_1 < \cdots < t_i < \cdots < t_n = t\}$ be any partition of the interval $[0,t]$ into n intervals and denote by

$$||\Pi_n|| = \max_{j=0,\ldots,n-1}(t_{j+1} - t_j)$$

the maximum step size of the partition Π_n, i.e. the mesh of the partition. The variation of X is defined as

$$V_t(X) = p - \lim_{||\Pi_n||\to 0} \sum_{k=0}^{n-1} |X(t_{k+1}) - X(t_k)| \,.$$

If X is differentiable, then $V_t(X) = \int_0^t |X'(u)|du$. If $V_t(X) < \infty$, then X is said to be of bounded variation on $[0,t]$. If this is true for all $t \geq 0$, then X is said to have bounded variation. The *quadratic variation* $[X,X]_t$ at time t of a process X is defined as

$$[X,X]_t = p - \lim_{||\Pi_n||\to 0} \sum_{k=0}^{n-1} |X(t_{k+1}) - X(t_k)|^2.$$

The limit exists for stochastic processes with continuous paths. In this case, the notation $<X,X>_t$ is usually adopted. The quadratic variation can also be introduced as

$$<X,X>_t = p - \lim_{n\to\infty} \sum_{k=1}^{2^n} \left(X_{t\wedge k/2^n} - X_{t\wedge(k-1)/2^n}\right)^2,$$

where $a \wedge b = \min(a,b)$. The second definition will be used in Chapter 3. If a process X is differentiable, then it has quadratic variation equal to zero. Moreover, total and quadratic variation are related by the following inequality

$$\sum_{k=0}^{n-1} |X(t_{k+1}) - X(t_k)| \geq \frac{\sum_{k=0}^{n-1} |X(t_{k+1}) - X(t_k)|^2}{\max_{\Pi_n} |X(t_{k+1}) - X(t_k)|}\,.$$

Therefore, if X is continuous and has finite quadratic variation, then its total variation is necessarily infinite. Note that $V_t(X)$ and $[X,X]_t$ are stochastic processes as well.

1.5.3 Moments, covariance, and increments of stochastic processes

The expected value and variance of a stochastic process are defined as

$$\mathbb{E}X_t = \int_\Omega X(t,\omega)\mathrm{d}P(\omega), \quad t \in [0,T],$$

and

$$\mathrm{Var}X_t = \mathbb{E}(X_t - \mathbb{E}X_t)^2, \quad t \in [0,T].$$

The kth moment of X_t, $k \geq 1$, is defined, for all $t \in [0,T]$, as $\mathbb{E}X_t^k$. These quantities are well-defined when the corresponding integrals are finite. The *covariance function* of the process for two time values s and t is defined as

$$\mathrm{Cov}(X_s, X_t) = \mathbb{E}\left\{(X_s - \mathbb{E}X_s)(X_t - \mathbb{E}X_t)\right\}.$$

The quantity $X_t - X_s$ is called the *increment* of the process from s to t, $s < t$.

These quantities are useful in the description of stochastic processes that are usually introduced to model evolution subject to some stochastic shocks. There are different ways to introduce processes based on the characteristics one wants to model. A couple of the most commonly used approaches are the modeling of increments and/or the choice of the covariance function.

1.5.4 Conditional expectation

The conditional probability of A given B is defined as $P(A|B) = P(A \cap B)/P(B)$ for $P(B) > 0$. In the same way, it is possible to introduce the conditional distribution of a random variable X with respect to the event B as

$$F_X(x|B) = \frac{P(X \leq x \cap B)}{P(B)}, \quad x \in \mathbb{R},$$

and the expectation with respect to this conditional distribution is naturally introduced as (see [163] for a similar treatise)

$$\mathbb{E}[X|B] = \frac{\mathbb{E}(X\mathbf{1}_B)}{P(B)},$$

where $\mathbf{1}_B$ is the indicator function of the set B, which means $\mathbf{1}_B(\omega) = 1$ if $\omega \in B$ and 0 otherwise. For discrete random variables, the conditional expectation takes the form

$$\mathbb{E}[X|B] = \sum_i x_i \frac{P(\{\omega : X(\omega) = x_i\} \cap B)}{P(B)} = \sum_i x_i P(X = x_i|B).$$

For continuous random variables with density f_X, we have

$$\mathbb{E}[X|B] = \frac{1}{P(B)} \int_{\mathbb{R}} x\mathbf{1}_B(x) f_X(x)\mathrm{d}x = \frac{1}{P(B)} \int_B x f_X(x)\mathrm{d}x.$$

Consider now a discrete random variable Y that takes distinct values in the sets A_i (i.e., $A_i = A_i(\omega) = \{\omega : Y(\omega) = y_i\}$, $i = 1, 2, \ldots$), and assume that all $P(A_i)$ are positive. Let $\mathbb{E}|X| < \infty$. Then a new random variable Z can be defined as follows:

$$Z(\omega) = \mathbb{E}[X|Y](\omega) = \mathbb{E}[X|A_i(\omega)] = \mathbb{E}[X|Y(\omega) = y_i], \quad \omega \in A_i\,.$$

For each fixed $\omega \in A_i$ the conditional expectation $\mathbb{E}[X|Y]$ coincides with $\mathbb{E}[X|A_i]$, but, as a whole, it is a random variable itself because it depends on the events generated by Y.

If instead of a single element A_i we consider a complete σ-algebra of events (for example, the one generated by the generic random variable Y), we arrive at the general definition of conditional expectation: let X be a random variable such that $\mathbb{E}|X| < \infty$.

A random variable Z is called the *conditional expectation* of X with respect to the σ-algebra $\mathcal{F}$ if:

i) Z is $\mathcal{F}$-measurable and
ii) Z is such that $\mathbb{E}(Z\mathbf{1}_A) = \mathbb{E}(X\mathbf{1}_A)$ for every $A \in \mathcal{F}$.

The conditional expectation is unique and will be denoted as $Z = \mathbb{E}[X|\mathcal{F}]$. With this notation, the equivalence above can be written as

$$\mathbb{E}(\,\mathbb{E}[X|\mathcal{F}]\,\mathbf{1}_A\,) = \mathbb{E}(X\mathbf{1}_A) \text{ for every } A \in \mathcal{F}\,. \tag{1.4}$$

As we noted, the conditional expectation is a random variable, and the equality is only true up to null-measure sets. Among the properties of the conditional expectation, we note only the following. Let X, X_1, X_2 be random variables and a, b two constants. Then,

$$\begin{aligned}
&\mathbb{E}[a \cdot X_1 + b \cdot X_2|\mathcal{F}] = a \cdot \mathbb{E}[X_1|\mathcal{F}] + b \cdot \mathbb{E}[X_2|\mathcal{F}],\\
&\mathbb{E}[X|\mathcal{F}_0] = \mathbb{E}X,
\end{aligned}$$

if $\mathcal{F}_0 = \{\Omega, \emptyset\}$. Moreover, if Y is $\mathcal{F}$-measurable, then

$$\mathbb{E}[Y \cdot X|\mathcal{F}] = Y \cdot \mathbb{E}[X|\mathcal{F}],$$

and choosing $X = 1$, it follows that

$$\mathbb{E}[Y|\mathcal{F}] = Y.$$

Finally, choosing $A = \Omega$ in (1.4), it follows that

$$\mathbb{E}\left\{\mathbb{E}[X|\mathcal{F}]\right\} = \mathbb{E}X.$$

If X is independent of $\mathcal{F}$, it follows that $\mathbb{E}[X|\mathcal{F}] = \mathbb{E}X$ and, in particular, if X and Y are independent, we have $\mathbb{E}[X|Y] = \mathbb{E}[X|\sigma(Y)] = \mathbb{E}X$, where $\sigma(Y)$ is the σ-algebra generated by the random variable Y.

1.5.5 Martingales

Given a probability space $(\omega, \mathcal{F}, P)$ and a filtration $\{\mathcal{F}_t, t \geq 0\}$ on $\mathcal{F}$, a *martingale* is a stochastic process $\{X_t, t \geq 0\}$ such that $\mathbb{E}|X_t| < \infty$ for all $t \geq 0$, it is adapted to a filtration $\{\mathcal{F}_t, t \geq 0\}$, and for each $0 \leq s \leq t < \infty$, it holds true that

$$\mathbb{E}[X_t|\mathcal{F}_s] = X_s,$$

i.e., X_s is the best predictor of X_t given $\mathcal{F}_s$. If in the definition above the equality "=" is replaced by "$\geq$", the process is called *submartingale*, and if it is replaced by "$\leq$", it is called *supermartingale*. From the properties of the expected value operator it follows that if X is a martingale, then

$$\begin{aligned}\mathbb{E}X_s &= (\text{by definition of martingale}) = \mathbb{E}\{\,\mathbb{E}[X_t|\mathcal{F}_s]\,\} \\ &= (\text{by measurability of } X_t \text{ w.r.t. } \mathcal{F}_s) = \mathbb{E}X_t,\end{aligned}$$

which means that martingales have a constant mean for all $t \geq 0$.

1.6 Brownian motion

The very basic ingredient of a model describing stochastic evolution is the so-called *Brownian motion* or *Wiener process.*[5] There are several alternative ways to characterize and define the Wiener process $W = \{W(t), t \geq 0\}$, and one is the following: it is a Gaussian process with continuous paths and with independent increments such that $W(0) = 0$ with probability 1, $\mathbb{E}W(t) = 0$, and $\text{Var}(W(t) - W(s)) = t - s$ for all $0 \leq s \leq t$. In practice, what is relevant for our purposes is that $W(t) - W(s) \sim N(0, t - s)$, for $0 \leq s \leq t$ and that on any two disjoint intervals, say (t_1, t_2), (t_3, t_4) with $t_1 \leq t_2 \leq t_3 \leq t_4$, the increments $W(t_2) - W(t_1)$ and $W(t_4) - W(t_3)$ are independent.

Simulation of the trajectory of the Brownian motion

Given a fixed time increment $\Delta t > 0$, one can easily simulate a trajectory of the Wiener process in the time interval $[0, T]$. Indeed, for $W_{\Delta t}$ it holds true that

$$W(\Delta t) = W(\Delta t) - W(0) \sim N(0, \Delta t) \sim \sqrt{\Delta t} \cdot N(0, 1),$$

and the same is also true for any other increment $W(t + \Delta t) - W(t)$; i.e.,

$$W(t + \Delta t) - W(t) \sim N(0, \Delta t) \sim \sqrt{\Delta t} \cdot N(0, 1)\,.$$

[5] This process was named in honor of the botanist Robert Brown. In 1827, Brown described the chaotic motion of a grain of pollen suspended on the water and repeatedly hit by water molecules, a motion well-modeled with Brownian motion. Louis Bachelier in 1900, and independently Albert Einstein, studied the details from the mathematical point of view. The Brownian motion is also a Wiener process (i.e., a continuous-time Gaussian process with independent increments).

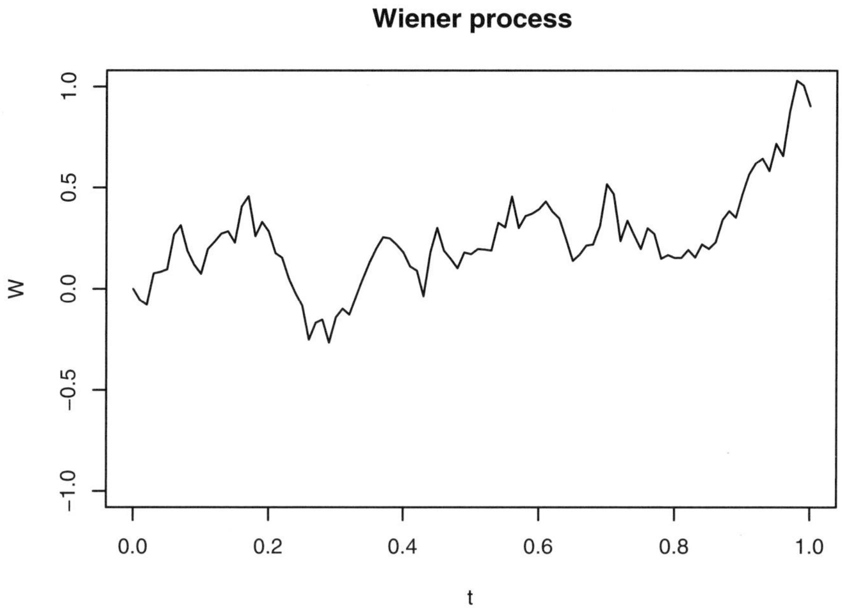

Fig. 1.2. A simulated path of the Wiener process.

Thus, we can simulate one such path as follows. Divide the interval $[0, T]$ into a grid such as $0 = t_1 < t_2 < \cdots < t_{N-1} < t_N = T$ with $t_{i+1} - t_i = \Delta t$. Set $i = 1$ and $W(0) = W(t_1) = 0$ and iterate the following algorithm.

1. Generate a (new) random number z from the standard Gaussian distribution.
2. $i = i + 1$.
3. Set $W(t_i) = W(t_{i-1}) + z \cdot \sqrt{\Delta t}$.
4. If $i \leq N$, iterate from step 1.

This method of simulation is valid only on the points of the grid, but in between any two points t_i and t_{i+1} the trajectory is usually approximated by linear interpolation. The algorithm above can be easily translated into some R code as follows.

```
> # ex1.06.R
> set.seed(123)
> N <- 100   # number of end-points of the grid including T
> T <- 1 # length of the interval [0,T] in time units
> Delta <- T/N # time increment
> W <- numeric(N+1) # initialization of the vector W
> t <- seq(0,T, length=N+1)
> for(i in 2:(N+1))
+         W[i] <- W[i-1] + rnorm(1) * sqrt(Delta)
> plot(t,W, type="l", main="Wiener process" , ylim=c(-1,1))
```

Interpolation between the points of the trajectory of W has been performed graphically by specifying the `type="l"` parameter in the `plot`[6] function (see Figure 1.2). But the code above might be considered as highly inefficient by R purists, as it implements an iteration using a `for` statement when this is not needed at all. In this simple case, the whole trajectory can be simulated in one line of R code as follows.

```
W <- c(0,cumsum( sqrt(Delta) * rnorm(N)))
```

due to the fact that the algorithm is just simulating a random walk with Gaussian increments of size $\sqrt{\Delta t} \cdot N(0,1)$. The reader can check that the two algorithms produce exactly the same trajectory by setting the seed of the random number generator to the same value by setting it with `set.seed(123)` (or any other value) before running each of the two R scripts.

Trajectories of Brownian motion can be obtained via other characterizations of the process. We review them just for completeness and to show some programming techniques.

1.6.1 Brownian motion as the limit of a random walk

One characterization of the Brownian motion says that it can be seen as the limit of a random walk in the following sense. Given a sequence of independent and identically distributed random variables $X_1, X_2, \ldots, X_n$, taking only two values $+1$ and -1 with equal probability and considering the partial sum,

$$S_n = X_1 + X_2 + \cdots + X_n\,.$$

Then, as $n \to \infty$,

$$P\left(\frac{S_{[nt]}}{\sqrt{n}} < x\right) \to P(W(t) < x)\,,$$

where $[x]$ is the integer part of the real number x. Please note that this result is a refinement of the central limit theorem that, in our case, asserts that $S_n/\sqrt{n} \to N(0,1)$. The following R code gives a graphical representation of how many random variables $X_1, X_2, \ldots, X_n$, we need to generate in order to obtain a good approximation.

```
> # ex1.07.R
> set.seed(123)
> n <- 10 # far from the CLT
> T <- 1
> t <- seq(0,T,length=100)
> S <- cumsum(2*(runif(n)>0.5)-1)
> W <- sapply(t, function(x) ifelse(n*x>0,S[n*x],0))
> W <- as.numeric(W)/sqrt(n)
```

6 One should notice that we have set `ylim=c(-1,1)` in the plot command. The interval $(-1,+1)$ is the 1-σ interval for the position of the Wiener process at time $t=1$; i.e., one should expect to find at least 68% of all possible realizations of the Wiener process up to time $t=1$, so this vertical axis limit specification allows us to have, with high probability, a display of the whole trajectory of W.

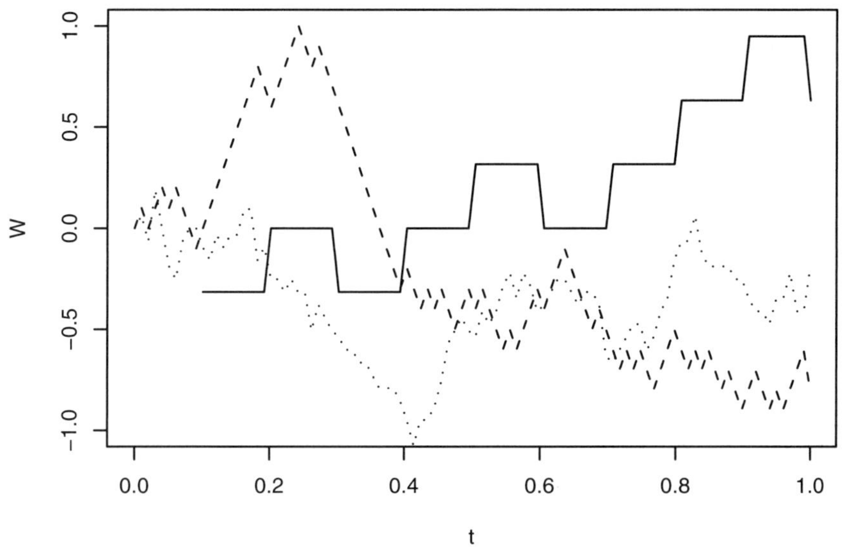

Fig. 1.3. Path of the Wiener process as the limit of a random walk; continuous line $n = 10$, dashed line $n = 100$, dotted line $n = 1000$.

```
> plot(t,W,type="l",ylim=c(-1,1))
> n <- 100 # closer to the CLT
> S <- cumsum(2*(runif(n)>0.5)-1)
> W <- sapply(t, function(x) ifelse(n*x>0,S[n*x],0))
> W <- as.numeric(W)/sqrt(n)
> lines(t,W,lty=2)
> n <- 1000 # quite close to the limit
> S <- cumsum(2*(runif(n)>0.5)-1)
> W <- sapply(t, function(x) ifelse(n*x>0,S[n*x],0))
> W <- as.numeric(W)/sqrt(n)
> lines(t,W,lty=3)
```

In the above, we have first simulated a sequence of random variates X_i taking values $+1$ and -1 with equal probability via the uniform distribution. The R command `runif(n)` generates n random numbers from the uniform distribution in $(0, 1)$, and `runif(n)>0.5` transforms these into a sequence of zeros and ones (actually `FALSE` and `TRUE`). Now, if x is either 0 or 1, the function $2 * x - 1$ maps 0 to -1 and 1 to 1. Thus, we now have a sequence of n equally distributed random numbers -1 and 1 and `cumsum` calculates S_n for us. Figure 1.3 shows the results of the approximation for $n = 10$, $n = 100$, and $n = 1000$.

1.6.2 Brownian motion as $L^2[0,T]$ expansion

Another characterization of the Wiener process quite useful in conjunction with empirical processes in statistics is the Karhunen-Loève expansion of W. The Karhunen-Loève expansion is a powerful tool that is nothing but an $L^2([0,T])$ expansion of random processes in terms of a sequence of independent random variables and coefficients. This is particularly useful for continuous-time processes that are a collection of uncountably many random variables (such as the Wiener process which is indeed a collection of uncountably many Gaussian variables). The Karhunen-Loève expansion is in fact a series of *only* countably many terms and is useful for representing a process on some fixed interval $[0,T]$. We recall that $L^2([0,T])$, or simply L^2, is the space of functions from $[0,T]$ to $\mathbb{R}$ defined as

$$L^2 = \{f : [0,T] \to \mathbb{R} : ||f||_2 < \infty\},$$

where

$$||f||_2 = \left(\int_0^T |f(t)|^2 \mathrm{d}t\right)^{\frac{1}{2}}.$$

Let us denote by $X(t)$ the trajectory of random process $X(t,\omega)$ for a given ω. The Wiener process $W(t)$ has trajectories belonging to $L^2([0,T])$ for almost all ω's, and the Karhunen-Loève expansion for it takes the form

$$W(t) = W(t,\omega) = \sum_{i=0}^{\infty} Z_i(\omega)\phi_i(t), \quad 0 \le t \le T,$$

with

$$\phi_i(t) = \frac{2\sqrt{2T}}{(2i+1)\pi} \sin\left(\frac{(2i+1)\pi t}{2T}\right).$$

The functions ϕ_i form a basis of orthogonal functions and Z_i a sequence of i.i.d. Gaussian random variables.

```
> # ex1.08.R
> set.seed(123)
> phi <- function(i,t,T){
+     (2*sqrt(2*T))/((2*i+1)*pi) * sin(((2*i+1)*pi*t)/(2*T))
+ }
> T <- 1
> N <- 100
> t <- seq(0,T,length=N+1)
> W <- numeric(N+1)
> n <- 10
> Z <- rnorm(n)
> for(i in 2:(N+1))
+     W[i] <- sum(Z*sapply(1:n, function(x) phi(x,t[i],T)))
> plot(t,W,type="l",ylim=c(-1,1))
> n <- 50
> Z <- rnorm(n)
> for(i in 2:(N+1))
+     W[i] <- sum(Z*sapply(1:n, function(x) phi(x,t[i],T)))
> lines(t,W,lty=2)
```

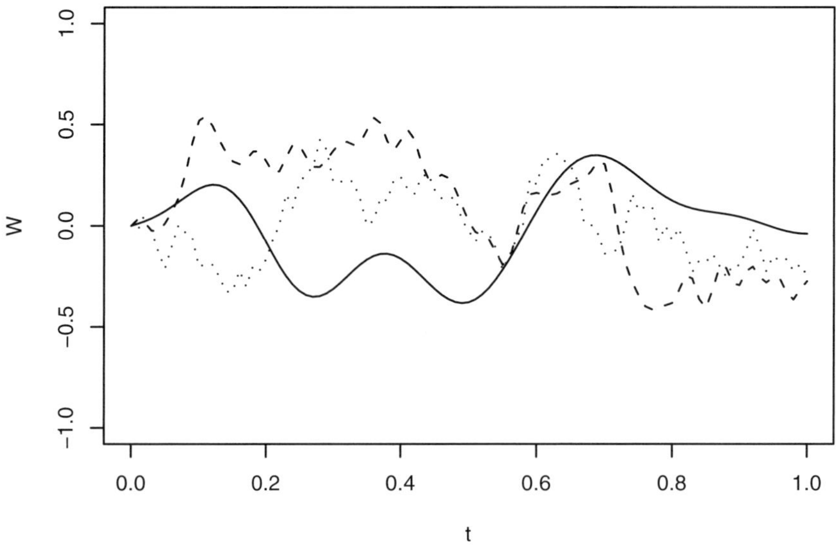

Fig. 1.4. Karhunen-Loève approximation of the path of the Wiener process with $n = 10$ (continuous line), $n = 50$ (dashed line), and $n = 100$ (dotted line) terms in the expansion.

```
> n <- 100
> Z <- rnorm(n)
> for(i in 2:(N+1))
+     W[i] <- sum(Z*sapply(1:n, function(x) phi(x,t[i],T)))
> lines(t,W,lty=3)
```

In Figure 1.4, three different approximations of the Wiener path based on Karhunen-Loève expansion are presented. The higher the number of terms, the better the approximation. Other L^2 methods, based on fractal theory and iterated function systems, have recently been introduced (see, e.g., [117]).

Which method to choose?

When one needs only to simulate the position of the Brownian motion at one fixed time point (which is quite common in finance in the evaluation of the payoff of contingent claims with a fixed exercise time), then the first method should be used, as it is accurate enough. The two other methods, in particular the last one, are useful if one needs more information and in particular if one needs information on the whole path of W (again in finance, for example in the evaluation of the payoff of Asian and barrier options, but also in physics when one needs the complete evolution of a system). The Karhunen-Loève expansion is of course more expensive in terms of CPU time, and its reliability also depends in principle on the reliability of the implementation of the trigonometric functions. But the Karhunen-Loève expansion method is also quite powerful for simulating paths of processes without independent increments or functionals of the Brownian motion (when the L^2 expansions are known). This is particularly useful in goodness of fit testing problems (see, e.g, [208]).

1.6.3 Brownian motion paths are nowhere differentiable

The Brownian motion has infinite simple variation (i.e., $\mathrm{V}_t(W) = \infty$) and covariance function $\mathrm{Cov}(W(s), W(t)) = t \wedge s$, where $s \wedge t = \min(s,t)$. Its paths are continuous but nowhere differentiable. To graphically view the weirdness of the trajectory of the Brownian motion and understand why it is nowhere differentiable, we can do the following experiment borrowed from [142]. We simulate the increments of the Brownian motion in two subsequent time points lagged by a time Δt, say $W(0.5)$ and $W(0.5+\Delta t)$, and we let $\Delta t \to 0$. Figure 1.5 shows the explosive behavior of $\lim_{\Delta t \to 0} |W(0.5+\Delta t) - W(0.5)|/\Delta t \to +\infty$. This happens because of the independence of the increments of the Brownian motion and also, more importantly, because the increments $W(t+\Delta t) - W(t)$ behave like $\sqrt{\Delta t}$ instead of Δt. Thus, in the limit one can expect

$$\lim_{\Delta t \to 0} \frac{|B(t+\Delta t) - B(t)|}{\Delta t} \simeq \lim_{\Delta t \to 0} \frac{|\sqrt{\Delta t}|}{\Delta t} = +\infty\,.$$

The next listing graphically proves this behavior. Notice that we choose the log scale for the y axis.

```
> # ex1.09.R
> set.seed(123)
> phi <- function(i,t,T){
+     (2*sqrt(2*T))/((2*i+1)*pi) * sin(((2*i+1)*pi*t)/(2*T))
+ }
> n <- 100
> Z <- rnorm(n)
> Delta <- seq(1e-7, 1e-2,length=30)
> W <- sum(Z*sapply(1:n, function(x) phi(x,0.5,T)))
> for(i in Delta)
+   Wh <- sum(Z*sapply(1:n, function(x) phi(x,0.5+i,T)))
> inc.ratio <- abs(Wh-W)/Delta
> plot(Delta,inc.ratio,type="l",log="y",xlab=expression(Delta*t),
+      ylab=expression(abs(W(0.5+Delta*t)-W(0.5))/Delta*t))
> max(inc.ratio,na.rm=T)
[1] 701496.4
```

The nice property of the Brownian motion is that its quadratic variation is finite. In particular, we have that $[W,W]_t = t$ for all $t \geq 0$.

1.7 Geometric Brownian motion

A process used quite often in finance to model the dynamics of some asset is the so-called *geometric Brownian motion*. This process has the property of

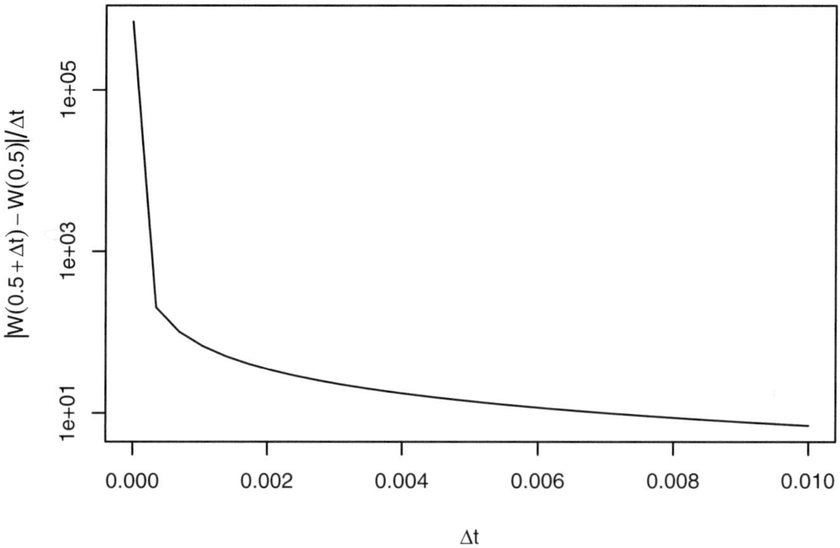

Fig. 1.5. Graphical evidence that the trajectory of the Brownian motion is nondifferentiable. The plot shows the behavior of $\lim_{\Delta t \to 0} |W(0.5+\Delta t) - W(0.5)|/\Delta t \to +\infty$.

having independent multiplicative increments and is defined as a function of the standard Brownian motion

$$S(t) = x \exp\left\{ \left(r - \frac{\sigma^2}{2} \right) t + \sigma W(t) \right\}, \quad t > 0, \tag{1.5}$$

with $S(0) = x$, $x \in \mathbb{R}$ is the initial value $\sigma > 0$ (interpreted as the volatility), and r (interpreted as the interest rate) two constants. It has been introduced in finance in [172]. One implementation to simulate a path of the geometric Brownian motion can be based on previous algorithms. Assuming that `W` contains a trajectory of W and `t` is the vector containing all the time points, a path can be drawn (see Figure 1.6) as follows.

```
> # ex1.10.R
> set.seed(123)
> r <- 1
> sigma <- 0.5
> x <- 10
> N <- 100    # number of end points of the grid including T
> T <- 1 # length of the interval [0,T] in time units
> Delta <- T/N # time increment
> W <- numeric(N+1) # initialization of the vector W
> t <- seq(0,T, length=N+1)
> for(i in 2:(N+1))
+         W[i] <- W[i-1] + rnorm(1) * sqrt(Delta)
```

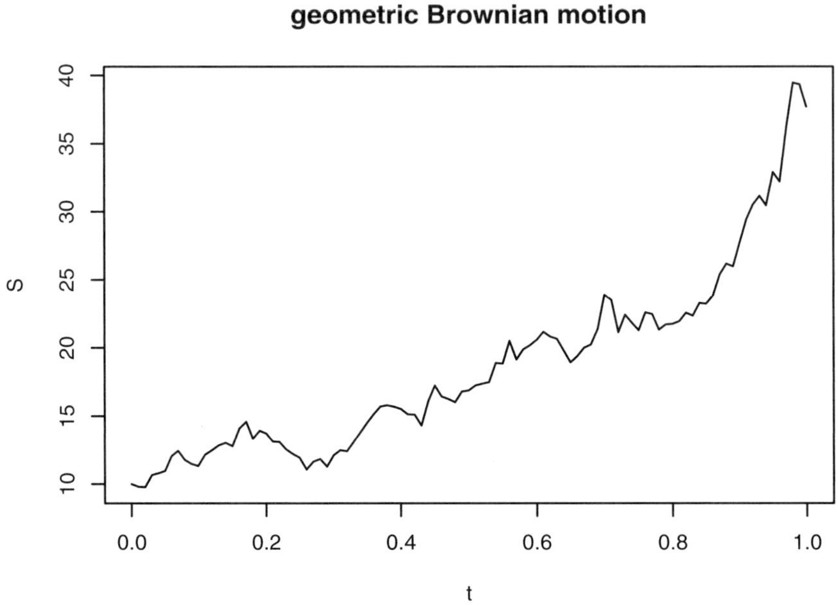

Fig. 1.6. A trajectory of the geometric Brownian motion obtained from the simulation of the path of the Wiener process.

```
> S <- x * exp((r-sigma^2/2)*t + sigma*W)
> plot(t,S,type="l",main="geometric Brownian motion")
```

An equivalent way of simulating a trajectory of the geometric Brownian motion is by simulating the increments of S. Indeed,

$$S(t+\Delta t) = S(t)e^{\left(r-\frac{\sigma^2}{2}\right)(t+\Delta t-t)+\sigma(W(t+\Delta t)-W(t))}, \tag{1.6}$$

which simplifies to

$$S(t+\Delta t) = S(t)\exp\left\{\left(r-\frac{\sigma^2}{2}\right)\Delta t + \sigma\sqrt{\Delta t}Z\right\} \tag{1.7}$$

with $Z \sim N(0,1)$. Formula (1.6), which we will derive formally later, is a particular case of the *generalized geometric Brownian motion*, which is a process starting from x at time s whose dynamic is

$$Z_{s,x}(t) = x\exp\left\{\left(r-\frac{\sigma^2}{2}\right)(t-s)+\sigma(W(t)-W(s))\right\}, \quad t \geq s\,. \tag{1.8}$$

Of course, $Z_{0,S(0)}(t) = S(t)$. In the same manner, we can consider the translated Brownian motion. Given a Brownian motion $W(t)$, we define a new process

$$W_{0,x}(t) = x + W(t)$$

with x a constant. Then $W_{0,x}(t)$ is a Brownian motion starting from x instead of 0. If we further want this to happen at some fixed time t_0 instead of at time 0, we need to translate the process further by $W(t_0)$. Thus,

$$W_{t_0,x}(t) = x + W(t) - W(t_0), \quad t \geq t_0\,, \tag{1.9}$$

is a Brownian motion starting at x at time t_0. More precisely, this is the process $W_{t_0,x} = \{W(t), t_0 \leq t \leq T | W(t_0) = x\}$ and, of course, $W_{0,W(0)}(t) = W(t)$. By the properties of the Brownian motion, $W_{t_0,x}(t)$ is equal in distribution to $x + W(t - t_0)$, and one way to simulate it is to simulate a standard Brownian motion, add the constant x, and then translate the time. The code in Listing 1.1 constructs such a trajectory.

1.8 Brownian bridge

Another useful and interesting manipulation of the Wiener process is the so-called *Brownian bridge*, which is a Brownian motion starting at x at time t_0 and passing through some point y at time T, $T > t_0$. It is defined as

$$W_{t_0,x}^{T,y}(t) = x + W(t - t_0) - \frac{t - t_0}{T - t_0} \cdot (W(T - t_0) - y + x)\,. \tag{1.10}$$

More precisely, this is the process $\{W(t), t_0 \leq t \leq T | W(t_0) = x, W(T) = y\}$. This process is easily simulated using the simulated trajectory of the Wiener process.

```
> # ex1.11.R
> set.seed(123)
> N <- 100    # number of end points of the grid including T
> T <- 1 # length of the interval [0,T] in time units
> Delta <- T/N # time increment
> W <- numeric(N+1) # initialization of the vector W
> t <- seq(0,T, length=N+1)
> for(i in 2:(N+1))
+          W[i] <- W[i-1] + rnorm(1) * sqrt(Delta)
> x <- 0
> y <- -1
> BB <- x + W - t/T * (W[N+1] - y +x)
> plot(t,BB,type="l")
> abline(h=-1, lty=3)
```

Figure 1.7 shows a simulated path of the Brownian bridge starting from $x = 0$ at time 0 and ending in $y = -1$ at time T. To conclude this section, we provide few functions to generate paths of the Wiener and other related processes. The functions, called `BM`, `GBM`, and `BBridge`, return an invisible[7] object of class `ts` (i.e., a regular time series object of the R language).

[7] Invisible objects returned by R functions are copies of the object that can be assigned to some other objects without being printed on the R console; i.e., `BM()` prints nothing on the console, assigning `X <- BM()`, typing `X` prints the whole simulated path on the R console, `plot(X)` plots the path on the graphic device, etc.

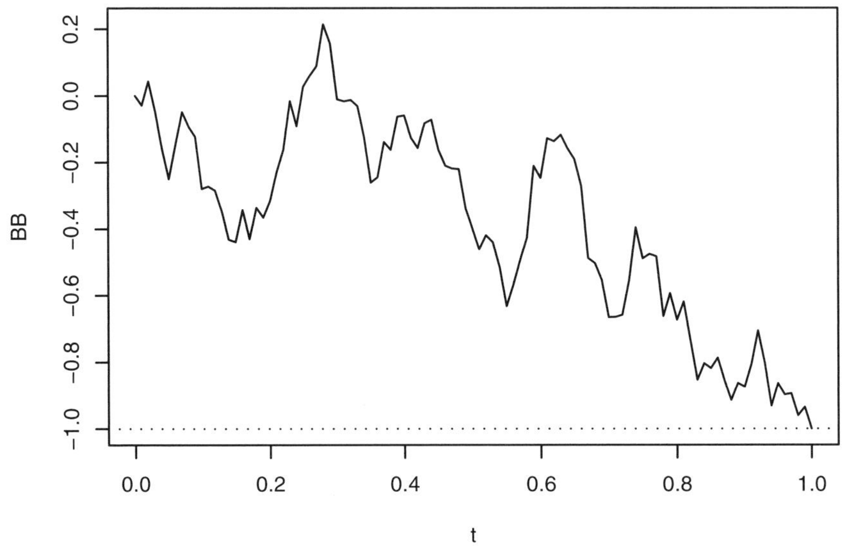

Fig. 1.7. A simulated trajectory of the Brownian bridge starting at x at time 0 and terminating its run at $y = -1$ at time T.

Listings 1.1, 1.2, and 1.3 construct respectively a trajectory of $W_{t_0,x} = \{W(t), t_0 \leq t \leq T | W(t_0) = x\}$, $W_{t_0,x}^{T,y}(t) = \{W(t), t_0 \leq t \leq T | W(t_0) = x, W(T) = y\}$, and the geometric Brownian motion. The functions `BM`, `GBM`, and `BBridge` are contained in the package `sde`. As the next listing shows, their use is very intuitive.

```
> # ex1.12.R
> require(sde)
>
> set.seed(123)
> plot(BM())
> plot(GBM(1,1,sqrt(0.5)))
> plot(BBridge(0,-1))
```

We will refine these functions in Chapter 2.

```
BM <- function(x=0, t0=0, T=1, N=100){
  if(T<= t0) stop("wrong times")
  dt <- (T-t0)/N
  t <- seq(t0,T, length=N+1)
  X <- ts(cumsum(c(x,rnorm(N)*sqrt(dt))),start=t0, deltat=dt)
  return(invisible(X))
}
```

Listing 1.1. Simulation $W_{s,x}(t) = \{W(t), t \geq s | W(s) = x\}$.

```
BBridge <- function(x=0, y=0, t0=0, T=1, N=100){
  if(T<= t0) stop("wrong times")
  dt <- (T-t0)/N
  t <- seq(t0, T, length=N+1)
  X <- c(0,cumsum( rnorm(N)*sqrt(dt)))
  BB <- x + X - (t-t0)/(T-t0)*(X[N+1]-y+x)
  X <- ts(BB, start=t0,deltat=dt)
  return(invisible(X))
}
```

Listing 1.2. Simulation of the Brownian bridge. $W_{t_0,x}^{T,y}(t) = \{W(t), t_0 \leq t \leq T | W(t_0) = x, W(T) = y\}$.

```
# x = starting point at time 0
# r = interest rate
# sigma = square root of the volatility
GBM <- function(x, r=0, sigma, T=1, N=100){
    tmp <- BM(T=T,N=N)
    S <- x * exp((r-sigma^2/2)*time(tmp) + sigma* as.numeric(tmp))
    X <- ts(S, start=0,deltat=1/N)
    return(invisible(X))
}
```

Listing 1.3. Simulation of the geometric Brownian motion.

1.9 Stochastic integrals and stochastic differential equations

Stochastic integrals and in particular Itô integrals are naturally introduced to correctly define a stochastic differential equation. We first present the heuristics behind the notion of stochastic differential equations with a classical example from mathematical finance. Let us suppose we have the quantity $S(t)$, $t \geq 0$, which represents the value of an asset at time t. Consider now the variation $\Delta S = S(t+\Delta t) - S(t)$ of S in a small time interval $[t, t+\Delta t)$. The returns of the asset for which S is the dynamics are defined as the ratio $\Delta S/S$. We can model the returns as

$$\frac{\Delta S}{S} = \text{deterministic contribution} + \text{stochastic contribution}.$$

The deterministic contribution might be assumed to be linked to the interest rate of non risky activities and thus proportional to time with some constant rate μ, thus

$$\text{deterministic contribution} = \mu \Delta t$$

(we will see later that μ can be made a function of either t or $S(t)$). The stochastic contribution is assumed to be related to the variation of some source of noise and to the natural variability of the market (the volatility). We denote by $\Delta X = X(t+\Delta t) - X(t)$ the variation of the noisy process (i.e., the shocks) and make it proportional to the market volatility σ; thus

$$\text{stochastic contribution} = \sigma \Delta X$$

(σ can also be made a function of t and/or S). The natural hypothesis is to assume Gaussian behavior of the noise (i.e., $\Delta X \sim N(0,1)$) which implies the assumption of X being the Wiener process if the shocks are, in addition, supposed to be independent. Finally, we have

$$\frac{\Delta S}{S} = \mu \Delta t + \sigma \Delta W.$$

Now, the evil temptation is to consider the difference equation above for infinitesimal time intervals (i.e., for $\Delta t \to 0$) in order to obtain a (stochastic) differential equation of the form

$$\frac{S'(t)}{S(t)} = \mu + \sigma W'(t), \text{ namely } S'(t) = \mu S(t) + \sigma S(t) W'(t),$$

which we can also write in differential form as

$$\mathrm{d}S(t) = \mu S(t)\mathrm{d}t + \sigma S(t)\mathrm{d}W(t). \tag{1.11}$$

The preceding equation is an example of a stochastic differential equation, but unfortunately this expression has no mathematical meaning, as we already mentioned that the variation of the Wiener process $\mathrm{d}W(t)$ is not finite and the Wiener process has continuous but nowhere differentiable paths. To make sense of (1.11), we switch to its integral form

$$S(t) = S(0) + \mu \int_0^t S(u)\mathrm{d}u + \sigma \int_0^t S(u)\mathrm{d}W(u). \tag{1.12}$$

Equation (1.12) introduces the *stochastic integral*

$$I(X) = \int_0^T X(u)\mathrm{d}W(u)$$

with respect to the Brownian motion. The definition of $I(X)$ is quite easy for *simple* (i.e., piecewise constant) processes X, but it requires more attention for generic processes. Even if we can't go into the details of the construction of the stochastic integral (the reader can refer to [170] or [130]), we will outline the basic steps in order to understand what $I(X)$ really means and find a way to simulate it. Given a generic integrand $f : [0,T] \times \Omega \to \mathbb{R}$, $I(f)$ is defined as the limit of the sequence of the integrals $I(f^{(n)})$, where $f^{(n)}$, called *simple processes*, are defined as

$$f^{(n)}(t,\omega) = f(t_j,\omega), \qquad t_j \le t < t_{j+1},$$

with $t_j \in \Pi_n([0,1])$ and such that $\Pi_n \to 0$ as $n \to \infty$. It is easy to show that $f^{(n)}$ converges to f in quadratic mean. Then $I\left(f^{(n)}\right)$ is defined as

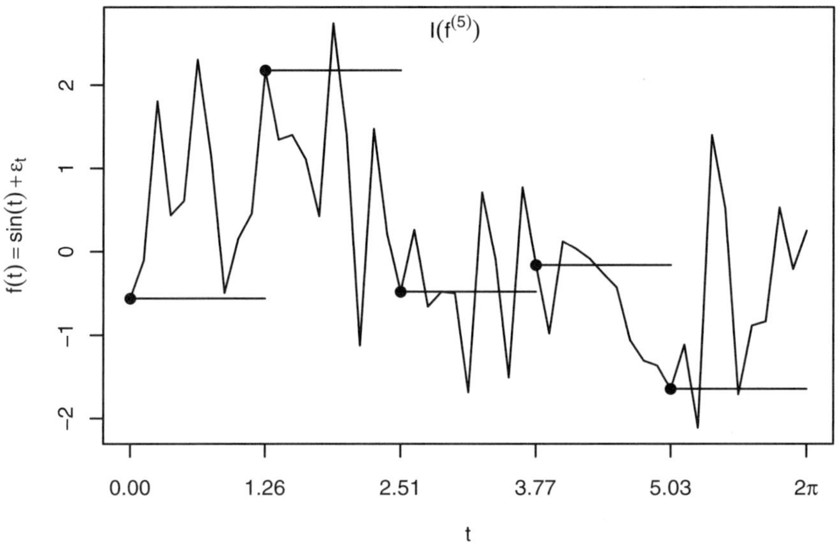

Fig. 1.8. The simple (piecewise constant) process $f^{(5)}(t)$ approximating $f(t) = \sin(t) + \varepsilon_t$ used to construct Itô integrals. Note that $f^{(n)}(t)$ is defined as right continuous.

$$
\begin{aligned}
I\left(f^{(n)}\right) &= \sum_{j=0}^{n-1} f^{(n)}\left(t_j\right)\left\{W\left(t_{j+1}\right) - W\left(t_j\right)\right\} \\
&= \sum_{j=0}^{n-1} f\left(t_j\right)\left\{W\left(t_{j+1}\right) - W\left(t_j\right)\right\}.
\end{aligned}
\tag{1.13}
$$

Equation (1.13) does not converge in the usual sense, as W does not have finite variation. On the contrary, if we consider the mean square convergence, the limit exists. Indeed, for every n, we have that

$$
\mathbb{E}\left\{I\left(f^{(n)}\right)\right\}^2 = \sum_{j=0}^{n-1} \mathbb{E}\left(f\left(t_j\right)\right)^2\left(t_{j+1} - t_j\right),
$$

from which it follows that $I\left(f^{(n)}\right) \to I(f)$ in quadratic mean, the limit being unique. Figure 1.8 is a representation of the simple process $f^{(5)}(t)$ approximating the target $f(t) = \sin(t) + \varepsilon_t$ (with ε_t a Gaussian noise). From the crude construction depicted above, few important things emerge as essential in the definition of $I(f)$. First of all, it is required that f be a process adapted to the natural filtration of the Wiener process; i.e., f is $\mathcal{F}_t$-measurable for every t. This is required in (1.13) in order to have a well-defined process and is the reason why, in the Itô sums of (1.13), the function is calculated at the beginning of the interval $[t_j, t_{j+1})$ instead of in the middle.[8] Moreover, the behavior of the integrand process needs to compensate for the weirdness of the path of the Brownian motion. This second fact implies the technical condition $\mathbb{E}\int_0^t X^2(u)\mathrm{d}u < \infty$. Figure 1.8 has been generated with the following lines of R code, which we report only for completeness.

```
> # ex1.13.R
> set.seed(123)
> n <- 5
> N <- n*10
> t <- seq(0, 2*pi, length=N+1)
> f <- sin(t)+rnorm(N+1)
> plot(t, f, type="l", axes=F,
+      ylab=expression(f(t)==sin(t)+epsilon[t]))
> idx <- seq(1, N+1, length=n+1)
> axis(1,t[idx], c(sprintf("%3.2f", t[idx[1:n]]),
+      expression(2*pi)))
> axis(2)
> box()
> for(i in 1:n){
+      lines( c(t[idx[i]],t[idx[i+1]]), c(f[idx[i]],f[idx[i]]) )
+      points( t[idx[i]], f[idx[i]] ,pch=19)
+ }
> text(t[idx[3]]+.5,f[idx[2]]+.5,expression(I(f^{(5)})))
```

1.9.1 Properties of the stochastic integral and Itô processes

Let $\{X(t), 0 \leq t \leq T\}$ be a stochastic process adapted to the filtration generated by the Brownian motion and such that $\int_0^T \mathbb{E}(X(s)^2)\mathrm{d}s < +\infty$. The stochastic integral of X is defined as

$$I_t(X) = \int_0^t X_s \mathrm{d}W_s = \lim_{||\Pi_n||\to 0} \sum_{i=0}^{n-1} X(t_i)(W(t_{i+1}) - W(t_i)),$$

where the convergence is in the quadratic mean and $t_i \in \Pi_n$. We will show in Section 1.13.1 how to simulate an Itô integral, but we summarize here, without proof, some nice properties of the Itô integral we will use at a later stage.

- If X is Itô integrable, then

$$\mathbb{E}\left(\int_0^T X(s)\mathrm{d}W(s)\right) = 0$$

and

---

[8] This alternative construction of the stochastic integral actually exists and leads to the Stratonovich stochastic integral, which unfortunately does not share the same properties of the Itô integral. In particular, it is not a martingale.

$$\text{Var}\left(\int_0^T X(s)\text{d}W(s)\right) = \int_0^T \mathbb{E}X^2(t)\text{d}t \quad \text{(Itô isometry)}.$$

- If X and Y are two Itô integrable processes and a and b two constants, then (*linearity*)

$$\int_0^T (aX(t) + bY(t))\text{d}W(t) = a\int_0^T X(t)\text{d}W(t) + b\int_0^T Y(t)\text{d}W(t).$$

- It follows from the linearity property above that

$$\int_0^T a\text{d}W(t) = a\int_0^T \text{d}W(t) = aW(T).$$

- As we will show later in Section 1.11, it can be proved that

$$\int_0^T W(t)\text{d}W(t) = \frac{1}{2}W^2(T) - \frac{1}{2}T. \tag{1.14}$$

- The process $M(t) = M(0) + \int_0^t X(s)\text{d}W(s)$ is a martingale with $M(0)$ a constant.

An Itô process $\{X_t, 0 \le t \le T\}$ is a stochastic process that can be written in the form

$$X_t = X_0 + \int_0^t g(s)\text{d}s + \int_0^t h(s)\text{d}W_s.$$

where $g(t,\omega)$ and $h(t,\omega)$ are two adapted and progressively measurable random functions such that

$$P\left\{\int_0^T |g(t,\omega)|\text{d}t < \infty\right\} = 1 \quad \text{and} \quad P\left\{\int_0^T h(t,\omega)^2\text{d}t < \infty\right\} = 1.$$

1.10 Diffusion processes

The class of processes that is considered in this book is that of diffusion process solutions to stochastic differential equations of the form

$$\text{d}X(t) = b(t, X(t))\text{d}t + \sigma(t, X(t))\text{d}W(t) \tag{1.15}$$

with some initial condition $X(0)$. As usual, (1.15) is interpreted in the Itô sense; i.e.,

$$X(t) = X(0) + \int_0^T b(u, X(u))\text{d}u + \int_0^T \sigma(u, X(u))\text{d}W(u). \tag{1.16}$$

The initial condition can be random or not. If random, say $X(0) = Z$, it should be independent of the σ-algebra generated by W and satisfy the condition $\mathbb{E}|Z|^2 < \infty$. The two deterministic functions $b(\cdot,\cdot)$ and $\sigma^2(\cdot,\cdot)$ are called respectively the *drift* and the *diffusion* coefficients of the stochastic differential equation (1.15). All over the text, even when not mentioned, they are supposed to be measurable and such that

$$P\left\{\int_0^T \sup_{|x|\leq R} (|b(t,x)| + \sigma^2(t,x))\mathrm{d}t < \infty\right\} = 1$$

for all $T, R \in [0,\infty)$ because (1.16) is an Itô process.

Assumption 1.1 (Global Lipschitz) *For all $x, y \in \mathbb{R}$ and $t \in [0,T]$, there exists a constant $K < +\infty$ such that*

$$|b(t,x) - b(t,y)| + |\sigma(t,x) - \sigma(t,y)| < K|x - y| \,. \tag{L}$$

Assumption 1.2 (Linear growth) *For all $x, y \in \mathbb{R}$ and $t \in [0,T]$, there exists a constant $C < +\infty$ such that*

$$|b(t,x)| + |\sigma(t,x)| < C(1 + |x|) \,. \tag{G}$$

The linear growth condition controls the behaviour of the solution so that X_t does not explode in a finite time.

Fact 1.2 (Existence and uniqueness, Theorem 5.2.1 [170]) *Under Assumptions 1.1 and 1.2, the stochastic differential equation (1.15) has a unique, continuous, and adapted strong solution such that*

$$\mathbb{E}\left\{\int_0^T |X_t|^2 \mathrm{d}t\right\} < \infty \,.$$

We call such a process X a *diffusion process*. The result above states that the solution X is of *strong* type. This essentially implies the pathwise uniqueness of the result. It is also possible to obtain *weak* solutions under different assumptions. In many cases in statistics, conditions for weak solutions are enough because they imply that any two weak solutions $X^{(1)}$ and $X^{(2)}$ are not necessarily pathwise identical, while their distributions are, and this is enough for likelihood inference. Of course, strong solutions, are also weak solutions but the contrary is not necessarily true.

The major part of this book will focus on the homogeneous version of the stochastic differential equation (1.15) with nonrandom initial condition, say $X(0) = x$. To keep the notation simpler, we will use the following notation throughout the text:

$$X_t = x + \int_0^t b(X_u)\mathrm{d}u + \int_0^t \sigma(X_u)\mathrm{d}W_u \tag{1.17}$$

and

$$\mathrm{d}X_t = b(X_t)\mathrm{d}t + \sigma(X_t)\mathrm{d}W_t\,. \tag{1.18}$$

The conditions above might be too restrictive in some cases, such as (see, e.g., [149])

$$\mathrm{d}X_t = -\theta X_t^3 \mathrm{d}t + \sigma \mathrm{d}W_t. \tag{1.19}$$

In many cases, local versions of the same conditions are enough.

Assumption 1.3 (Local Lipschitz) *For any $N < \infty$, $|x|, |y| \le N$, there exists a constant $L_N > 0$ such that*

$$|b(x) - b(y)| + |\sigma(x) - \sigma(y)| \le L_N |x - y|$$

and

$$2xb(x) + \sigma^2(x) \le B(1 + x^2)\,. \tag{1.20}$$

Indeed, these conditions are satisfied in (1.19). For the class of ergodic diffusion processes that will be discussed later, it is usually true that $xb(x) < 0$. Hence (1.20) is just a condition on the growth of the diffusion coefficient.

Assumption 1.4 *Let $b(\cdot)$ be locally bounded, $\sigma^2(\cdot)$ continuous, and positive, and for some A the following condition holds:*

$$xb(x) + \sigma^2(x) \le A(1 + x^2)\,. \tag{\textbf{B}}$$

Fact 1.3 (See [73]) *Under Assumption 1.4, the stochastic differential equation (1.17) has a unique weak solution.*

For a discussion about strong and weak solutions, the reader can consider references [150], [219], [130], and [149].

1.10.1 Ergodicity

Diffusion processes possess the Markov property and may or may not be ergodic. The ergodic property implies that, for any measurable function $h(\cdot)$, the following result holds with probability 1:

$$\frac{1}{T}\int_0^T h(X_t)\mathrm{d}t \to \int_{-\infty}^{+\infty} h(x)\pi(x)\mathrm{d}x = \mathbb{E}h(\xi),$$

where $\pi(\cdot)$ is called the *invariant* or *stationary* density of the diffusion process and ξ is some random variable with $\pi(\cdot)$ as density. Diffusion processes have the nice property that the stationary distribution, when it exists, can be expressed in terms of the *scale measure* and the *speed measure*,[9] defined respectively as

$$s(x) = \exp\left\{-2\int_{x_0}^{x} \frac{b(y)}{\sigma^2(y)}\mathrm{d}y\right\} \tag{1.21}$$

[9] Sometimes they are both termed “measure” or “density.”

and

$$m(x) = \frac{1}{\sigma^2(x)s(x)} \,. \tag{1.22}$$

In particular, the density of the invariant distribution $\pi(\cdot)$ is proportional, up to a normalizing constant, to the speed measure $m(\cdot)$; i.e.,

$$\pi(x) = \frac{m(x)}{M} \,, \tag{1.23}$$

where $M = \int m(x)\mathrm{d}x$.

Assumption 1.5 *Let (l, r), with $-\infty \leq l \leq r \leq +\infty$, be the state space of the diffusion process X solution to (1.18), and assume that*

$$\int_l^r m(x)\mathrm{d}x < \infty \,.$$

Let x^* be an arbitrary point in the state space of X such that

$$\int_{x^*}^r s(x)\mathrm{d}x = \int_l^{x^*} s(x)\mathrm{d}x = \infty \,.$$

If one or both of the integrals above are finite, the corresponding boundary is assumed to be instantaneously reflecting.

Under Assumption 1.5, the process X is ergodic and has an invariant distribution function.

1.10.2 Markovianity

From the Markovian property of the diffusion, it is also possible to define the transition density from value x at time s to value y at time t by $p(t, y|s, x)$ or, when convenient, as $p(t - s, y|x)$. For parametric models, we will later use the notation $p(t, y|s, x; \theta)$ or $p_\theta(t, y|s, x)$ and $p(t - s, y|x; \theta)$ or $p_\theta(t - s, y|x)$, respectively. The transition density satisfies the *Kolmogorov forward equation*

$$\frac{\partial}{\partial t}p(t, y|s, x) = -\frac{\partial}{\partial y}(b(y)p(t, y|s, x)) + \frac{1}{2}\frac{\partial^2}{\partial y^2}(\sigma^2(y)p(t, y|s, x)) \tag{1.24}$$

and *Kolmogorov backward equation*

$$-\frac{\partial}{\partial s}p(t, y|s, x) = b(x)\frac{\partial}{\partial x}p(t, y|s, x) + \frac{1}{2}\sigma^2(x)\frac{\partial^2}{\partial x^2}p(t, y|s, x); \tag{1.25}$$

see, e.g, [170]. Letting $t \to -\infty$ in the Kolmogorov forward equation (1.24), it is possible to obtain

$$\frac{\mathrm{d}^2}{\mathrm{d}x^2}(\sigma^2(x)\pi(x)) = 2\frac{\mathrm{d}}{\mathrm{d}x}(b(x)\pi(x)) \,, \tag{1.26}$$

where $\pi(x)$ is the stationary density. Equation (1.26) establishes a relationship between the drift $b(\cdot)$, the diffusion coefficient $\sigma(\cdot)$, and the invariant density $\pi(\cdot)$. Hence, in principle, given either of the two, one can obtain the third. For example, by integrating (1.26), we obtain (see, e.g., [19])

$$b(x) = \frac{1}{2\pi(x)}\frac{\mathrm{d}}{\mathrm{d}x}(\sigma^2(x)\pi(x)),$$

and integrating again, one gets (see, e.g., [3])

$$\sigma^2(x) = \frac{2}{\pi(x)}\int_0^x b(u)\pi(u)\mathrm{d}u\,.$$

All these facts and relationships will be useful for both the simulation algorithms and in the chapters on parametric and nonparametric inference that follow.

1.10.3 Quadratic variation

It can be seen [170] that the quadratic variation of a diffusion process solution to (1.15) is given by

$$<X,X>_t = \int_0^t \sigma^2(u, X_u)\mathrm{d}u\,. \tag{1.27}$$

1.10.4 Infinitesimal generator of a diffusion process

Given a diffusion process X solution to $\mathrm{d}X_t = b(X_t)\mathrm{d}t + \sigma(X_t)\mathrm{d}W_t$, $X_0 = x$, a differential operator $\mathcal{L}$ of the form

$$(\mathcal{L}f)(x) = \frac{\sigma^2(x)}{2}f''(x) + b(x)f'(x) \tag{1.28}$$

with f two times differentiable is called the *infinitesimal generator* of the diffusion process X.

1.10.5 How to obtain a martingale from a diffusion process

If Z is a process defined as

$$Z(t) = f(X_t) - \int_0^t (\mathcal{L}f)(X_s)\mathrm{d}s\,,$$

where X is a diffusion process and $f(\cdot) \in C_0^2(\mathbb{R})$, then Z is a martingale with respect to the Brownian motion. Moreover, we have that

$$\mathbb{E}f(X_t) = f(x) + \mathbb{E}\left(\int_0^t (\mathcal{L}f)(X_s)\mathrm{d}s\right).$$

1.11 Itô formula

An important tool of stochastic calculus that is also useful in simulations is the Itô formula. This formula can be seen as the stochastic version of a Taylor expansion of $f(X)$ stopped at the second order, where X is a diffusion process. Itô's lemma says that if $f(t,x)$ is a twice differentiable function on both t and x, then

$$\begin{aligned} f(t, X_t) = f(0, X_0) &+ \int_0^t f_t(u, X_u)\mathrm{d}u + \int_0^t f_x(u, X_u)\mathrm{d}X_u \\ &+ \frac{1}{2}\int_0^t f_{xx}(u, X_u)(\mathrm{d}X_u)^2, \end{aligned}$$

where

$$\frac{\partial f(t,x)}{\partial t} = f_t(t,x), \quad \frac{\partial f(t,x)}{\partial x} = f_x(t,x), \quad \frac{\partial^2 f(t,x)}{\partial x^2} = f_{xx}(t,x),$$

or, in differential form,

$$\mathrm{d}f(t,X_t) = f_t(t,X_t)\mathrm{d}t + f_x(t,X_t)\mathrm{d}X_t + \frac{1}{2}f_{xx}(t,X_t)(\mathrm{d}X_t)^2.$$

If X_t is the Brownian motion, this simplifies to the following

$$f(t,W_t) = f(0,0) + \int_0^t \left(f_t(u,W_u) + \frac{1}{2}f_{xx}(u,W_u)\right)\mathrm{d}u + \int_0^t f_x(u,W_u)\mathrm{d}W_u$$

or, in differential form,

$$\mathrm{d}f(t,W_t) = \left(f_t(t,W_t) + \frac{1}{2}f_{xx}(t,W_t)\right)\mathrm{d}t + f_x(t,W_t)\mathrm{d}W_t.$$

As an example, consider the function $f(t,x) = f(x) = x^2$. The Itô formula applied to $f(W_t)$ then leads to

$$W_t^2 = 0^2 + \int_0^t 2W_s\mathrm{d}W_s + \frac{1}{2}\int_0^t 2\mathrm{d}s,$$

and therefore

$$\int_0^t W_s\mathrm{d}W_s = \frac{1}{2}W_t^2 - \frac{1}{2}t$$

as already mentioned (see formula (1.14)).

1.11.1 Orders of differentials in the Itô formula

Despite the apparent simplicity of the Itô formula, terms such as $(\mathrm{d}X_t)^2$ are not easy to derive in many concrete cases without some additional knowledge about the stochastic integral. In particular, from the point of view of application of the Itô formula, one should keep in mind that $(\mathrm{d}t \cdot \mathrm{d}W_t)$ and $(\mathrm{d}t)^2$ are of order $O(\mathrm{d}t)$, which means that after developing the term $(\mathrm{d}X_t)^2$, all terms in the formula for which the differential part is either $(\mathrm{d}t \cdot \mathrm{d}W_t)$ or $(\mathrm{d}t)^2$ can be disregarded. Moreover, terms of order $(\mathrm{d}W_t)^2$ behave like $\mathrm{d}t$ for the properties of the Brownian motion. Hence, the differential part $(\mathrm{d}W_t)^2$ can be replaced everywhere simply by $\mathrm{d}t$.

1.11.2 Linear stochastic differential equations

Direct application of the Itô formula helps in finding out the solution of stochastic differential equations. Consider the stochastic differential equation

$$\mathrm{d}X_t = b_1(t)X_t\mathrm{d}t + \sigma_1(t)X_t\mathrm{d}W_t\,. \tag{1.29}$$

This equation is called a stochastic differential equation with *multiplicative noise*. Choosing $f(t,x) = \log x$, one obtains

$$X_t = X_0 \exp\left\{\int_0^t \left(b_1(s) - \frac{1}{2}\sigma_1(s)\right)\mathrm{d}s + \int_0^t \sigma_1(s)\mathrm{d}W_s\right\}. \tag{1.30}$$

Consider now the nonhomogeneous version of equation (1.29)

$$\mathrm{d}X_t = (b_1(t)X_t + b_2(t))\mathrm{d}t + (\sigma_1(t)X_t + \sigma_2(t))\mathrm{d}W_t, \tag{1.31}$$

and let Y_t be as in (1.30) with $Y_0 = 1$. Then the solution of (1.31) is

$$X_t = Y_t\left(X_0 + \int_0^t (b_2(s) - \sigma_1(s)\sigma_2(s))Y_s^{-1}\mathrm{d}s + \int_0^t \sigma_2(s)Y_s^{-1}\mathrm{d}W_s\right).$$

A simple derivation of this fact can be found in [163] Section 3.3.3. It is clear that geometric Brownian motion is a particular case of (1.29) with constant coefficients $b_1(t) = \mu$ and $\sigma_1(t) = \sigma$.

1.11.3 Derivation of the SDE for the geometric Brownian motion

It is now easy to derive the stochastic differential equation for the geometric Brownian motion,

$$S_t = S_0 \exp\left\{\left(r - \frac{\sigma^2}{2}\right)t + \sigma W_t\right\}, \quad t > 0, \tag{1.32}$$

by choosing

$$f(t,x) = S_0 \exp\left\{\left(r - \frac{\sigma^2}{2}\right)t + \sigma x\right\}.$$

Thus, $f(t, W_t) = S_t$ and

$$f_t(t,x) = \left(r - \frac{\sigma^2}{2}\right) f(t,x) \qquad f_x(t,x) = \sigma f(t,x) \qquad f_{xx}(t,x) = \sigma^2 f(t,x).$$

Hence

$$\begin{aligned} \mathrm{d}S_t &= \mathrm{d}f(t,W_t) \\ &= \left(f_t(t,W_t) + \frac{1}{2} f_{xx}(t,W_t)\right) \mathrm{d}t + f_x(t,W_t)\mathrm{d}W(t) \\ &= \left(\left(r - \frac{\sigma^2}{2}\right) S_t + \frac{1}{2}\sigma^2 S_t\right) \mathrm{d}t + \sigma S_t \mathrm{d}W_t \\ &= rS_t \mathrm{d}t + \sigma S_t \mathrm{d}W_t, \end{aligned} \tag{1.33}$$

which justifies (1.11).

1.11.4 The Lamperti transform

There is one particular application of the Itô formula that is of interest in many of the simulation and estimation methods we are going to describe in the next chapters (for example, this transform has been used in [85], [207], and [6]). Suppose we have the stochastic differential equation

$$\mathrm{d}X_t = b(t,X_t)\mathrm{d}t + \sigma(X_t)\mathrm{d}W_t,$$

where the diffusion coefficient depends only on the state variable. Such a stochastic differential equation can always be transformed into one with a unitary diffusion coefficient by applying the *Lamperti transform*,

$$Y_t = F(X_t) = \int_z^{X_t} \frac{1}{\sigma(u)} \mathrm{d}u\,. \tag{1.34}$$

Here z is any arbitrary value in the state space of X. Indeed, the process Y_t solves the stochastic differential equation

$$\mathrm{d}Y_t = b_Y(t,Y_t)\mathrm{d}t + \mathrm{d}W_t,$$

where

$$b_Y(t,y) = \frac{b(t,F^{-1}(y))}{\sigma(F^{-1}(y))} - \frac{1}{2}\sigma_x(F^{-1}(y)),$$

which we can also write as

$$\mathrm{d}Y_t = \left(\frac{b(t,X_t)}{\sigma(X_t)} - \frac{1}{2}\sigma_x(X_t)\right) \mathrm{d}t + \mathrm{d}W_t\,. \tag{1.35}$$

To obtain the result one, should use the Itô formula with

$$f(t,x) = \int_z^x \frac{1}{\sigma(u)}\mathrm{d}u, \quad f_t(t,x) = 0, \quad f_x(t,x) = \frac{1}{\sigma(x)}, \quad f_{xx}(t,x) = -\frac{\sigma_x(x)}{\sigma^2(x)}.$$

Therefore

$$
\begin{aligned}
\mathrm{d}f(t,x) &= 0\cdot \mathrm{d}t + f_x(t,X_t)\mathrm{d}X_t + \frac{1}{2} f_{xx}(t,X_t)(\mathrm{d}X_t)^2 \\
&= \frac{b(t,X_t)\mathrm{d}t + \sigma(X_t)\mathrm{d}W_t}{\sigma(X_t)} - \frac{1}{2}\frac{\sigma_x(t,X_t)}{\sigma^2(X_t)}(b(t,X_t)\mathrm{d}t + \sigma(X_t)\mathrm{d}W_t)^2 \\
&= \frac{b(t,X_t)}{\sigma(X_t)}\mathrm{d}t + \mathrm{d}W_t - \frac{1}{2}\frac{\sigma_x(t,X_t)}{\sigma^2(X_t)} \\
&\quad \times \left(b(t,X_t)(\mathrm{d}t)^2 + 2\cdot b(t,X_t)\sigma(X_t)\mathrm{d}t\cdot \mathrm{d}W_t + \sigma^2(x)(\mathrm{d}W_t)^2\right).
\end{aligned}
$$

Now recall that terms $(\mathrm{d}t)^2$ and $(\mathrm{d}t \cdot \mathrm{d}W_t)$ can be discarded and $(\mathrm{d}W_t)^2$ replaced by $\mathrm{d}t$. Hence the result is obtained.

1.12 Girsanov's theorem and likelihood ratio for diffusion processes

Girsanov's theorem is a change-of-measure theorem for stochastic processes. In inference for diffusion processes, this is used to obtain the likelihood ratio on which likelihood inference is based. There are different versions of this theorem, and here we will give one useful in statistics for diffusion processes [149]. Consider the three stochastic differential equations

$$
\begin{aligned}
\mathrm{d}X_t &= b_1(X_t)\mathrm{d}t + \sigma(X_t)\mathrm{d}W_t, && X_0^{(1)}, && 0\le t\le T,\\
\mathrm{d}X_t &= b_2(X_t)\mathrm{d}t + \sigma(X_t)\mathrm{d}W_t, && X_0^{(2)}, && 0\le t\le T,\\
\mathrm{d}X_t &= \sigma(X_t)\mathrm{d}W_t, && X_0, && 0\le t\le T,
\end{aligned}
$$

and denote by P_1, P_2, and P the three probability measures induced by the solutions of the three equations.

Fact 1.4 (see, e.g., [159] or [124]) *Assume that the coefficients satisfy Assumptions 1.1 and 1.2 or Assumption 1.4 or any other set of conditions that guarantee the existence of the solution of each stochastic differential equation. Assume further that the initial values are either random variables with densities $f_1(\cdot)$, $f_2(\cdot)$, and $f(\cdot)$ with the same common support or nonrandom and equal to the same constant. Then the three measures P_1, P_2, and P are all equivalent and the corresponding Radon-Nikodým derivatives are*

$$
\frac{\mathrm{d}P_1}{\mathrm{d}P}(X) = \frac{f_1(X_0)}{f(X_0)} \exp\left\{\int_0^T \frac{b_1(X_s)}{\sigma^2(X_s)}\mathrm{d}X_s - \frac{1}{2}\int_0^T \frac{b_1^2(X_s)}{\sigma^2(X_s)}\mathrm{d}s\right\} \tag{1.36}
$$

and

$$
\begin{aligned}
\frac{\mathrm{d}P_2}{\mathrm{d}P_1}(X) = \frac{f_2(X_0)}{f_1(X_0)} \exp\Bigg\{ &\int_0^T \frac{b_2(X_s) - b_1(X_s)}{\sigma^2(X_s)}\mathrm{d}X_s \\
&- \frac{1}{2}\int_0^T \frac{b_2^2(X_s) - b_1^2(X_s)}{\sigma^2(X_s)}\mathrm{d}s\Bigg\}.
\end{aligned} \tag{1.37}
$$

A version of (1.36) for parametric families $P_1 = P_1(\theta)$ with $b_1(x) = b_1(x;\theta)$ is usually adopted as the likelihood ratio to find maximum likelihood estimators. To understand why (1.37) (or (1.36)) can be assimilated to a likelihood, consider the following example inspired by the one by M. Keller-Ressel and T. Steiner.[10] Suppose we have a standard Brownian motion W_t on $[0,1]$ rescaled by a constant σ,

$$X_t = \sigma W_t,$$

and the same rescaled Brownian motion with drift

$$Y_t = \nu t + \sigma W_t, \quad 0 \le t \le 1,$$

with $\nu < 0$ and $\sigma > 0$. We apply Girsanov's formula (1.37) to weight the trajectories of the process Y_t, and in our case $b_1(x) = \nu$ for the process Y_t, $b_2(x) = 0$ for X_t, and $\sigma(x) = \sigma$. We have that

$$\begin{aligned}
\frac{\mathrm{d}P_2}{\mathrm{d}P_1}(Y) &= \exp\left\{\int_0^1 \frac{0-\nu}{\sigma^2}\mathrm{d}Y_s - \frac{1}{2}\int_0^1 \frac{0^2-\nu^2}{\sigma^2}\mathrm{d}t\right\} \\
&= \exp\left\{\frac{-\nu}{\sigma^2}\int_0^1 (\nu\mathrm{d}s + \sigma\mathrm{d}W_s) + \frac{1}{2}\frac{\nu^2}{\sigma^2}\right\} \\
&= \exp\left\{-\left(\frac{\nu}{\sigma}\right)^2 + \frac{-\nu}{\sigma}W_1 + \frac{1}{2}\frac{\nu^2}{\sigma^2}\right\} \\
&= \exp\left\{-\frac{\nu W_1}{\sigma} - \frac{1}{2}\left(\frac{\nu}{\sigma}\right)^2\right\}.
\end{aligned}$$

The previous weights depend on the final value of W_1. Of course, the lower and more negative W_1 is, the higher is the likelihood that an observed trajectory Y really comes from the model Y_t because $\nu < 0$. The following code simulates 30 paths of X_t and Y_t with $\nu = -0.7$ and $\sigma = 0.5$ and calculates the weights using Girsanov's formula. Finally, it plots the paths of Y_t and the same paths with a color proportional to the weights: the darker trajectories are more likely to come from the true model Y_t. The result is shown in Figure 1.9.

```
> # ex1.14.R
> set.seed(123)
> par("mar"=c(3,2,1,1))
> par(mfrow=c(2,1))
> npaths <- 30
> N <- 1000
> sigma <- 0.5
> nu <- -0.7
> X <- sde.sim(drift=expression(0),sigma=expression(0.5),
+            pred=F, N=N,M=npaths)
> Y <- X + nu*time(X)
> girsanov <- exp(0.25 * (-nu/sigma*X[N,] - 0.5*(nu/sigma)^2))
> girsanov <- (girsanov - min(girsanov)) / diff(range(girsanov))
> col.girsanov <- gray(girsanov)
> matplot(time(X),Y,type="l",lty=1, col="black",xlab="t")
> matplot(time(X),Y,type="l",lty=1,col=col.girsanov,xlab="t")
```

In the preceding R code, the weights are rescaled a bit to enhance visual contrast and are further normalized to one.

[10] See Wikipedia at http://commons.wikimedia.org/wiki/Image:Girsanov.png.

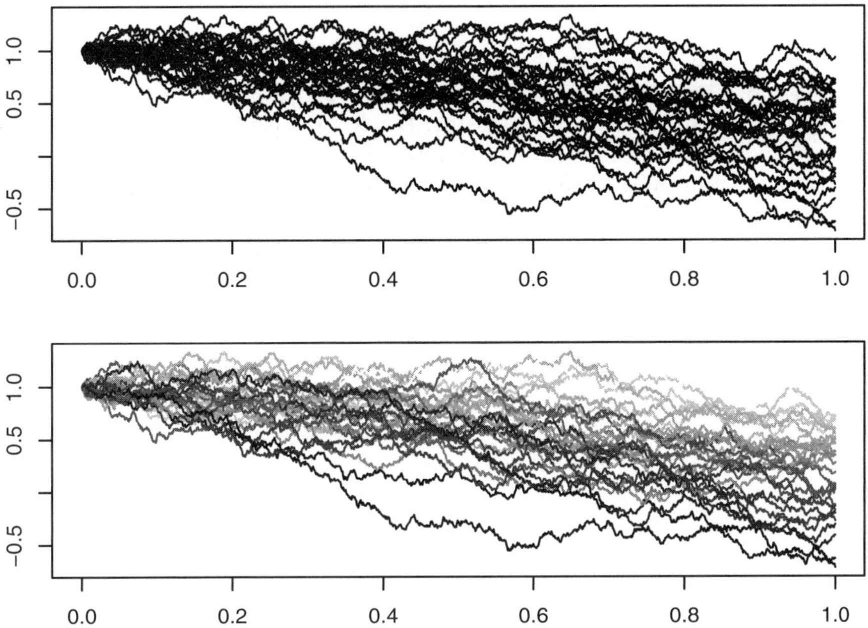

Fig. 1.9. Simulated paths of Brownian motion with drift (top panel) and the same paths (bottom panel) colored using Girsanov's weights. The darker trajectories are more likely to come from the model of Brownian motion with drift.

1.13 Some parametric families of stochastic processes

In this section, we present some of the well-known and widely used stochastic process solutions to the general stochastic differential equation

$$\mathrm{d}X_t = b(X_t)\mathrm{d}t + \sigma(X_t)\mathrm{d}W_t \tag{1.38}$$

with a quick review of their main properties. When possible, we will describe each process in terms of its first moments and covariance function, in terms of the stationary density $\pi(x)$ (1.23), and in terms of its conditional distribution ($p(t-s, y|x; \theta)$ or $p_\theta(t-s, y|x)$) of X_t given some previous state of the process $X_s = x_s$. In some cases, it will be simpler to express the stationary density $\pi(x)$ of a diffusion in terms of the scale measure $s(\cdot)$ (1.21) and the speed measure $m(\cdot)$ (1.22) of the diffusion.

1.13.1 Ornstein-Uhlenbeck or Vasicek process

The Ornstein-Uhlenbeck [224] or Vasicek [225] process is the unique solution to the following stochastic differential equation

$$dX_t = (\theta_1 - \theta_2 X_t)dt + \theta_3 dW_t, \qquad X_0 = x_0, \tag{1.39}$$

with $\theta_3 \in \mathbb{R}_+$ and $\theta_1, \theta_2 \in \mathbb{R}$. The model ($\theta_1 = 0$) was originally proposed by Ornstein and Uhlenbeck [224] in the context of physics and then generalized by Vasicek [225] to model interest rates. For $\theta_2 > 0$, this is a *mean reverting* process, which means that the process tends to oscillate around some equilibrium state. Another interesting property of the Ornstein-Uhlenbeck process is that, contrary to the Brownian motion, it is a process with finite variance for all $t \geq 0$. Another parametrization of the Ornstein-Uhlenbeck process more common in finance modeling is

$$dX_t = \theta(\mu - X_t)dt + \sigma dW_t, \qquad X_0 = x_0, \tag{1.40}$$

where σ is interpreted as the volatility, μ is the long-run equilibrium value of the process, and θ is the speed of reversion. As an application of the Itô lemma, we can show the explicit solution of (1.39) by choosing $f(t, x) = xe^{\theta_2 t}$. Indeed,

$$f_t(t,x) = \theta_2 f(t,x), \qquad f_x(t,x) = e^{\theta_2 t}, \qquad f_{xx}(t,x) = 0\,.$$

Therefore,

$$\begin{aligned} X_t e^{\theta_2 t} &= f(t, X_t) = f(0, X_0) + \int_0^t \theta_2 X_u e^{\theta_2 u} du + \int_0^t e^{\theta_2 u} dX_u \\ &= x_0 + \int_0^t \theta_2 X_u e^{\theta_2 u} du + \int_0^t e^{\theta_2 u} \{(\theta_1 - \theta_2 X_u)du + \theta_3 dW_u\} \\ &= x_0 + \frac{\theta_1}{\theta_2}\left(e^{\theta_2 t} - 1\right) + \theta_3 \int_0^t e^{\theta_2 u} dW_u, \end{aligned}$$

from which we obtain

$$X_t = \frac{\theta_1}{\theta_2} + \left(x_0 - \frac{\theta_1}{\theta_2}\right) e^{-\theta_2 t} + \theta_3 \int_0^t e^{-\theta_2 (t-u)} dW_u$$

or, in the second parametrization,

$$X_t = \mu + (x_0 - \mu)e^{-\theta t} + \sigma \int_0^t e^{-\theta(t-u)} dW_u.$$

The invariant law

For $\theta_2 > 0$, the process is also ergodic, and its invariant law is the Gaussian density with mean θ_1/θ_2 and variance $\theta_3^2/2\theta_2$,

$$X_t \sim N\left(\frac{\theta_1}{\theta_2}, \frac{\theta_3^2}{2\theta_2}\right).$$

The covariance function is $\text{Cov}(X_t, X_s) = \theta_3^2 e^{-\theta_2|t-s|}/2\theta_2$. Sometimes it is convenient to rewrite the process as the scaled time-transformed Wiener process

$$X_t = \frac{\theta_1}{\theta_2} + \frac{\theta_3 e^{-\theta_2 t}}{\sqrt{2\theta_2}} W\left(e^{2\theta_2 t}\right).$$

This relationship comes directly from the properties of the Brownian motion.

The conditional law

For any $t \geq 0$, the Ornstein-Uhlenbeck process has a Gaussian transition (or conditional) density $p_\theta(t, X_t|X_0 = x_0)$; i.e., the density of the distribution of X_t given $X_0 = x_0$, with mean and variance respectively

$$m(t,x) = \mathbb{E}_\theta(X_t|X_0 = x_0) = \frac{\theta_1}{\theta_2} + \left(x_0 - \frac{\theta_1}{\theta_2}\right) e^{-\theta_2 t} \tag{1.41}$$

and

$$v(t,x) = \text{Var}_\theta(X_t|X_0 = x_0) = \frac{\theta_3^2\left(1 - e^{-2\theta_2 t}\right)}{2\theta_2}. \tag{1.42}$$

The conditional covariance function is

$$\text{Cov}(X_s, X_t|X_0 = x_0) = \frac{\theta_3^2}{2\theta_2} e^{-\theta_2(s+t)} \left(e^{2\theta_2(s\wedge t)} - 1\right),$$

and its scaled time-transformed Wiener representation is

$$X_t = \frac{\theta_1}{\theta_2} + \left(x_0 - \frac{\theta_1}{\theta_2}\right) e^{-\theta_2 t} + \frac{\theta_3}{\sqrt{2\theta_2}} W\left(e^{2\theta_2 t} - 1\right) e^{-\theta_2 t}.$$

Example of simulation of the stochastic integral

It can be seen that for $\theta_1 = 0$ the trajectory of X_t is essentially a negative exponential perturbed by the stochastic integral. One way of simulating trajectories of the Ornstein-Uhlenbeck process is indeed via the simulation of the stochastic integral. As promised in Section 1.9, we now show how to simulate a stochastic integral, but the next chapter will cover other different (and preferred) ways to simulate paths from the Ornstein-Uhlenbeck process and others. The following script, given without further comments, essentially simulates Itô sums. The result is shown in Figure 1.10.

```
> # ex1.15.R
> W <- BM()
> t <- time(W)
> N <- length(t)
> x <- 10
> theta <- 5
> sigma <- 3.5
> X <- numeric(N)
> X[1] <- x
> ito.sum <- c(0, sapply(2:N, function(x) {
+     exp(-theta*(t[x]-t[x-1])) * (W[x]-W[x-1])} ) )
> X <- sapply(1:N, function(x)  {X[1]*exp(-theta*t[x]) +
+      sum(ito.sum[1:x])} )
> X <- ts(X,start=start(W), deltat=deltat(W))
> plot(X,main="Ornstein-Uhlenbeck process")
```

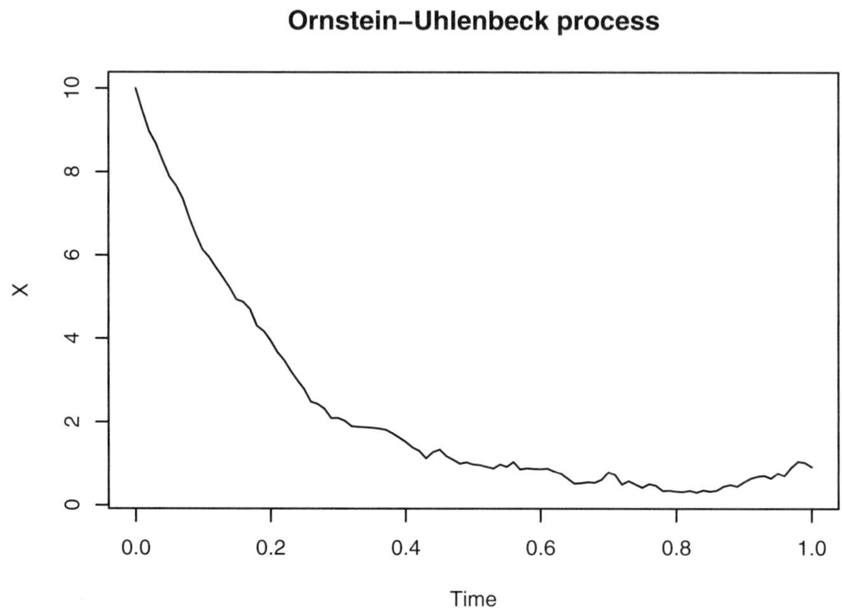

Fig. 1.10. Simulated path of the Ornstein-Uhlenbeck process $\mathrm{d}X_t = -\theta X_t + \sigma \mathrm{d}W_t$ with $X(0) = 10$, $\theta = 5$, and $\sigma = 3.5$.

1.13.2 The Black-Scholes-Merton or geometric Brownian motion model

We have already presented the derivation of the geometric Brownian motion in Section 1.7, so here we just recall the basic facts. This process is sometimes called the Black-Scholes-Merton model after its introduction in the finance context to model asset prices (see [36], [162]). The process is the solution to the stochastic differential equation

$$\mathrm{d}X_t = \theta_1 X_t \mathrm{d}t + \theta_2 X_t \mathrm{d}W_t, \quad X_0 = x_0,$$

with $\theta_2 > 0$. The parameter θ_1 is interpreted as the constant interest rate and θ_2 as the volatility of risky activities. The explicit solution is

$$X_t = x_0 e^{(\theta_1 - \frac{1}{2}\theta_2^2)t + \theta_2 W_t}.$$

The conditional density function is log-normal with the mean and variance of its logarithm transform (i.e., the log-mean and log-variance) given by

$$\mu = \log(x_0) + \left(\theta_1 - \frac{1}{2}\theta_2^2\right) t, \qquad \sigma^2 = \theta_2^2 t, \tag{1.43}$$

with mean and variance

$$m(t,x_0) = e^{\mu+\frac{1}{2}\sigma^2} = x_0 e^{\theta_1 t}, \tag{1.44}$$

$$v(t,x_0) = e^{2\mu+\sigma^2}\left(e^{\sigma^2}-1\right) = x_0^2 e^{2\theta_1 t}\left(e^{\theta_2^2 t}-1\right). \tag{1.45}$$

Hence

$$\begin{aligned} p_\theta(t,y|x_0) &= \frac{1}{y\sigma\sqrt{2\pi}}\exp\left\{-\frac{(\log y-\mu)^2}{2\sigma^2}\right\} \\ &= \frac{1}{y\theta_2\sqrt{2\pi t}}\exp\left\{-\frac{(\log y-(\log x_0+(\theta_1-\frac{1}{2}\theta_2^2)t))^2}{2\theta_2^2 t}\right\}. \end{aligned}$$

1.13.3 The Cox-Ingersoll-Ross model

Another interesting family of parametric models is that of the Cox-Ingersoll-Ross process. This model was introduced by Feller as a model for population growth (see [82] and [83]) and became quite popular in finance after Cox, Ingersoll, and Ross proposed it to model short-term interest rates [59]. It was recently adopted to model nitrous oxide emission from soil by Pedersen in [181] and to model the evolutionary rate variation across sites in molecular evolution (see [157]). The CIR process is the solution to the stochastic differential equation

$$\mathrm{d}X_t = (\theta_1 - \theta_2 X_t)\mathrm{d}t + \theta_3\sqrt{X_t}\mathrm{d}W_t, \quad X_0 = x_0 > 0\,, \tag{1.46}$$

sometimes parametrized as

$$\mathrm{d}X_t = \theta(\beta - X_t)\mathrm{d}t + \sigma\sqrt{X_t}\mathrm{d}W_t, \quad X_0 = x_0 > 0\,, \tag{1.47}$$

where $\theta_1, \theta_2, \theta_3 \in \mathbb{R}_+$. If $2\theta_1 > \theta_3^2$, the process is strictly positive otherwise it is nonnegative, which means that it can reach the state 0. Of course, the last case is not admitted in finance when the CIR process is used to model interest rates. The stochastic differential equation (1.46) has the explicit solution

$$X_t = \left(X_0 - \frac{\theta_1}{\theta_2}\right)e^{-\theta_2 t} + \theta_3 e^{-\theta_2 t}\int_0^t e^{\theta_2 u}\sqrt{X_u}\mathrm{d}W_u\,.$$

The conditional distribution

Under the hypothesis $2\theta_1 > \theta_3^2$, the process is also stationary. Moreover, the conditional transition density exists in explicit form. The process $Y_t = 2cX_t$, with $c = 2\theta_2/(\theta_3^2(1-e^{-\theta_2 t}))$, has a conditional distribution $Y_t|Y_0 = y_0$, which follows the law of the noncentral chi-squared distribution with $\nu = 4\theta_1/\theta_3^2$ degrees of freedom and noncentrality parameter $y_0 e^{-\theta_2 t}$. The transition density of the original process (i.e., the distribution of $X_t|X_0 = x_0$) can be easily obtained from the distribution of $Y_t|Y_0 = y_0$, and the result is

$$p_\theta(t,y|x_0) = ce^{-(u+v)}\left(\frac{u}{v}\right)^{q/2} I_q(2\sqrt{uv})\,, \tag{1.48}$$

with $u = cx_0e^{-\theta_2 t}$, $v = cy$, $q = 2\theta_1/\theta_3^2 - 1$, where $I_q(\cdot)$ is the modified Bessel function of the first kind of order q (see, e.g., [1])

$$I_q(x) = \sum_{k=0}^{\infty}\left(\frac{x}{2}\right)^{2k+q}\frac{1}{k!\Gamma(k+q+1)}, \qquad x \in \mathbb{R},$$

and $\Gamma(\cdot)$ is the gamma function (i.e., $\Gamma(z) = \int_0^\infty x^{z-1}e^{-x}\mathrm{d}x$, $z \in \mathbb{R}_+$). These derivations can be found, for example, in [82].

Behavior of the conditional distribution

From the point of view of likelihood inference, it should be noted that even if the non-central chi-squared distribution is a special yet well-known probability distribution, its actual computation is not an easy task. Indeed, as $t \to 0$, both u and v explode, which means that when the distribution has to be calculated for very large arguments, it implies the evaluation of the chi-squared distribution in the right tail. The contrary is also true. If $t \to \infty$, $u \to 0$, and $u \to 2\theta_2 x/\theta_3^2$, when the left tail of the distribution is interested in the calculations (see, e.g., [72]). Being that the chi-squared distribution is based on the modified Bessel function of the first kind $I_q(\cdot)$, the problem is essentially in the evaluation of $I_q(\cdot)$ for very small and large arguments. We will discuss the numerical implementation of the CIR density later in Section 2.3.2.

For the Cox-Ingersoll-Ross process, the mean of the conditional density is that of the Ornstein-Uhlenbeck process

$$m(t,x_0) = \frac{\theta_1}{\theta_2} + \left(x_0 - \frac{\theta_1}{\theta_2}\right)e^{-\theta_2 t}, \tag{1.49}$$

and the variance is

$$v(t,x_0) = x_0\frac{\theta_3^2(e^{-\theta_2 t} - e^{-2\theta_2 t})}{\theta_2} + \frac{\theta_1\theta_3^2(1-e^{-2\theta_2 t})}{2\theta_2^2}\,. \tag{1.50}$$

The covariance function is given by

$$\mathrm{Cov}(X_s,X_t) = x_0\frac{\theta_3^2}{\theta_2}\left(e^{-\theta_2 t} - e^{-\theta_2(s+t)}\right) + \frac{\theta_1\theta_3^2}{2\theta_2^2}\left(e^{-\theta_2(t-s)} - e^{-\theta_2(t+s)}\right).$$

The stationary law

The stationary distribution of the CIR process is a Gamma law with shape parameter $2\theta_1/\theta_3^2$ and scale parameter $\theta_3^2/2\theta_2$. Hence the stationary law has mean equal to $\frac{\theta_1}{\theta_2}$ and variance $\theta_1\theta_3^2/(2\theta_2^2)$. The covariance function in the stationary case is given by

$$\mathrm{Cov}(X_s,X_t) = \frac{\theta_1\theta_3^2}{2\theta_2^2}e^{-\theta_2(t-s)}\,.$$

1.13.4 The CKLS family of models

The Chan-Karolyi-Longstaff-Sanders (CKLS) family of models (see [49]) is a class of parametric stochastic differential equations widely used in many finance applications, in particular to model interest rates or asset prices. The CKLS process solves the stochastic differential equation

$$\mathrm{d}X_t = (\theta_1 + \theta_2 X_t)\mathrm{d}t + \theta_3 X_t^{\theta_4}\mathrm{d}W_t\,. \tag{1.51}$$

This CKLS model is a further extension of the Cox-Ingersoll-Ross model and hence embeds all previous models. Table 1.4 presents some cases. The CKLS model does not admit an explicit transition density unless $\theta_1 = 0$ or $\theta_4 = \frac{1}{2}$. It takes values in $(0, +\infty)$ if $\theta_1, \theta_2 > 0$, and $\theta_4 > \frac{1}{2}$. In all cases, θ_3 is assumed to be positive.

1.13.5 The modified CIR and hyperbolic processes

Another example of an application of the Lamperti transform (see, e.g., [56]) is the *modified* Cox-Ingersoll-Ross process solution to

$$\mathrm{d}X_t = -\theta_1 X_t \mathrm{d}t + \theta_2\sqrt{1 + X_t^2}\mathrm{d}W_t \tag{1.52}$$

with $\theta_1 + \theta_2^2/2 > 0$ (this is needed to make the process positive recurrent). This process has a stationary distribution whose density $\pi(\cdot)$ is proportional to

$$\pi(x) \propto \frac{1}{(1+x^2)^{1+\theta_1/\theta_2^2}}\,.$$

Setting $\nu = 1 + 2\theta_1/\theta_2^2$, then $X_t \sim t(\nu)/\sqrt{\nu}$, where t is the Student t distribution with ν degrees of freedom. Applying the Lamperti transform

$$F(x) = \int_0^x \frac{1}{\theta_2\sqrt{1+u^2}}\mathrm{d}u = \frac{\operatorname{arcsinh}(x)}{\theta_2} = \frac{1}{\theta_2}\log\left(x + \sqrt{1+x^2}\right)$$

Table 1.4. The family of CKLS process and its embedded elements under different parametric specifications. In all cases, $\theta_3 > 0$.

	θ_1	θ_2	θ_4	mean reverting	see
Merton	any	0	0	no	[162]
Vasicek or Ornstein-Uhlenbeck	any	any	0	yes	[225]
CIR or Square Root process	any	any	1/2	yes	[59]
Dothan	0	0	1	no	[71]
Geometric BM or B&S	0	any	1	yes	[36]
Brennan and Schwartz	any	any	1	yes	[42]
CIR VR	0	0	3/2	no	[58]
CEV	0	any	any	yes	[57]

to (1.52), we have that $Y_t = F(X_t)$ satisfies the stochastic differential equation

$$\mathrm{d}F(X_t) = -(\theta_1/\theta_2 + \theta_2/2)\frac{X_t}{\sqrt{1+X_t^2}}\mathrm{d}t + \mathrm{d}W_t, \tag{1.53}$$

which we can rewrite in terms of the Y_t process as

$$\mathrm{d}Y_t = -(\theta_1/\theta_2 + \theta_2/2)\tanh(\theta_2 Y_t) + \mathrm{d}W_t \tag{1.54}$$

with $Y_0 = F(X_0)$.

1.13.6 The hyperbolic processes

A process X satisfying

$$\mathrm{d}X_t = -\frac{\theta X_t}{\sqrt{1+X_t^2}}\mathrm{d}t + \mathrm{d}W_t \tag{1.55}$$

with $\theta > 0$ is called the *hyperbolic process*. It is a special case of the general hyperbolic diffusion introduced in [21] and solution of

$$\mathrm{d}X_t = \frac{\sigma^2}{2}\left[\beta - \gamma\frac{X_t}{\sqrt{\delta^2 + (X_t-\mu)^2}}\right]\mathrm{d}t + \sigma \mathrm{d}W_t.$$

Its invariant density is

$$\pi(x) = \frac{\sqrt{\gamma^2-\beta^2}}{2\gamma\delta K_1(\delta\sqrt{\gamma^2-\beta^2})}\exp\left\{-\gamma\sqrt{\delta^2+(x-\mu)^2}+\beta(x-\mu)\right\},$$

where $K_\nu(x)$ is the modified Bessel function of second kind of index ν (see, e.g., [1]). The parameters $\gamma > 0$ and $0 \le |\beta| < \gamma$ determine the shape of the distribution, and $\delta \ge 0$, and μ are, respectively, the scale and location parameters of the distribution. This process was used to model log-returns of assets prices in stock markets in [77] and [32]. A further generalization of the hyperbolic process (1.55) is the solution to the stochastic differential equation

$$\mathrm{d}X_t = -\frac{\alpha X_t}{\sqrt{1+X_t^2}}\mathrm{d}t + \frac{\beta + X_t^2}{1+X_t^2}\mathrm{d}W_t \tag{1.56}$$

with $X_0 = 0$, $\alpha > 0$, and $\beta \ge 1$. This process, introduced in [161], is also ergodic.

1.13.7 The nonlinear mean reversion Aït-Sahalia model

This model satisfies the nonlinear stochastic differential equation

$$\mathrm{d}X_t = (\alpha_{-1}X_t^{-1} + \alpha_0 + \alpha_1 X_t + \alpha_2 X_t^2)\mathrm{d}t + \beta_1 X_t^\rho \mathrm{d}W_t\,. \tag{1.57}$$

In general, there are no exact distributional results although, as explained in Chapter 3, approximate transition densities can be obtained via Hermite polynomial expansion. This model was proposed by Aït-Sahalia to model interest rates in [7], [8]. The same author [4] proposed a further generalization that includes more structure in the diffusion coefficient. The second model is of the form

$$dX_t = (\alpha_{-1}X_t^{-1} + \alpha_0 + \alpha_1 X_t + \alpha_2 X_t^2)dt + \sqrt{\beta_0 + \beta_1 X_t + \beta_2 X_t^{\beta_3}}dW_t\,.$$

Some natural constraints on the parameters are needed in order to have a meaningful specification of the model. Moreover, Markovianity is granted only under additional constraints. The next table summarizes relations between coefficients. For the proof of these facts, see the Appendix in [4].

condition	when
$\beta_0 \geq 0$	*always*
$\beta_1 > 0$	$\beta_0 = 0$ and $\beta_3 > 1$
$\beta_1 > 0$	$0 < \beta_3 < 1$ or $\beta_2 = 0$
$\beta_2 > 0$	$\beta_0 = 0$ and $0 < \beta_3 < 1$
$\beta_2 > 0$	$\beta_3 > 1$ or $\beta_1 = 0$
$\beta_3 > 1$	$\beta_2 > 0$
$\alpha_2 \leq 0$ and $\alpha_1 < 0$	$\alpha_2 = 0$, $\alpha_{-1} > 0$, and $2\alpha_{-1} \geq \beta_0 \geq 0$
$\alpha_2 \leq 0$ and $\alpha_1 < 0$	$\alpha_3 = 0$, $\alpha_0 > 0$, $\beta_0 = 0$, $\beta_3 > 1$, and $2\alpha_0 \geq \beta_1 > 0$

It is clear that the CKLS model is just a particular case of the Aït-Sahalia model.

1.13.8 Double-well potential

This model is interesting because of the fact that its density has a bimodal shape. The process satisfies the stochastic differential equation

$$dX_t = (X_t - X_t^3)dt + dW_t\,.$$

This model is challenging in the sense that the standard Euler approximation that we discuss in Section 2.1 could not be expected to work due to the high nonlinearity of the stochastic differential equation and high non-Gaussianity of its finite-dimensional distributions.

1.13.9 The Jacobi diffusion process

The Jacobi diffusion process is the solution to the stochastic differential equation

$$dX_t = -\theta\left(X_t - \frac{1}{2}\right)dt + \sqrt{\theta X_t(1-X_t)}dW_t \qquad (1.58)$$

for $\theta > 0$. It has an invariant distribution that is uniform on $(0,1)$. The peculiar thing is that, given any twice differentiable distribution function F, the transformed diffusion $Y_t = F^{-1}(X_t)$ has an invariant density $\pi(\cdot)$ that is the density of F; i.e., $\pi = F'$.

1.13.10 Ahn and Gao model or inverse of Feller's square root model

A further generalization in the direction of nonlinear and mean reversion is the model proposed in [2], which is a suitable transformation of the Cox-Ingersoll-Ross model. The process is the solution to the stochastic differential equation

$$dX_t = X_t(\theta_1 - (\theta_3^3 - \theta_1\theta_2)X_t)dt + \theta_3 X_t^{\frac{3}{2}} dW_t\,.$$

The conditional distribution of this process is related to that of the Cox-Ingersoll-Ross model as

$$p_\theta(t, y|x_0) = \frac{1}{y^2} p_\theta^{CIR}\left(t, \frac{1}{y}\middle|\frac{1}{x_0}\right),$$

where p_θ^{CIR} is the conditional density (1.48).

1.13.11 Radial Ornstein-Uhlenbeck process

The radial Ornstein-Uhlenbeck process is the solution to the stochastic differential equation

$$dX_t = (\theta X_t^{-1} - X_t)dt + dW_t$$

for $\theta > 0$. This model is a special case of (1.57) but still interesting because some distributional results are known. In particular, the conditional distribution has the explicit form

$$p_\theta(t, y|x_0) = \frac{\left(\frac{y}{x}\right)^\theta \sqrt{xy}e^{-y^2+(\theta+\frac{1}{2})t}}{\sinh(t)} \exp\left\{-\frac{x^2+y^2}{e^{2t}-1}\right\} I_{\theta-\frac{1}{2}}\left(\frac{xy}{\sinh(t)}\right),$$

where I_ν is the modified Bessel function of order ν.

1.13.12 Pearson diffusions

A class that further generalizes the Ornstein-Uhlenbeck and Cox-Ingersoll-Ross processes is the class of *Pearson diffusion*. This class of models and their statistical properties have been analyzed in [89], and its name is due to the fact that, when a stationary solution exists for this model, its invariant density belongs to the Pearson system [177]. The Pearson system allows for a big variety of distributions which can take positive and/or negative values, and can be bounded, symmetric or skewed, and heavy or light tailed. In particular, the type IV Pearson distribution is a kind of skewed t distribution that seems to fit financial time series particularly well. Pearson diffusions have also been studied just from the probabilistic point of view in [34] and [230]. Using the parametrization in [89], these diffusions solve the stochastic differential equation

$$\mathrm{d}X_t = -\theta(X_t - \mu)\mathrm{d}t + \sqrt{2\theta(aX_t^2 + bX_t + c)}\mathrm{d}W_t \qquad (1.59)$$

with $\theta > 0$ and a, b, and c such that the diffusion coefficient is well-defined (i.e., the square root can be extracted) for all the values of the state space of X_t. Pearson diffusions are characterized as having a mean reverting linear drift and a squared diffusion coefficient that is a second-order polynomial of the state of the process. A further nice property of these models is that they are closed under translation and scale transformations: if X_t is an ergodic Pearson diffusion then also $\tilde{X}_t = \gamma X_t + \delta$ is a Pearson diffusion satisfying the stochastic differential equation (1.59) with parameters $\tilde{a} = a$, $\tilde{b} = b\gamma - 2a\delta$, $\tilde{c} = c\gamma^2 - b\gamma\delta + a\delta^2$, $\tilde{\theta} = \theta$ and $\tilde{\mu} = \gamma\mu + \delta$. The parameter γ may also be negative, and in that case the state space of $\tilde{X}_t$ will change its sign. The scale and the speed measures of these processes have the forms

$$s(x) = \exp\left\{\int_{x_0}^{x} \frac{u - \mu}{au^2 + bu + c}\mathrm{d}u\right\}$$

and

$$m(x) = \frac{1}{2\theta s(x)(ax^2 + bx + c)},$$

where x_0 is some value such that $ax_0^2 + bx_0 + c > 0$. Let (l, r) be an interval such that for $x \in (l, r)$ we have $ax^2 + bx + c > 0$. As seen before, a unique ergodic invariant distribution with values in (l, r) exists only if $\int_{x_0}^{r} s(x)\mathrm{d}x = \infty$, $\int_{l}^{x_0} s(x)\mathrm{d}x = \infty$, and $M = \int_l^r m(x)\mathrm{d}x < \infty$. In this case, the invariant distribution has density $m(x)/M$. The relation between this and the Pearson family of distributions is that $m(x)$ solves the following differential equation that characterizes the system of Pearson distributions:

$$\frac{\mathrm{d}m(x)}{\mathrm{d}x} = -\frac{(2a+1)x - \mu + b}{ax^2 + bx + c} m(x)\,.$$

Some interesting cases are included in this family.[11] In particular:

1. When the diffusion coefficient $\sigma^2(x) = 2\theta$, we recover the Ornstein-Uhlenbeck process.
2. For $\sigma^2(x) = 2\theta x$ and $0 < \mu \leq 1$, we obtain the Cox-Ingersoll-Ross process (under some additional reflecting conditions), and if $\mu > 1$ the invariant distribution is a Gamma law with scale parameter 1 and shape parameter μ.
3. For $a > 0$ and $\sigma^2(x) = 2\theta a(x^2 + 1)$, the invariant distribution always exists on the real line, and for $\mu = 0$ the invariant distribution is a scaled t distribution with $\nu = 1 + a^{-1}$ degrees of freedom and scale parameter $\nu^{-\frac{1}{2}}$, while for $\mu \neq 0$ the distribution is a form of skewed t distribution that is called Pearson type IV distribution.

[11] For an extensive description of all cases, refer to [89].

4. For $a > 0$, $\mu > 0$, and $\sigma^2(x) = 2\theta a x^2$, the distribution is defined on the positive half line and it is an inverse Gamma distribution with shape parameter $1 + a^{-1}$ and scale parameter a/μ.
5. For $a > 0$, $\mu \geq a$, and $\sigma^2(x) = 2\theta a x(x+1)$, the invariant distribution is the scaled F distribution with $2\mu/a$ and $2/a + 2$ degrees of freedom and scale parameter $\mu/(1+a)$. For $0 < \mu < 1$, some reflecting conditions on the boundaries are also needed.
6. If $a < 0$ and $\mu > 0$ are such that $\min(\mu, 1-\mu) \geq -a$ and $\sigma^2(x) = 2\theta a x(x-1)$, the invariant distribution exists on the interval $(0,1)$ and is a Beta distribution with parameters $-\mu/a$ and $(\mu-1)/a$. The Jacobi diffusion solution to (1.58) is a particular case of this class of Pearson diffusion, which is in fact called the general Jacobi diffusion. If $F(x) = e^x/(1+e^x)$ is the logistic distribution, the transformed process $Y_t = F^{-1}(X_t)$ with $F^{-1}(y) = \log(y/(1-y))$ has an invariant density, which is the generalized logistic distribution [22],
$$\pi(x) = \frac{e^{\alpha x}}{(1+e^x)^{\alpha+\beta} B(\alpha,\beta)},$$
with $\alpha = (\mu-1)/a$, $\beta = \mu/a$, and $B(\cdot,\cdot)$ the Beta function.

When the second moment of the distribution exists, then the autocorrelation function takes the form
$$\mathrm{Cor}(X_s, X_{s+t}) = e^{-\theta t}.$$
Moments $\mathbb{E}|X_t|^k$ exist if and only if $a < (k-1)^{-1}$. Hence, if $a \leq 0$, all the moments exist, while for $a > 0$ only the moments up to order $k < a^{-1} + 1$ exist. If $\mathbb{E}(X_t^{2n}) < \infty$, then the moments satisfy the recurrent formula
$$\mathbb{E}(X_t^n) = a_n^{-1}\left\{b_n \mathbb{E}(X_n^{n-1}) + c_n \mathbb{E}(X_t^{n-2})\right\},$$
where $a_n = n(1-(n-1)a)\theta$, $b_n = n(\mu+(n-1)b)\theta$, $c_n = n(n-1)c\theta$ for $n = 0, 1, 2, \ldots$ with initial conditions $\mathbb{E}(X_t^0) = 1$ and $\mathbb{E}(X_t) = \mu$ (see again [89]). Finally, for these processes, it is possible to derive the eigenfunction of the infinitesimal operator of the diffusions that are useful to construct estimating functions, as will be explained in Section 3.5.6.

1.13.13 Another classification of linear stochastic systems

In the social sciences, the quantity of interest is often the stationary distribution of the stochastic process that describes the equilibrium of some dynamics. In most of the cases (see, e.g., [23], [54]), the diffusion coefficient is chosen to model some kind of variability related to the state of the process. For these disciplines, the interest is in the shape of the stationary distributions and the statistical indexes related to them (mean, mode, etc.). The stochastic differential equations used to model social processes [55] usually have a linear drift of the form $b(x) = r(\theta - x)$ with $r > 0$. For this reason, the models are called *linear feedback models.* The diffusion coefficient $\sigma^2(x)$ may be constant, linearly depending on x, or of polynomial type, leading, respectively, to Gaussian, Gamma, or Beta stationary distributions.[12] Of course, all of these models fall into one or more of the previous classes, but we collect some of their properties in the following.

Type N model

For this model, we have:

drift coefficient: $b(x) = r(\theta - x)$, $r > 0$.

diffusion coefficient: $\sigma^2(x) = \epsilon$, $\epsilon > 0$.

s.d.e.: $\mathrm{d}X_t = r(\theta - X_t)\mathrm{d}t + \sqrt{\epsilon}\mathrm{d}W_t$.

stationary density: $\pi(x) = \dfrac{1}{2\pi\delta}\exp\left\{-\dfrac{(x-\theta)^2}{2\delta}\right\}$, $\delta = \dfrac{\epsilon}{r}$.

statistics: mean, mode $= \theta$, variance $= \delta$.

Type G model

For this model, we have:

drift coefficient: $b(x) = r(\theta - x)$, $r > 0$.

diffusion coefficient: $\sigma^2(x) = \epsilon x$, $\epsilon > 0$.

s.d.e.: $\mathrm{d}X_t = r(\theta - X_t)\mathrm{d}t + \sqrt{\epsilon X_t}\mathrm{d}W_t$.

stationary density: $\pi(x) = \left(\dfrac{x}{\delta}\right)^{-1+\frac{\theta}{\delta}} \dfrac{e^{-\frac{x}{\delta}}}{\Gamma\left(\frac{\theta}{\delta}\right)}$, $\delta = \dfrac{\epsilon}{r}$.

statistics: mean $= \theta$, mode $= \theta - \delta$, variance $= \delta\theta$.

Type B model

For this model, we have:

[12] These correspond to the following Pearson family of distributions: Gamma = type III, Beta = type I, and Gaussian = limit of type I, III, IV, V, or VI.

drift coefficient: $b(x) = r(\theta - x)$, $r > 0$.

diffusion coefficient: $\sigma^2(x) = \epsilon x(1-x)$, $\epsilon > 0$.

s.d.e.: $\mathrm{d}X_t = r(\theta - X_t)\mathrm{d}t + \sqrt{\epsilon X_t(1-X_t)}\mathrm{d}W_t$.

stationary density: $\pi(x) = \dfrac{\Gamma\left(\frac{1}{\delta}\right)}{\Gamma\left(\frac{\theta}{\delta}\right)\Gamma\left(\frac{1-\theta}{\delta}\right)} x^{-1+\frac{\theta}{\delta}}(1-x)^{-1+\frac{1-\theta}{\delta}}$, $\delta = \dfrac{\epsilon}{r}$.

statistics: mean $= \theta$, mode $= \dfrac{\theta - \delta}{1 - 2\delta}$, variance $= \dfrac{\theta(1-\theta)}{1+\delta}$.

The Type B model is used as a model of political polarization, where X_t is the political persuasion of a subject. The persuasion is measured on the hypothetical axis of conviction, say $X_t = 0$ "liberal" and $X_t = 1$ "conservative". The choice of $\sigma(x)$ in this model is motivated by the fact that people who hold extreme views are much less subject to random fluctuations than those near the center. The drift explains the move toward the average persuasion of the whole population. The ratio δ controls the extent of polarization. If μ is the mean and σ^2 is the variance, the model can be reparametrized accordingly and $\delta = \sigma^2/(\mu(1-\mu) - \sigma^2)$.

1.13.14 One epidemic model

The mathematical theory of epidemiology [17] contains many types of models, both deterministic and stochastic which have been adapted and expanded in the social sciences [23]. We just mention here a simple epidemic model. Let X_t be the fraction of a population that has an infectious disease at time t. If the disease does not confer immunity (e.g., gonorrhea), then the rate of change usually follows the nonstochastic differential equation

$$\mathrm{d}x_t = a x_t(1-x_t)\mathrm{d}t - b x_t \mathrm{d}t + c(1-x_t)\mathrm{d}t\,,$$

where $a > 0$ is the rate of person-to-person transmission, $b > 0$ is the rate of recovery (or forgetting), and $c > 0$ is the rate of transmission from an external source. The transformation of the deterministic model into a stochastic one motivates the choice of the diffusion coefficient as in Type B models. Therefore, the corresponding stochastic differential equation is

$$\mathrm{d}X_t = \mu(X_t)\mathrm{d}t + \sigma(X_t)\mathrm{d}W_t\,,$$

where

$$\mu(x) = ax(1-x) - bx + c(1-x)\,,$$
$$\sigma^2(x) = \epsilon x(1-x), \quad \epsilon > 0\,.$$

In this case, the density of the stationary distribution takes the form

$$\pi(x) = \frac{M}{x(1-x)} \exp\left\{\int_0^x \left(\frac{a}{\epsilon} - \frac{b}{\epsilon(1-u)} + \frac{c}{\epsilon u}\right) \mathrm{d}u\right\},$$

where M is the normalizing constant over the state space $[0, 1]$. This distribution may be bimodal depending on the number of real solutions to $d \pm \sqrt{d^2 - (\epsilon - c)/a}$, where $d = (a - b - c + 2\epsilon)/2a$. When there are two positive real solutions, the smaller is the epidemic threshold and the larger is the size of the epidemic. This is a model of stochastic threshold when $0 < c < \epsilon$. The epidemic is unlikely to happen if the model (the number of infected people in the population) stays below the lowest mode. When the model is above it the epidemic size is around the second mode. When $c > \epsilon$ (high infection rate from external sources), the epidemic is guaranteed.

1.13.15 The stochastic cusp catastrophe model

Contrary to what the name suggests, the catastrophe theory has little or nothing to do with disasters [187] but more with the classification of nondegenerate singularities (minima, maxima, saddle points) and also degenerate singularities (e.g., when the second derivative vanishes). From the dynamical systems point of view, the dynamics of a catastrophe model is usually written in terms of a potential function: the system behaves as though it moves toward the point of lowest potential. If $V(x)$ denotes the potential, the dynamics is then $\mathrm{d}x/\mathrm{d}t = -\partial V/\partial x$. The corresponding stochastic differential equation is of the form

$$\mathrm{d}X_t = -\frac{\partial V}{\partial x}(X_t)\mathrm{d}t + \sqrt{\epsilon}\mathrm{d}W_t$$

with stationary density given by

$$\pi(x) = M e^{\frac{V(x)}{\epsilon}},$$

with M the normalizing constant. Singularities of the deterministic system are the solutions of $\partial V/\partial x = 0$, and the equilibrium of the system is stable or unstable according to whether $\partial^2 V/\partial x^2$ is positive or negative, while catastrophe points are those values of x for which $\partial^2 V/\partial x^2 = 0$. The potential V is usually approximated via polynomials. The "cusp" or Zeeman model [232] considers a third-order polynomial $\mathrm{d}x/\mathrm{d}t = -(a_1 + a_2 x + a_3 x^2 + a_4 x^3)$, which is usually parametrized in the so-called *standard form*

$$\mathrm{d}X_t = r(\alpha + \beta(X_t - \lambda) - (X_t - \lambda)^3)\mathrm{d}t + \sqrt{\epsilon}\mathrm{d}W_t$$

where α and β are called the *normal* and *splitting* factors respectively. The stationary density has the form

$$\pi(x) = M \exp\left\{\frac{\alpha(x-\lambda) + \frac{1}{2}\beta(x-\lambda)^2 - \frac{1}{4}(x-\lambda)^4}{\delta}\right\}, \quad \delta = \frac{\epsilon}{r}.$$

The four parameters $(\alpha, \beta, \lambda, \delta)$ characterize the cusp density via Cardan's discriminant $C = 27\alpha^2 - 4\beta^3$. In particular, according to Zeeman's terminology:

- *Asymmetry* α: If $C < 0$, the cusp density is bimodal and α is the relative height of the two modes. If $C > 0$, the cusp density is unimodal and α measures the skewness.
- *Bifurcation* β: If $C < 0$, then β determines the separation of the two modes; if $C > 0$, then β measures the kurtosis.
- *Location* λ : The cusp catastrophe points are located at $x = \lambda$ with $\alpha = 0$ and $\beta = 0$. Moving λ shifts the cusp density horizontally on the x axis without changing the shape.
- *Dispersion* δ: The parameter determines the variation about the two modes of a bimodal cusp density like the variance does in the Gaussian law, but it is not a scale parameter.

1.13.16 Exponential families of diffusions

Exponential families of diffusion processes have been introduced in [216] and are solutions to the stochastic differential equation

$$\mathrm{d}X_t = \sum_{i=1}^{p} \beta_i b_i(X_t)\mathrm{d}t + \lambda v(X_t)\mathrm{d}W_t\,,$$

with $\lambda > 0$, and $v(\cdot) > 0$, and $b_i(\cdot)$ functions such that a unique weak solution exists. These models are not exponential families of stochastic processes in the sense of [146]. By using the reparametrization $\theta_i = \beta_i/\lambda^2$, and the functions

$$T_i(x) = 2\int_{x_0}^{x} \frac{b_i(y)}{v^2(y)}\mathrm{d}y$$

for some x_0 in the state space of X, the scale measure has the following representation

$$s(x) = \exp\left\{-\sum_{i=1}^{p} \theta_i T_i(x)\right\}$$

and the speed measure is

$$m(x) = v(x)^{-2}\exp\left\{\sum_{i=1}^{p} \theta_i T_i(x)\right\}.$$

If we denote by $\phi(\theta)$ the normalizing constant

$$\phi(\theta) = \int m(x)\mathrm{d}x\,,$$

the invariant distribution of this class of models takes the canonical form of an exponential family, i.e.

$$\pi(x) = \exp\left\{\sum_{i=1}^{p} \theta_i T_i(x) - 2\log v(x) - \phi(\theta)\right\}.$$

These models are ergodic (and hence the invariant density $\pi(x)$ exists) under proper conditions on the parameter θ_i, $i = 1, \ldots, p$ (see [216] for details).

1.13.17 Generalized inverse gaussian diffusions

The exponential family of diffusions also encompasses some peculiar interesting models like the generalized inverse Gaussian diffusions which are solutions to the stochastic differential equation

$$\mathrm{d}X_t = (\beta_1 X_t^{2\alpha-1} - \beta_2 X_t^{2\alpha} + \beta_3 X_t^{2(\alpha-1)})\mathrm{d}t + \lambda X_t^{\alpha}\mathrm{d}W_t, \quad X_0 > 0\,.$$

With the reparametrization $\theta_i = 2\beta_i\lambda^{-2}, i = 1, 2, 3$, the scale measure takes the form

$$s(x) = x^{-\theta_1}\exp\left(\theta_2 x + \theta_3 x^{-1}\right), \quad x > 0$$

and the speed measure is

$$m(x) = x^{\theta_1 - 2\alpha}\exp\left(-\theta_2 x - \theta_3 x^{-1}\right)\,.$$

The invariant density, under suitable conditions on the parameters, takes the form of the generalized inverse Gaussian distribution introduced in [98] (see as well [129] for an extensive statistical analysis).

2

Numerical Methods for SDE

This chapter covers basic and new material on the subject of simulation of solutions of stochastic differential equations. The chapter reviews results from the well-known reference [142] but also covers new results such as, but not only, exact sampling (see [29]). Other related references mentioned in this chapter are, for example, [156] and [125].

There are two main objectives in the simulation of the trajectory of a process solution of a stochastic differential equation: either interest is in the whole trajectory or in the expected value of some functional of the process (moments, distributions, etc) which usually are not available in explicit analytical form. Variance reduction techniques for stochastic differential equations can be borrowed from standard variance reduction techniques of the Monte Carlo method (see Section 1.4) and clearly only apply when interest is in the functionals of the process.

Simulation methods are usually based on discrete approximations of the continuous solution to a stochastic differential equation. The methods of approximation are classified according to their different properties. Mainly two criteria of optimality are used in the literature: the *strong* and the *weak* (orders of) convergence.

Strong order of convergence

A time-discretized approximation Y_δ of a continuous-time process Y, with δ the maximum time increment of the discretization, is said to be of general *strong* order of convergence γ to Y if for any fixed time horizon T it holds true that

$$\mathbb{E}|Y_\delta(T) - Y(T)| \leq C\delta^\gamma, \quad \forall\, \delta < \delta_0\,,$$

with $\delta_0 > 0$ and C a constant not depending on δ. This kind of criterion is similar to the one used in the approximation of the trajectories of nonstochastic dynamical systems.

S.M. Iacus, *Simulation and Inference for Stochastic Differential Equations*,
doi: 10.1007/978-0-387-75839-8_2,

Weak order of convergence

Along with the strong convergence, the *weak* convergence can be defined. Y_δ is said to converge *weakly* of order β to Y if for any fixed horizon T and any $2(\beta+1)$ continuous differentiable function g of polynomial growth, it holds true that

$$|\mathbb{E}g(Y(T)) - \mathbb{E}g(Y_\delta(T))| \leq C\delta^\beta, \quad \forall\, \delta < \delta_0\,,$$

with $\delta_0 > 0$ and C a constant not depending on δ.

Schemes of approximation of some order that strongly converge usually have a higher order of weak convergence. This is the case with the Euler scheme, which is strongly convergent of order $\gamma = \frac{1}{2}$ and weakly convergent of order $\beta = 1$ (under some smoothness conditions on the coefficients of the stochastic differential equation). While the schemes have their own order of convergence, it is usually the case that, for some actual specifications of the stochastic differential equations, they behave better.

2.1 Euler approximation

One of the most used schemes of approximation is the *Euler* method, originally used to generate solutions to deterministic differential equations. We implicitly used this method in Chapter 1 several times. The idea is the following: given an Itô process $\{X_t, 0 \leq t \leq T\}$ solution of the stochastic differential equation

$$\mathrm{d}X_t = b(t, X_t)\mathrm{d}t + \sigma(t, X_t)\mathrm{d}W_t,$$

with initial deterministic value $X_{t_0} = X_0$ and the discretization $\Pi_N = \Pi_N([0,T])$ of the interval $[0,T]$, $0 = t_0 < t_1 < \cdots < t_N = T$. The *Euler approximation* of X is a continuous stochastic process Y satisfying the iterative scheme

$$Y_{i+1} = Y_i + b(t_i, Y_i)(t_{i+1} - t_i) + \sigma(t_i, Y_i)(W_{i+1} - W_i), \tag{2.1}$$

for $i = 0, 1, \ldots, N-1$, with $Y_0 = X_0$. We have simplified the notation setting $Y(t_i) = Y_i$ and $W(t_i) = W_i$. Usually the time increment $\Delta t = t_{i+1} - t_i$ is taken to be constant (i.e., $\Delta t = 1/N$). In between any two time points t_i and t_{i+1}, the process can be defined differently. One natural approach is to consider linear interpolation so that $Y(t)$ is defined as

$$Y(t) = Y_i + \frac{t - t_i}{t_{i+1} - t_i}(Y_{i+1} - Y_i), \quad t \in [t_i, t_{i+1})\,.$$

From (2.1), one can see that to simulate the process Y one only needs to simulate the increment of the Wiener process, and we already know how to do this from Chapter 1. As mentioned, the Euler scheme has order $\gamma = \frac{1}{2}$ of strong convergence.

2.1.1 A note on code vectorization

Consider for example the Ornstein-Uhlenbeck process (see Section 1.13.1) solution of

$$\mathrm{d}X_t = (\theta_1 - \theta_2 X_t)\mathrm{d}t + \theta_3 \mathrm{d}W_t\,.$$

For this process, $b(t,x) = (\theta_1 - \theta_2 x)$ and $\sigma(t,x) = \theta_3$. Suppose we fix an initial value $X_0 = x$ and the set of parameters $(\theta_1, \theta_2, \theta_3) = (0, 5, 3.5)$. The following algorithm can be used to simulate the trajectory of the process using the Euler algorithm instead of the simulation of the stochastic integral as in Section 1.13.1:

```
> # ex 2.01.R
> set.seed(123)
> N <- 100
> T <- 1
> x <- 10
> theta <- c(0, 5, 3.5)
> Dt <- 1/N
> Y <- numeric(N+1)
> Y[1] <- x
> Z <- rnorm(N)
> for(i in 1:N)
+  Y[i+1] <- Y[i] + (theta[1] - theta[2]*Y[i])*Dt + theta[3]*sqrt(Dt)*Z[i]
> Y <- ts(Y,start=0, deltat=1/N)
> plot(Y)
```

At first glance, this algorithm appears not to be efficient from the R point of view. We can try to optimize this code by replacing the `for` loop with some `*apply` function. Indeed, noticing that

$$Y(2) = Y(1)(1 - \theta_2 \Delta t) + \theta_3 * \sqrt{\Delta t} Z(1)$$

and

$$\begin{aligned} Y(3) &= Y(2)(1 - \theta_2 \Delta t) + \theta_3 \sqrt{\Delta t} Z(2) \\ &= \{Y(1)(1 - \theta_2 \Delta t) + \theta_3 \sqrt{\Delta t} Z(1)\}(1 - \theta_2 \Delta t) + \theta_3 \sqrt{\Delta t} Z(2) \\ &= Y(1)(1 - \theta_2 \Delta t)^2 + \theta_3 \sqrt{\Delta t} Z(1)(1 - \theta_2 \Delta t) + \theta_3 \sqrt{\Delta t} Z(2) \end{aligned}$$

by iterative substitution, we get the general formula for the kth step of the Euler scheme for the Ornstein-Uhlenbeck process,

$$Y(k) = Y(1) \cdot (1 - \theta_2 \Delta t)^{k-1} + \sum_{j=1}^{k-1} (\theta_3 * \sqrt{\Delta t} * Z(j)) * (1 - \theta_2 \Delta t)^{(k-j-1)},$$

which leads to the following algorithm that uses linear algebra instead of `for` loops.

```
> # ex 2.02.R
> set.seed(123)
> theta <- c(0, 5, 3.5)
> N <- 100
> T <- 1
> x <- 10
```

```
> Z <- rnorm(N)
> Dt <- 1/N
> A <- theta[3]*sqrt(Dt)*Z
> P <- (1-theta[2]*Dt)^(0:(N-1))
> X0 <- x
> X <- sapply(2:(N+1), function(x) X0*(1-theta[2]*Dt)^(x-1) +
+     A[1:(x-1)] %*% P[(x-1):1])
> Y <- ts(c(X0,X),start=0, deltat=1/N)
> plot(Y)
```

But, this one-line code is not at all better than the one implying the `for` loop. In fact, it is even worse, as it implies more calculations than needed. To show this, we embed the two codes into two functions `OU`, and `OU.vec`, to measure their performance in terms of CPU time.[1]

```
> # ex 2.03.R
> OU <- function(a,b, x, N=1000){
+   Y <- numeric(N+1)
+   Y[1] <- x
+   Z <- rnorm(N)
+   Dt <- 1/N
+   for(i in 1:N)
+     Y[i+1] <- Y[i] - a*Y[i]*Dt + b*sqrt(Dt)*Z[i]
+    invisible(Y)
+ }

> OU.vec <- function(a, b, x, N=1000){
+   Dt <- 1/N
+   Z <- rnorm(N)
+   A <- b*sqrt(Dt)*Z
+   P <- (1-a*Dt)^(0:(N-1))
+    X0 <- x
+    X <- c(X0, sapply(2:(N+1),
+       function(x) X0*(1-a*Dt)^(x-1) +
+       sum(A[1:(x-1)] * P[(x-1):1])))
+  invisible(X)
+ }
```

Using the `system.time` function, we test the two implementations

```
> set.seed(123)
> system.time(OU(10,5,3.5))
[1] 0.037 0.001 0.044 0.000 0.000

> set.seed(123)
> system.time(OU.vec(10,5,3.5))
[1] 0.198 0.024 0.261 0.000 0.000
```

which shows that the vectorized version is much slower than the naive one. The moral of this example is that `for` loops in R should be replaced by vectorized code *only if the number of calculations does not increase*, as the case above shows. Vectorization can really reduce simulation/estimation time, but only under the conditions above. As vectorization can become a nightmare for the average R programmer, the reader has been warned.

To convince the reader that it is worth using vectorized code when it is appropriate, we report without comments the example from Chapter 1 for generating paths of Brownian motion with two levels of optimization in the R code.

---

[1] Times may vary on the reader's machine.

```
# ex 2.04.R
> BM.1 <- function(N=10000){ # brutal code
+  W <- NULL
+  for(i in 2:(N+1))
+         W <- c(W, rnorm(1) / sqrt(N))
+ }

> BM.2 <- function(N=10000){ # smarter
+  W <- numeric(N+1)
+  Z <- rnorm(N)
+  for(i in 2:(N+1))
+         W[i] <- W[i-1] + Z[i-1] / sqrt(N)
+ }

> BM.vec <- function(N=10000) # awesome !
+  W <- c(0,cumsum(rnorm(N)/sqrt(N)))

> set.seed(123)
> system.time(BM.1())
[1] 1.354 1.708 3.856 0.000 0.000

> set.seed(123)
> system.time(BM.2())
[1] 0.281 0.011 0.347 0.000 0.000

> set.seed(123)
> system.time(BM.vec())
[1] 0.008 0.001 0.010 0.000 0.000
```

2.2 Milstein scheme

The Milstein scheme[2] [164] makes use of Itô's lemma to increase the accuracy of the approximation by adding the second-order term. Denoting by σ_x the partial derivative of $\sigma(t,x)$ with respect to x, the Milstein approximation looks like

$$\begin{aligned} Y_{i+1} =& Y_i + b(t_i, Y_i)(t_{i+1} - t_i) + \sigma(t_i, Y_i)(W_{i+1} - W_i) \\ &+ \frac{1}{2}\sigma(t_i, Y_i)\sigma_x(t_i, Y_i)\left\{(W_{i+1} - W_i)^2 - (t_{i+1} - t_i)\right\} \end{aligned} \tag{2.2}$$

or, in more symbolic form,

$$Y_{i+1} = Y_i + b\Delta t + \sigma \Delta W_t + \frac{1}{2}\sigma\sigma_x \left\{(\Delta W_t)^2 - \Delta t\right\}.$$

This scheme has strong and weak orders of convergence equal to one. Let us consider once again the Ornstein-Uhlenbeck process solution of (1.39). For this process, $b(t,x) = \theta_1 - \theta_2 \cdot x$ and $\sigma(t,x) = \theta_3$, and thus $\sigma_x(t,x) = 0$ and the Euler and Milstein schemes coincide. This is one case in which the Euler scheme is of strong order of convergence $\gamma = 1$.

[2] Actually, Milstein proposed two schemes of approximation. The one presented in this section corresponds to the one most commonly known as the "Milstein scheme" and has the simplest form. Another Milstein scheme of higher-order approximation, will be presented later in this chapter.

The geometric Brownian motion

A more interesting case is that of geometric Brownian motion, presented in Section 1.7, solving the stochastic differential equation

$$\mathrm{d}X_t = \theta_1 X_t \mathrm{d}t + \theta_2 X_t \mathrm{d}W_t.$$

For this process, $b(t,x) = \theta_1 \cdot x$, $\sigma(t,x) = \theta_2 \cdot x$, and $\sigma_x(t,x) = \theta_2$. The Euler discretization for this process looks like

$$Y_{i+1}^E = Y_i^E(1 + \theta_1 \cdot \Delta t) + \theta_2 Y_i^E \Delta W_t,$$

and the Milstein scheme reads

$$\begin{aligned} Y_{i+1}^M &= Y_i^M + \theta_1 \cdot Y_i^M \Delta t + \theta_2 Y_i^M \Delta W_t + \frac{1}{2}\theta_2^2 Y_i^M \left\{(\Delta W_t)^2 - \Delta t\right\} \\ &= Y_i^M \left(1 + \left(\theta_1 - \frac{1}{2}\theta_2^2\right)\Delta t\right) + \theta_2 Y_i^M \Delta W_t + \frac{1}{2}\theta_2^2 Y_i^M (\Delta W_t)^2. \end{aligned}$$

Recall that $\Delta W_t \sim \sqrt{\Delta t} Z$ with $Z \sim N(0,1)$. Thus

$$Y_{i+1}^E = Y_i^E(1 + \theta_1 \cdot \Delta t + \theta_2 \sqrt{\Delta t} Z)$$

and

$$\begin{aligned} Y_{i+1}^M &= Y_i^M \left(1 + \left(\theta_1 - \frac{1}{2}\theta_2^2\right)\Delta t\right) + \theta_2 Y_i^M \sqrt{\Delta t} Z + \frac{1}{2}\theta_2^2 Y_i^M \Delta t Z^2 \\ &= Y_i^M \left(1 + \left(\theta_1 + \frac{1}{2}\theta_2^2 (Z^2 - 1)\right)\Delta t + \theta_2 \sqrt{\Delta t} Z\right). \end{aligned}$$

Looking at the exact solution in (1.7), the Milstein scheme makes the expansion exact up to order $O(\Delta t)$. Indeed, formal Taylor expansion leads to

$$\begin{aligned} X_{t+\Delta t} &= X_t \exp\left\{\left(\theta_1 - \frac{\theta_2^2}{2}\right)\Delta t + \theta_2 \sqrt{\Delta t} Z\right\} \\ &= X_t \left\{1 + \left(\theta_1 - \frac{\theta_2^2}{2}\right)\Delta t + \theta_2 \sqrt{\Delta t} Z + \frac{1}{2}\theta_2^2 \Delta t Z^2 + O(\Delta t)\right\} \\ &= Y_{i+1}^M. \end{aligned}$$

2.3 Relationship between Milstein and Euler schemes

Following [125], we now show a result on transformations of stochastic differential equations and the two schemes of approximation. Given the generic stochastic differential equation

$$\mathrm{d}X_t = b(t, X_t)\mathrm{d}t + \sigma(t, X_t)\mathrm{d}W_t, \tag{2.3}$$

the Milstein scheme for it is

$$\Delta X = X_{i+1} - X_i = \left(b(t_i, X_i) - \frac{1}{2}\sigma(t_i, X_i)\sigma_x(t_i, X_i)\right)\Delta t \\ + \sigma(t_i, X_i)\sqrt{\Delta t}Z + \frac{1}{2}\sigma(t_i, X_i)\sigma_x(t_i, X_i)\Delta t Z^2 \tag{2.4}$$

with $Z \sim N(0,1)$. Consider now the transformation $y = F(x)$ and its inverse $x = G(y)$. Then (2.3) becomes by Itô's lemma

$$\mathrm{d}Y_t = \left(F'(X_t)b(t, X_t) + \frac{1}{2}F''(X_t)\sigma^2(t, X_t)\right)\mathrm{d}t + F'(X_t)\sigma(t, X_t)\mathrm{d}W_t \tag{2.5}$$

with $Y_t = F(X_t)$. Now choose F as the Lamperti transform (1.34) so that

$$F'(x) = \frac{1}{\sigma(t,x)}, \qquad F''(x) = -\frac{\sigma_x(t,x)}{\sigma^2(t,x)}.$$

We know from (1.35) that (2.5) becomes

$$\mathrm{d}Y_t = \left(\frac{b(t, X_t)}{\sigma(t, X_t)} - \frac{1}{2}\sigma_x^2(t, X_t)\right)\mathrm{d}t + \mathrm{d}W_t\,.$$

We remark again that the Lamperti transform is such that the multiplicative factor in front of the Wiener process no longer depends on the state of the process. Thus the Euler scheme for $Y_t = F(X_t)$ reads

$$\Delta Y = \left(\frac{b(t_i, X_i)}{\sigma(t_i, X_i)} - \frac{1}{2}\sigma_x(t_i, X_i)\right)\Delta t + \sqrt{\Delta t}Z\,.$$

The next step is to calculate the Taylor expansion of the inverse transformation in order to obtain some comparable expression.

$$G(Y_i + \Delta Y) = G(Y_i) + G'(Y_i)\Delta Y + \frac{1}{2}G''(Y_i)(\Delta Y_i)^2 + O\left(\Delta Y^3\right).$$

Notice that

$$G'(y) = \frac{\mathrm{d}}{\mathrm{d}y}F^{-1}(y) = \frac{1}{F'(G(y))} = \sigma(t, G(y))$$

and

$$G''(y) = G'(y)\sigma_x(t, G(y)) = \sigma(t, G(y))\sigma_x(t, G(y)),$$

which finally leads to

$$G(Y_i + \Delta Y) - G(Y_i) = \left(b(t_i, X_i) - \frac{1}{2}\sigma(t_i, X_i)\sigma_x(t_i, X_i)\right)\Delta t \\ + \sigma(t_i, X_i)\sqrt{\Delta t}Z + \frac{1}{2}\sigma(t_i, X_i)\sigma_x(t_i, X_i)\Delta t Z^2 \\ + O\left(\Delta t^{\frac{3}{2}}\right).$$

Thus the Milstein scheme on the original process (2.4) and the Euler scheme on the transformed process are equal up to and including the order Δt. So, in principle, whenever possible, one should use the Euler scheme on the transformed process.

In general, if F (not necessarily the Lamperti transformation) eliminates the interactions between the state of the process and the increments of the Wiener process, this transformation method is probably always welcome because it reduces instability in the simulation process. A detailed account on this matter can be found in [70]. We now show a couple of applications of the transformation method just presented.

2.3.1 Transform of the geometric Brownian motion

The first example is on the geometric Brownian motion. If we use $F(x) = \log(x)$, then from (1.32) and Itô's lemma, we obtain

$$\mathrm{d}\log X_t = \left(\theta_1 - \frac{1}{2}\theta_2^2\right)\mathrm{d}t + \theta_2 \mathrm{d}W_t.$$

Thus the Euler scheme for the transformed process is

$$\Delta \log X = \left(\theta_1 - \frac{1}{2}\theta_2^2\right)\Delta t + \theta_2\sqrt{\Delta t}Z\,.$$

Now, using the Taylor expansion on the inverse transform $G(y) = x^y$, we get the Milstein scheme.

2.3.2 Transform of the Cox-Ingersoll-Ross process

One more interesting application of the Lamperti transform concerns the Cox-Ingersoll-Ross process introduced in Section 1.13.3. The dynamics of the process is

$$\mathrm{d}X_t = (\theta_1 - \theta_2 X_t)\mathrm{d}t + \theta_3\sqrt{X_t}\mathrm{d}W_t, \tag{2.6}$$

and the Milstein scheme for it reads as

$$\Delta X = \left((\theta_1 - \theta_2 X_i) - \frac{1}{4}\theta_3^2\right)\Delta t + \theta_3\sqrt{X_i}\sqrt{\Delta t}Z + \frac{1}{4}\theta_3^2\Delta t Z^2. \tag{2.7}$$

Now, using the transformation $y = \sqrt{x}$, we obtain the transformed stochastic differential equation

$$\mathrm{d}Y_t = \frac{1}{2Y_t}\left((\theta_1 - \theta_2 Y_t^2) - \frac{1}{4}\theta_3^2\right)\mathrm{d}t + \frac{1}{2}\theta_3\mathrm{d}W_t\,,$$

for which the Euler scheme is

$$\Delta Y = \frac{1}{2Y_i}\left((\theta_1 - \theta_2 Y_i^2) - \frac{1}{4}\theta_3^2\right)\Delta t + \frac{1}{2}\theta_3\sqrt{\Delta t}Z\,.$$

Since $G(y) = x^2$, we obtain

$$\begin{aligned} G(Y_i + \Delta Y) - G(Y_i) &= (Y_i + \Delta Y)^2 - Y_i^2 = (\Delta Y)^2 + 2Y_i\Delta Y \\ &= \frac{1}{4}\theta_3^2\Delta t Z^2 + O(\Delta t^2) \\ &\quad + \left((\theta_1 - \theta_2 Y_i^2) - \frac{1}{4}\theta_3^2\right)\Delta t + Y_i\theta_3\sqrt{\Delta t}Z\,, \end{aligned}$$

which is exactly (2.7) given that $Y_i = \sqrt{X_i}$.

2.4 Implementation of Euler and Milstein schemes: the sde.sim function

We now show generic implementations of both the Euler and Milstein schemes.

```
sde.sim <- function(t0=0, T=1, X0=1, N=100, delta,
         drift, sigma, sigma.x,
         method=c("euler","milstein")){

  if(missing(drift) )
   stop("please specify at least the drift coefficient of the SDE")

  if(missing(sigma))
   sigma <- expression(1)

  if(!is.expression(drift) | !is.expression(sigma))
   stop("coefficients must be expressions in `t' and `x'")

  method <- match.arg(method)
  needs.sx <- FALSE

  if(method=="milstein") needs.sx <- TRUE

   if(needs.sx & missing(sigma.x)){
    cat("sigma.x not provided, attempting symbolic derivation.\n")
    sigma.x <- D(sigma,"x")
  }

  d1 <- function(t,x) eval(drift)
  s1 <- function(t,x) eval(sigma)
  sx <- function(t,x) eval(sigma.x)

  if(t0<0 | T<0)
   stop("please use positive times!")

  if(missing(delta)){
    t <- seq(t0,T, length=N+1)
  } else {
   t <- c(t0,t0+cumsum(rep(delta,N)))
   T <- t[N+1]
   warning("T set to =",T,"\n")
  }

  Z <- rnorm(N)
  X <- numeric(N+1)
```

```
  Dt <- (T-t0)/N
  sDt <- sqrt(Dt)
  X[1] <- X0

  if(method=="euler"){
   for(i in 2:(N+1))
     X[i] <- X[i-1] + d1(t[i-1],X[i-1])*Dt +
                s1(t[i-1],X[i-1])*sDt*Z[i-1]
  }
  if(method=="milstein"){
   for(i in 2:(N+1)){
    X[i] <- X[i-1] + d1(t[i-1],X[i-1])*Dt +
                s1(t[i-1],X[i-1])*sDt*Z[i-1] +
                0.5*s1(t[i-1],X[i-1])* sx(t[i-1],X[i-1]) *
                (Dt*Z[i-1]^2-Dt)
   }
  }
  X <- ts(X,start=t0 ,deltat=Dt)
  invisible(X)
}
```

The `sde.sim` function above can be used to simulate paths of solutions to generic stochastic differential equations. The function can simulate trajectories with either the "`euler`" or "`milstein`" scheme. The function accepts the two coefficients `drift` and `sigma` and eventually the partial derivative of the diffusion coefficient `sigma.x` for the Milstein scheme. If `sigma.x` is not provided by the user, the function tries to provide one itself using the R function `D`. Coefficients must be objects of class `expression` with arguments named `t` and `x`, respectively, interpreted as time and space (i.e., the state of the process). If the diffusion coefficient `sigma` is not specified, it is assumed to be unitary (i.e., identically equal to one). The user can specify the initial value X_0 (defaulted to 1), the interval $[t_0, T]$ (defaulted to $[0,1]$), the Δ step `delta`, and the number `N` of values of the process to be generated. The function always returns a `ts` (time series) object of length $N+1$; i.e., the initial value X_0 and the N new simulated values of the trajectory. If Δ is not specified, $\Delta = (T-t_0)/N$. If Δ is specified, then N new observations are generated at time increments of Δ and the time horizon is adjusted accordingly as $T = \Delta \cdot N$. We have added the opportunity to specify the Δ step directly because this will be relevant in Chapter 3.

2.4.1 Example of use

The following examples use the `sde.sim` function for some of the processes introduced earlier. The reader doesn't need to write the code for the `sde.sim` function because it is included in the CRAN package called `sde`. Moreover, this function is going to evolve during this chapter as new methods are introduced.

The Ornstein-Uhlenbeck process

$$dX_t = (\theta_1 - \theta_2 X_t)dt + \theta_3 dW_t, \qquad X_0 = 10\,,$$

with $(\theta_1, \theta_2, \theta_3) = (0, 5, 3.5)$.

```
> # ex 2.05.R
> # Ornstein-Uhlenbeck process
> require(sde)
[1] TRUE
> set.seed(123)
> d <- expression(-5 * x)
> s <- expression(3.5)
> sde.sim(X0=10,drift=d, sigma=s) -> X
sigma.x not provided, attempting symbolic derivation.
> plot(X,main="Ornstein-Uhlenbeck")
```

The Cox-Ingersoll-Ross process

$$\mathrm{d}X_t = (\theta_1 - \theta_2 X_t)\mathrm{d}t + \theta_3\sqrt{X_t}\mathrm{d}W_t, \qquad X_0 = 10\,,$$

with $(\theta_1, \theta_2, \theta_3) = (6, 3, 2)$.

```
> # ex 2.06.R
> # Cox-Ingersoll-Ross (CIR-1)
> set.seed(123)
> d <- expression( 6-3*x )
> s <- expression( 2*sqrt(x) )
> sde.sim(X0=10,drift=d, sigma=s) -> X
sigma.x not provided, attempting symbolic derivation.
> plot(X,main="Cox-Ingersoll-Ross")
```

The Cox-Ingersoll-Ross process with Milstein scheme

We need the partial derivative with respect to variable x of the coefficient $\sigma(\cdot,\cdot)$,

$$\sigma_x(t,x) = \frac{\partial}{\partial x} 2\sqrt{x} = \frac{1}{\sqrt{x}}\,.$$

Therefore,

```
> # ex 2.07.R
> # Cox-Ingersoll-Ross (CIR-2)
> d <- expression( 6-3*x )
> s <- expression( 2*sqrt(x) )
> s.x <- expression( 1/sqrt(x) )
> set.seed(123)
> sde.sim(X0=10, drift=d, sigma=s, sigma.x=s.x,
+      method="milstein") -> X
> plot(X,main="Cox-Ingersoll-Ross")
```

The Cox-Ingersoll-Ross process with Euler scheme on the transformed process $Y_t = \sqrt{X_t}$

$$\mathrm{d}Y_t = \frac{1}{2Y_t}\left(\theta_1 - \theta_2 Y_t^2 - \frac{1}{4}\theta_3^2\right)\mathrm{d}t + \frac{1}{2}\theta_3 \mathrm{d}W_t, \qquad Y_0 = \sqrt{10}\,,$$

with $(\theta_1, \theta_2, \theta_3) = (6, 3, 2)$.

```
> # ex 2.08.R
> # Cox-Ingersoll-Ross (CIR-3)
> set.seed(123)
> d <- expression( (6-3*x^2 - 1)/(2*x) )
> s <- expression( 1 )
> sde.sim(X0=sqrt(10),drift=d, sigma=s) -> Y
> plot(Y^2,main="Cox-Ingersoll-Ross")
```

The reader can verify that the Milstein scheme CIR-2 returns the same path as CIR-3 but on a different scale (i.e., $Y^2 = X$).

The geometric Brownian motion process

$$\mathrm{d}X_t = \theta_1 X_t \mathrm{d}t + \theta_2 X_t \mathrm{d}W_t, \qquad X_0 = 10,$$

with $(\theta_1, \theta_2) = \left(1, \frac{1}{2}\right)$.

```
> # ex 2.09.R
> # geometric Brownian Motion
> set.seed(123)
> d <- expression( x )
> s <- expression( 0.5*x )
> sde.sim(X0=10,drift=d, sigma=s) -> X
> plot(X,main="geometric Brownian Motion")
```

2.5 The constant elasticity of variance process and strange paths

The constant elasticity of variance (CEV) process introduced in finance in option pricing (see [201] and [25]) is another particular member of the CKLS family of models (see Section 1.13.4) and is the solution of the stochastic differential equation

$$\mathrm{d}X_t = \mu X_t \mathrm{d}t + \sigma X_t^\gamma \mathrm{d}W_t, \quad \gamma \geq 0. \tag{2.8}$$

This process is quite useful in modeling a skewed implied volatility. In particular, for $\gamma < 1$, the skewness is negative, and for $\gamma > 1$ the skewness is positive. For $\gamma = 1$, the CEV process is a particular version of the geometric Brownian motion. Even if this process is assumed to be positive, the discretized version of it can reach negative values. To understand why this could happen and what to do with the paths that cross zero, one should go back to the theoretical properties of the process. For example, in [38] it is shown that for $\gamma < \frac{1}{2}$ there is a positive probability for this process to be absorbed in zero. Hence, if in one simulation the path crosses the zero line, one should stop the simulation and consider this path as actually absorbed in 0.

2.6 Predictor-corrector method

Both schemes of discretization consider the coefficients b and σ as not varying during the time interval Δt, which is of course untrue for a generic stochastic differential equation, as b and σ can depend on both the time t and the state of the process X_t. One way to recover the varying nature of these coefficients is to average their values in some way. Since the coefficients depend on X_t and we are simulating X_t, the method we present here just tries to approximate the states of the process first. This method is of weak convergence order 1. The predictor-corrector algorithm is as follows. First consider the simple approximation (the *predictor*)

$$\tilde{Y}_{i+1} = Y_i + b(t_i, Y_i)\Delta t + \sigma(t_i, Y_i)\sqrt{\Delta t}Z.$$

Then choose two weighting coefficients α and η in $[0,1]$, and calculate the *corrector* as

$$\begin{aligned} Y_{i+1} = Y_i &+ \left(\alpha\tilde{b}(t_{i+1}, \tilde{Y}_{i+1}) + (1-\alpha)\tilde{b}(t_i, Y_i)\right)\Delta t \\ &+ \left(\eta\sigma(t_{i+1}, \tilde{Y}_{i+1}) + (1-\eta)\sigma(t_i, Y_i)\right)\sqrt{\Delta t}Z \end{aligned}$$

with

$$\tilde{b}(t_i, Y_i) = b(t_i, Y_i) - \eta\sigma(t_i, Y_i)\sigma_x(t, Y_i)\,.$$

The next code shows an implementation of the predictor-corrector method. With a (small) loss of efficiency, the new `sde.sim` function can replace the old one. Note that the predictor-corrector method falls back to the standard Euler method for $\alpha = \eta = 0$. The function by default implements the predictor-corrector method with $\alpha = \eta = \frac{1}{2}$. We only report here the modification to previous code. As usual, the complete version of `sde.sim` can be found in the `sde` package.

```
sde.sim <- function(t0=0, T=1, X0=1, N=100, delta,
         drift, sigma, sigma.x,
         method=c("euler","milstein"),
         alpha=0.5, eta=0.5, pred.corr=T){

# (...)

  if(pred.corr==F){
   alpha <- 0
   eta <- 0
   sigma.x <- NULL
  }

# (...)

  if(method=="milstein" | (method=="euler" & pred.corr==T))
   needs.sx <- TRUE

  d1.t <- function(t,x) d1(t,x) - eta * s1(t,x) * sx(t,x)

  if(method=="euler"){
   for(i in 2:(N+1)){
     Y[i] <- X[i-1] + d1(t[i-1],X[i-1])*Dt +
              s1(t[i-1],X[i-1])*sDt*Z[i-1]
     if(pred.corr==T)
      X[i] <- X[i-1] + (alpha*d1.t(t[i],Y[i]) +
              (1-alpha)* d1.t(t[i],X[i-1]))*Dt +
              (eta * s1(t[i],Y[i]) +
              (1-eta)*s1(t[i-1],Y[i-1]))*sDt*Z[i-1]
     else
      X[i] <- Y[i]
   }
  }
# (...)
}
```

There are different predictor-corrector methods available that can be applied to discretization schemes of order greater than 1. The reader should look, for example, at Section 5.3 in [142].

2.7 Strong convergence for Euler and Milstein schemes

To show in practice how strong convergence takes place in discretization schemes, we reproduce here nice empirical evidence from [125]. The objective of this section is to show how the Milstein scheme outperforms the Euler scheme in convergence in the simple case of geometric Brownian motion. As already mentioned, the theory says that convergence is in the limit of the discretization step as $\Delta t \to 0$. This experiment proceeds as follows:

1. We first simulate trajectories of the Brownian motion with an increased level of refinement (i.e., with a decreasing value of Δt). This is done iteratively using the Brownian bridge.
2. We then construct the trajectory of the geometric Brownian motion with both Euler and Milstein schemes using the path of the Wiener process available.
3. We then compare the values of the process X at time T in the two cases.

Figure 2.1 is the result of step 1, and we now explain the rationale behind it. On the top-left corner is depicted the trajectory of a Brownian bridge starting from 1 at time $t_0 = 0$ and ending at 1 at time $T = 1$ using $N = 2$ intervals, and indeed we have three points of the trajectory at times 0, $\frac{1}{2}$, and 1. The top-right picture has been generated using two Brownian bridges. The first Brownian bridge starts at 1 at time 0 and ends at $B(\frac{1}{2})$ at time $t = \frac{1}{2}$. The second Brownian bridge starts at $B(\frac{1}{2})$ at time $t = \frac{1}{2}$ and ends up at 1 at time 1. This trajectory has five points at times 0, $\frac{1}{4}$, $\frac{1}{2}$, $\frac{3}{4}$, 1. In the next step, the procedure is iterated splitting each interval $[0, \frac{1}{4}], \ldots, [\frac{3}{4}, 1]$ into two parts up to the final bottom-right picture, consisting of $2^{14}+1$ points of the trajectory of the Wiener process. The figure is generated with the following code, which we show because this dyadic algorithm is also a constructive way of building a process with continuous but nowhere differentiable path (i.e., the Wiener process).

```
> # ex 2.10.R
> set.seed(123)
> W <- vector(14,mode="list")
> W[[1]] <- BBridge(1,1,0,1,N=2)
>
> for(i in 1:13){
+  cat(paste(i,"\n"))
+  n <- length(W[[i]])
+  t <- time(W[[i]])
+  w <- as.numeric(W[[i]])
+  tmp <- w[1]
+  for(j in 1:(n-1)){
+     tmp.BB <- BBridge(w[j],w[j+1],t[j],t[j+1],N=2)
+     tmp <- c(tmp, as.numeric(tmp.BB[2:3]))
+   }
+   W[[i+1]] <- ts(tmp,start=0,deltat=1/(2^(i+1)))
+ }
> min.w <- min(unlist(W))-0.5
> max.w <- max(unlist(W))+0.5
> opar <- par(no.readonly = TRUE)
> par(mfrow=c(7,2),mar=c(3,0,0,0))
> for(i in 1:14){
```

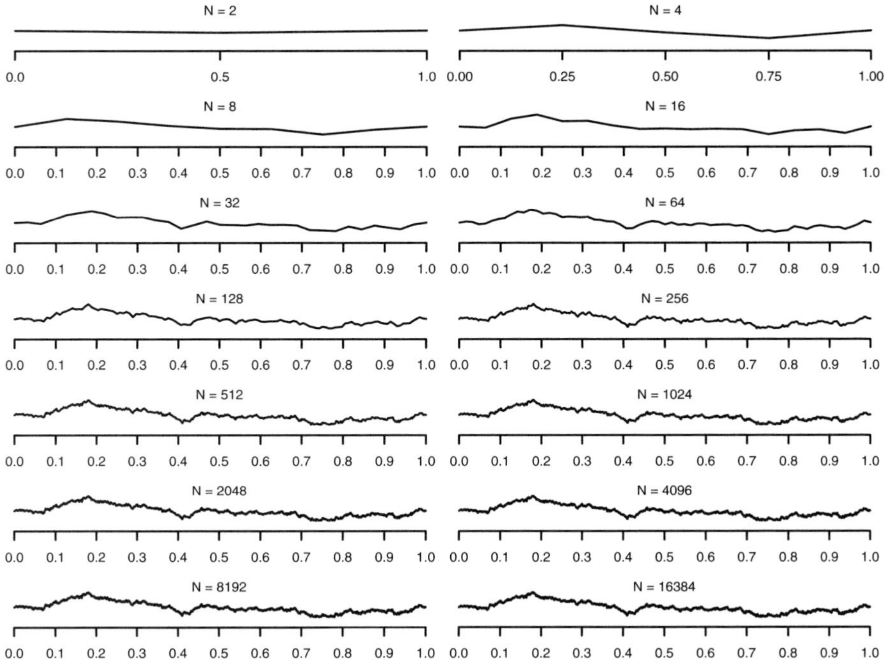

Fig. 2.1. A simulated path of the Wiener process for increased levels of discretization, N being the number of subintervals of [0,1].

```
+   plot(W[[i]], ylim=c(min.w, max.w),axes=F)
+   if(i==1)
+     axis(1,c(0,0.5,1))
+   if(i==2)
+   axis(1,c(0,0.25,0.5,0.75,1))
+   if(i>2)
+     axis(1,c(0,0.1,0.2,0.3,0.4,0.5,0.6,0.7,0.8,0.9,1))
+   text(0.5,2.2,sprintf("N = %d",2^i))
+ }
> par(opar)
```

In the R code above, we make use of the `BBridge` function of the `sde` package (see also Listing 1.2).

The next step is to simulate the trajectory of the geometric Brownian motion using both the Euler and Milstein schemes and calculate the value $X(T)$ with the two schemes. The following script does the calculations and plots both values against the true value $X(T)$ for the given Wiener process path (i.e., the Wiener process ending in 1 at time 1, which is a sort of Brownian bridge) $X(1) = \exp\{\theta_1 - \frac{1}{2}\theta_2^2\}$. We have chosen $\theta_1 = 1$ and $\theta_2 = \frac{1}{2}$.

```
> # ex 2.11.R
> S0 <- 1
> theta <- c(1, 0.5)
> euler <- NULL
> milstein <- NULL
> for(i in 1:14){
```

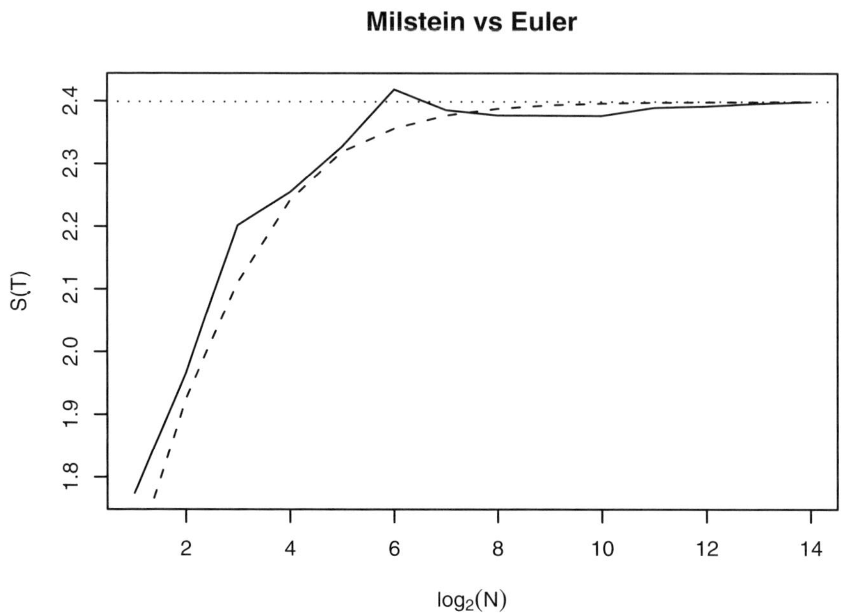

Fig. 2.2. Speed of convergence of Euler and Milstein schemes (Euler = solid line, Milstein = dashed line) to the true value (dotted line) as a function of the discretization step $\Delta t = 1/N$.

```
+     n <- length(W[[i]])
+     Dt <- 1/n
+     sDt <- sqrt(Dt)
+     E <- numeric(n)
+     E[1] <- S0
+     M <- numeric(n)
+     M[1] <- S0
+     for(j in 2:n){
+         Z <- W[[i]][j]-W[[i]][j-1]
+         E[j] <-  E[j-1] * (1 + theta[1] * Dt + theta[2] * Z)
+         M[j] <-  M[j-1] * (1 + (theta[1] - 0.5* theta[2]^2) * Dt +
+          theta[2] * Z + 0.5 * theta[2]^2 * Z^2)
+     }
+     cat(paste(E[n],M[n],"\n"))
+     euler <- c(euler, E[n])
+     milstein <- c(milstein, M[n])
+ }

> plot(1:14,euler,type="l",main="Milstein vs Euler",
+   xlab=expression(log[2](N)), ylab="S(T)")
> lines(1:14,milstein,lty=2)
> abline(h=exp(theta[1]-0.5*theta[2]^2),lty=3)
```

Figure 2.2 shows the speed of convergence of both schemes (Euler = solid line, Milstein = dashed line) to the true value (dotted line) as a function of $\Delta t = 1/N$.

2.8 KPS method of strong order $\gamma = 1.5$

By adding more terms to the Itô-Taylor expansion, one can achieve a strong order γ higher than 1 (for a detailed review and implementation, see [142]). In particular, the following scheme[3] (see, e.g., [143]) has strong order $\gamma = 1.5$:

$$\begin{aligned} Y_{i+1} = Y_i &+ b\Delta t + \sigma \Delta W_t + \frac{1}{2}\sigma\sigma_x \left\{(\Delta W_t)^2 - \Delta t\right\} \\ &+ \sigma b_x \Delta U_t + \frac{1}{2}\left\{bb_x + \frac{1}{2}\sigma^2 b_{xx}\right\}\Delta t^2 \\ &+ \left\{b\sigma_x + \frac{1}{2}\sigma^2\sigma_{xx}\right\}\{\Delta W_t \Delta t - \Delta U_t\} \\ &+ \frac{1}{2}\sigma(\sigma\sigma_x)_x \left\{\frac{1}{3}(\Delta W_t)^2 - \Delta t\right\}\Delta W_t\,, \end{aligned}$$

where

$$\Delta U_t = \int_{t_0}^{t_{i+1}} \int_{t_i}^{s} \mathrm{d}W_u \mathrm{d}s$$

is a Gaussian random variable with zero mean and variance $\frac{1}{3}\Delta t^3$ and such that $\mathbb{E}(\Delta U_t \Delta W_t) = \frac{1}{2}\Delta t^2$. All the pairs $(\Delta W_t, \Delta U_t)$ are mutually independent for all t_i's. To implement this scheme, additional partial derivatives of the drift and diffusion coefficient are required and the algorithm in `sde.sim` that generates Gaussian variates must be changed to allow for bivariate Gaussian variates for the pairs $(\Delta W_t, \Delta U_t)$. With this aim, we use the function `mvrnorm` in the `MASS` package, which implements the algorithm described in [195].

Note that the Euler scheme is not of strong order $\gamma = 1.5$ for the Ornstein-Uhlenbeck process, as there is the additional term $\sigma b_x \Delta U_t$ in the expansion. The next code implements the strong order scheme above and, as usual, we just report the additional part of the `sde.sim` function.

```
sde.sim <- function(t0=0, T=1, X0=1, N=100, delta,
         drift, sigma, drift.x, sigma.x, drift.xx, sigma.xx,
         method=c("euler","milstein","KPS"),
         alpha=0.5, eta=0.5, pred.corr=T){

# (...)

  if(method == "KPS") {
   needs.sx <- TRUE
   needs.dx <- TRUE
   needs.sxx <- TRUE
   needs.dxx <- TRUE
  }

# (...)

  if(method=="euler")
   X <- sde.sim.euler(X0, t0, Dt, N, d1, s1, s1.x, alpha, eta, pred.corr)
```

[3] We name this scheme KPS after the authors who proposed it.

```
  if(method=="milstein")
   X <- sde.sim.milstein(X0,  t0, Dt, N, d1, s1, s1.x)

  if(method=="KPS"){
    require(MASS)
    Sigma <- matrix(c(Dt, 0.5*Dt^2, 0.5*Dt^2, 1/3*Dt^3),2,2)
    tmp <- mvrnorm(N, c(0,0), Sigma)
    Z <- tmp[,1]
    U <- tmp[,2]
    X <- sde.sim.KPS(X0,  t0, Dt, N, d1, d1.x, d1.xx,
               s1, s1.x, s1.xx, Z, U)
  }

# (...)

}
```

Listing 2.1. Simulation of paths of processes governed by stochastic differential equations.

In the code of Listing 2.1, we have separated into three functions the different schemes of simulation: `sde.sim.euler`, `sde.sim.milstein`, and `sde.sim.KPS`. The reason for this approach is twofold. On the one hand, the `sde.sim` became just an interface for different schemes, hence allowing even for generalization toward the implementation of new schemes. On the other hand, separating functions allows for an implementation in C code to speed up the execution (this will be appreciated in Monte Carlo experiments). The R code corresponding to the functions above can be found in Listings 2.2, 2.3, and 2.4. Of their C counterparts we only present the Milstein scheme, in Listing 2.5, while the complete source code can be found as part of the R package `sde` available through the CRAN repository. The C code will be called in R using the following lines.

```
sde.sim.milstein <- function(X0,  t0, Dt, N, d1, s1, s1.x){
  return( .Call("sde_sim_milstein",  X0,  t0, Dt, as.integer(N), d1,
           s1, s1.x, .GlobalEnv) )
}
```

What is interesting from the reading of C code is the use of the `feval()` function. This function, defined in Listing 2.6, evaluates an R function directly from the C code and uses the result in the internal loops, speeding up the whole simulation scheme. We found that these C routines are about two times faster than their R counterparts but no more. Indeed, we are paying the cost of flexibility by allowing the `sde.sim` function to accept any sort of specification of drift and diffusion coefficients. Of course, for intensive simulation studies on a specific model, writing the complete code in C might be worth trying.

```
sde.sim.euler <- function(X0, t0, Dt, N, d1, s1, s1.x,
   alpha, eta, pred.corr){
  X <- numeric(N+1)
  Y <- numeric(N+1)
  sDt <- sqrt(Dt)
  Z <- rnorm(N, sd=sDt)
  X[1] <- X0
  Y[1] <- X0
```

```
d1.t <- function(t,x)  d1(t,x) - eta * s1(t,x) * s1.x(t,x)
  if(pred.corr==TRUE){
   for(i in 2:(N+1)){
     Y[i] <- X[i-1] + d1(t[i-1],X[i-1])*Dt + s1(t[i-1],X[i-1])*Z[i-1]
     X[i] <- X[i-1] + (alpha*d1.t(t[i],Y[i]) +
              (1-alpha)* d1.t(t[i],X[i-1]))*Dt +
              (eta * s1(t[i],Y[i]) +
              (1-eta)*s1(t[i-1],Y[i-1]))*Z[i-1]
   }
  } else {
   for(i in 2:(N+1))
    X[i] <- X[i-1] + d1(t[i-1],X[i-1])*Dt + s1(t[i-1],X[i-1])*Z[i-1]
  }
  return(X)
}
```

Listing 2.2. R code for Euler simulation scheme.

```
sde.sim.milstein <- function(X0,  t0, Dt, N, d1, s1, s1.x){
  X <- numeric(N+1)
  Y <- numeric(N+1)
  sDt <- sqrt(Dt)
  Z <- rnorm(N, sd=sDt)

  X[1] <- X0
  Y[1] <- X0

  for(i in 2:(N+1)){
    X[i] <- X[i-1] + d1(t[i-1],X[i-1])*Dt +
             s1(t[i-1],X[i-1])*Z[i-1] +
             0.5*s1(t[i-1],X[i-1])* s1.x(t[i-1],X[i-1]) *
             (Z[i-1]^2-Dt)
  }
 return(X)
}
```

Listing 2.3. R code for Milstein simulation scheme.

```
sde.sim.KPS <- function(X0,  t0, Dt, N, d1, d1.x, d1.xx,
   s1, s1.x, s1.xx, Z, U){
  X <- numeric(N+1)
  Y <- numeric(N+1)
  Dt <- (T-t0)/N
  sDt <- sqrt(Dt)
  X[1] <- X0
  Y[1] <- X0

   for(i in 2:(N+1)){
    D1 <- d1(t[i-1],X[i-1])
    D1.x <- d1.x(t[i-1],X[i-1])
    D1.xx <- d1.xx(t[i-1],X[i-1])
    S1 <- s1(t[i-1],X[i-1])
    S1.x <- s1.x(t[i-1],X[i-1])
    S1.xx <- s1.xx(t[i-1],X[i-1])

    X[i] <- X[i-1] + D1 * Dt + S1 * Z[i-1] +
         0.5 * S1 * S1.x * (Z[i-1]^2-Dt) +
    S1 * D1.x * U[i-1] +
    0.5 * (D1 * D1.x + 0.5 * S1^2 * D1.xx) * Dt^2 +
    (D1 * S1.x + 0.5 * S1^2 * S1.xx) * (Z[i-1] * Dt - U[i-1]) +
     0.5 * S1 * (S1.x^2 + S1*S1.xx) * (1/3*Z[i-1]^2 - Dt) * Z[i-1]
   }
  return(X)
```

```
}
```

Listing 2.4. R code for KPS simulation scheme.

```
SEXP sde_sim_milstein(SEXP x0, SEXP t0, SEXP delta, SEXP N,
                      SEXP d, SEXP s, SEXP sx, SEXP rho)
{
  double T, DELTA;
  double sdt, Z, tmp, D, S, Sx;
  int i, n;
  SEXP X;

  if(!isNumeric(x0)) error("`x0' must be numeric");
  if(!isNumeric(t0)) error("`t0' must be numeric");
  if(!isNumeric(delta)) error("`delta' must be numeric");
  if(!isInteger(N)) error("`N' must be integer");
  if(!isFunction(d)) error("`d' must be a function");
  if(!isFunction(s)) error("`s' must be a function");
  if(!isFunction(sx)) error("`sx' must be a function");
  if(!isEnvironment(rho)) error("`rho' must be an environment");

  PROTECT(x0 = AS_NUMERIC(x0));
  PROTECT(delta = AS_NUMERIC(delta));
  PROTECT(t0 = AS_NUMERIC(t0));
  PROTECT(N = AS_INTEGER(N));

  T = *NUMERIC_POINTER(t0);
  n = *INTEGER_POINTER(N);
  DELTA = *NUMERIC_POINTER(delta);

  PROTECT(X = NEW_NUMERIC(n+1));
  REAL(X)[0] = *NUMERIC_POINTER(x0);
  sdt = sqrt(DELTA);

  GetRNGstate();
  for(i=1; i<= n+1; i++){
   Z = rnorm(0,sdt);
   T = T + DELTA;
   tmp = REAL(X)[i-1];
   D = feval(T,tmp,d,rho);
   S = feval(T,tmp,s,rho);
   Sx = feval(T,tmp,sx,rho);
   REAL(X)[i] = tmp + D*DELTA + S*Z + 0.5*S*Sx*(Z*Z-DELTA);
  }
  PutRNGstate();

  UNPROTECT(5);
  return(X);
}
```

Listing 2.5. C code for Milstein simulation scheme.

```
/*
   t : time variable
   x : space variable
   f : a SEXP to a R function
   rho : the environment `f' is going to be evaluated

   on return:   the value of f(t,x)
*/
double feval(double t, double x, SEXP f, SEXP rho)
{
 double val= 0.0;
 SEXP R_fcall, tpar, xpar;
```

```
  PROTECT(tpar = allocVector(REALSXP, 1));
  PROTECT(xpar = allocVector(REALSXP, 1));
  REAL(tpar)[0] = t;
  REAL(xpar)[0] = x;

  PROTECT(R_fcall = allocList(3));
  SETCAR(R_fcall, f);
  SET_TYPEOF(R_fcall, LANGSXP);
  SETCADR(R_fcall, tpar);
  SETCADDR(R_fcall, xpar);

  val = *NUMERIC_POINTER(eval(R_fcall, rho));
  UNPROTECT(3);
  return(val);
}
```

Listing 2.6. C code for the `feval` function, which allows for the calculation of R functions directly in the C code.

2.9 Second Milstein scheme

In [164], Milstein proposed both (2.2) and the approximation

$$
\begin{aligned}
Y_{i+1} = Y_i &+ \left(b - \frac{1}{2}\sigma\sigma_x\right)\Delta t + \sigma Z\sqrt{\Delta t} + \frac{1}{2}\sigma\sigma_x \Delta t Z^2 \\
&+ \Delta t^{\frac{3}{2}}\left(\frac{1}{2}b\sigma_x + \frac{1}{2}b_x\sigma + \frac{1}{4}\sigma^2\sigma_{xx}\right) Z \\
&+ \Delta t^2\left(\frac{1}{2}bb_x + \frac{1}{4}b_{xx}\sigma^2\right).
\end{aligned}
\tag{2.9}
$$

This method has weak second-order convergence in contrast to the weak first-order convergence of the Euler scheme. This method requires partial (first and second) derivatives of both drift and diffusion coefficients. Listing 2.7 contains the code corresponding to the approximation (2.9). The function `sde.sim` needs to be modified as follows.

```
 if(method == "KPS" | method == "milstein2") { # added "milstein2" method
  needs.sx <- TRUE
  needs.dx <- TRUE
  needs.sxx <- TRUE
  needs.dxx <- TRUE
 }
# (...)

 if(method=="milstein2") # added a call to the sde.sim.milstein2
  X <- sde.sim.milstein2(X0, t0, Dt, N, d1, d1.x, d1.xx, s1, s1.x, s1.xx)
```

```
sde.sim.milstein2 <- function(X0, t0, Dt, N, d1, d1.x, d1.xx,
      s1, s1.x, s1.xx){
  X <- numeric(N+1)
  Y <- numeric(N+1)
  sDt <- sqrt(Dt)
```

```
  Z <- rnorm(N, sd=sDt)
  X[1] <- X0
  Y[1] <- X0

  for(i in 2:(N+1)){
    X[i] <- X[i-1] + d1(t[i-1],X[i-1])*Dt +
              s1(t[i-1],X[i-1])*Z[i-1] +
              0.5*s1(t[i-1],X[i-1])* s1.x(t[i-1],X[i-1]) *(Z[i-1]^2-Dt) +
          Dt^(3/2)*(0.5*d1(t[i-1],X[i-1])*s1.x(t[i-1],X[i-1]) +
          0.5*d1.x(t[i-1],X[i-1])*s1(t[i-1],X[i-1])+
          0.25*s1(t[i-1],X[i-1])^2 * s1.xx(t[i-1],X[i-1]))*Z[i-1] +
          Dt^2*(0.5* d1(t[i-1],X[i-1])*d1.x(t[i-1],X[i-1])+
          0.25*d1.xx(t[i-1],X[i-1])*s1(t[i-1],X[i-1])^2)
  }
 return(X)
}
```

Listing 2.7. R code for the second Milstein simulation scheme.

2.10 Drawing from the transition density

All the methods presented so far are based on the discretized version of the stochastic differential equation. In the case where a transition density of X_t given some previous value X_s, $s < t$, is known in explicit form, direct simulation from this can be done. Unfortunately, the transition density is known for very few processes, and these cases are the ones for which exact likelihood inference can be done, as will be discussed in Chapter 3. In these fortunate cases, the algorithm for simulating processes is very easy to implement. We suppose that a random number generator is available for the transition density for the process $p_\theta(\Delta, y|x) = Pr(X_{t+\Delta} \in \mathrm{d}y|X_t = x)$. If this generator is not available one can always use one of the standard methods to draw from known densities, such as the rejection method. We are not going to discuss this approach here and assume this random number generator exists in the form of an R function that accepts the number `n` of pseudo random numbers to draw, a vector of length `n` of values `x` (this will play the role of the X_t's), the time lags `Dt`, and a vector of parameters `theta`. The fact that we allow for `n` random numbers to be generated will be useful whenever one wants to simulate multiple trajectories of the same process in a way we discuss at the end of the chapter. The next function generates a complete path of a stochastic process for which the random number generator `rcdist` is known. We assume that the corresponding model is parametrized through a vector of parameters `theta`.

```
sde.sim.cdist <- function(X0=1, t0=0, Dt=0.1, N, rcdist=NULL, theta=NULL){
   X <- numeric(N+1)
   X[1] <- X0
   for(i in 2:(N+1)){
    X[i] <- rcdist(1, Dt, X[i-1], theta)
   }
 ts(X, start=t0, deltat=Dt)
}
```

The function `sde.sim.cdist` just iterates calls to the random number generator `rcdist`, assigning to `X[i]` the pseudo random number generated from the law of $X_{t+\Delta t}|X_t =$ `X[i-1]`. We now write the random number generators for the processes for which the conditional distribution is known.

2.10.1 The Ornstein-Uhlenbeck or Vasicek process

Recall that the Ornstein-Uhlenbeck or Vasicek process solution to

$$\mathrm{d}X_t = (\theta_1 - \theta_2 X_t)\mathrm{d}t + \theta_3 \mathrm{d}W_t, \qquad X_0 = x_0,$$

has a known Gaussian transition density $p_\theta(\Delta, y|X_t = x)$ with mean and variance as in (1.41) and (1.42), respectively. Hence we can just make use of the `rnorm` function to build our random number generator.

```
rcOU <- function(n=1, t, x0, theta){
  Ex <- theta[1]/theta[2]+(x0-theta[1]/theta[2])*exp(-theta[2]*t)
  Vx <- theta[3]^2*sqrt((1-exp(-2*theta[2]*t))/(2*theta[2]))
  rnorm(n, mean=Ex, sd = sqrt(Vx))
}
```

The functions `[rpdq]cOU` in the `sde` package provide interfaces to random number generation, cumulative distribution function, density function, and quantile calculations, respectively, for the *conditional* law of the Ornstein-Uhlenbeck process. Similarly, the functions `[rpdq]sOU` provide the same functionalities for the *stationary* law of the process.

2.10.2 The Black and Scholes process

The Black and Scholes or geometric Brownian motion process solution of

$$\mathrm{d}X_t = \theta_1 X_t \mathrm{d}t + \theta_2 X_t \mathrm{d}W_t$$

has a log-normal transition density $p_\theta(\Delta, y|x)$, where the log-mean and log-variance are given in (1.43). The following code implements the random number generator from the conditional law.

```
rcBS <- function(n=1, Dt, x0, theta){
   lmean <- log(x0) + (theta[1]-0.5*theta[2]^2)*Dt
   lsd <- sqrt(Dt)*theta[2]
   rlnorm(n, meanlog = lmean, sdlog = lsd)
}
```

The package `sde` provides the functions `[rpdq]cBS` for random number generation, cumulative distribution function, density function, and quantile calculations of the conditional law of $X_{t+\Delta}|X_t$.

2.10.3 The CIR process

The conditional density of $X_{t+\Delta}|X_t = x$ for the Cox-Ingersoll-Ross process solution of

$$\mathrm{d}X_t = (\theta_1 - \theta_2 X_t)\mathrm{d}t + \theta_3\sqrt{X_t}\mathrm{d}W_t$$

is a noncentral chi-squared distribution (see Section 1.13.3). In particular, we have shown in (1.48) that $p_\theta(\Delta, y|x)$ can be rewritten in terms of the transition density of $Y_t = 2cX_t$, which has a chi-squared distribution with $\nu = 4\theta_1/\theta_3^2$ degrees of freedom and noncentrality parameter $Y_s e^{-\theta_2 t}$, where $c = 2\theta_2/(\theta_3^2(1-e^{-\theta_2 t}))$. So we just need to simulate a value of y from $Y_t|Y_s = 2cx_s$ and return $y/(2c)$. The following code does the job.

```
rcCIR <- function(n=1, Dt, x0, theta){
   c <- 2*theta[2]/((1-exp(-theta[2]*t))*theta[3]^2)
   ncp <- 2*c*x0*exp(-theta[2]*Dt)
   df <- 4*theta[1]/theta[3]^2
   rchisq(n, df=df, ncp=ncp)/(2*c)
}
```

Also, for the Cox-Ingersoll-Ross process, the package `sde` provides the functions `[rpdq]cCIR` and `[rpdq]sCIR` for random number generation, cumulative distribution function, density function, and quantile calculations, respectively, for the conditional and stationary laws.

2.10.4 Drawing from one model of the previous classes

Given that for the Cox-Ingersoll-Ross, Ornstein-Uhlenbeck, and geometric Brownian motion processes the transition densities are known in explicit form, we can add a more flexible interface in `sde.sim` to simulate these models. In fact, we add the switch `model`, which can be one between "`OU`," "`BS`," and "`CIR`" (with the obvious meanings) and a vector of parameter `theta`. So, for example,

```
sde.sim(model="CIR", theta=c(3,2,1))
```

simulates a path of the Cox-Ingersoll-Ross process solution of $\mathrm{d}X_t = (3 - 2X_t)\mathrm{d}t + \sqrt{X_t}\mathrm{d}W_t$ with initial value $X_0 = 1$ (the default value in `sde.sim`). Below we show the relevant code change to the `sde.sim` function.

```
sde.sim <- function (t0 = 0, T = 1, X0 = 1, N = 100, delta, drift, sigma,
    drift.x, sigma.x, drift.xx, sigma.xx, drift.t, method = c("euler",
       "milstein", "KPS", "milstein2", "cdist"),
    alpha = 0.5, eta = 0.5, pred.corr = T, rcdist = NULL, theta = NULL,
    model = c("CIR", "VAS", "OU", "BS"))
{
    method <- match.arg(method)
    if(!missing(model)){
     model <- match.arg(model)
     method <- "model"
        }

    if (missing(drift)){
     if (method == "cdist" || !missing(model))
      drift <- expression(NULL)
     else
        stop("please specify al least the drift coefficient of the SDE")
    }

# (...)
```

```
    if(method == "model"){
     if(is.null(theta))
       stop("please provide a vector of parameters for the model")
    if(model == "CIR")
       X <- sde.sim.cdist(X0, t0, Dt, N, rcCIR, theta)
    if(model == "OU")
       X <- sde.sim.cdist(X0, t0, Dt, N, rcOU, theta)
    if(model == "BS")
       X <- sde.sim.cdist(X0, t0, Dt, N, rcBS, theta)
     }
# (...)
}
```

2.11 Local linearization method

The local linearization method consists in approximating locally the drift of the stochastic differential equation with a linear function. The main idea behind this technique is that a linear approximation is better than the simple constant approximation made by the Euler method (see, e.g., [24], [13]). The method has been proposed in the context of stochastic differential equations by Ozaki and developed by him and his co-authors (see [173], [174], [175], [204], [206], [207]).

2.11.1 The Ozaki method

The first approach we present is the Ozaki method, and it works for homogeneous stochastic differential equations. Consider the stochastic differential equation

$$\mathrm{d}X_t = b(X_t)\mathrm{d}t + \sigma \mathrm{d}W_t\,, \tag{2.10}$$

where σ is supposed to be constant. The construction of the method starts from the corresponding deterministic dynamical system

$$\frac{\mathrm{d}x_t}{\mathrm{d}t} = b(x_t)\,,$$

where x_t has to be a smooth function of t in the sense that it is two times differentiable with respect to t. Then, with a little abuse of notation, we have

$$\frac{\mathrm{d}^2 x_t}{\mathrm{d}t^2} = b_x(x_t)\frac{\mathrm{d}x_t}{\mathrm{d}t}\,.$$

Suppose now that $b_x(x)$ is constant in the interval $[t, t+\Delta t)$, and hence by iterated integration of both sides of the equation above, first from t to $u \in [t, t+\Delta t)$ and then from t to $t+\Delta t$, we obtain the difference equation

$$x_{t+\Delta t} = x_t + \frac{b(x_t)}{b_x(x_t)}\left(e^{b_x(x_t)\Delta t} - 1\right)\,. \tag{2.11}$$

Now we translate the result above back to the stochastic dynamical system in (2.10). So, suppose $b(x)$ is approximated by the linear function $K_t x$, where K_t is constant in the interval[4] $[t, t+\Delta t)$. The solution to the stochastic differential equation is

$$X_{t+\Delta t} = X_t e^{K_t \Delta t} + \sigma \int_t^{t+\Delta t} e^{K_t(t+\Delta t - u)} \mathrm{d}W_u \,.$$

Now what remains to be done is to determine the constant K_t. The main assumption is that the conditional expectation of $X_{t+\Delta t}$ given X_t,

$$\mathbb{E}(X_{t+\Delta t}|X_t) = X_t e^{K_t \Delta t},$$

coincides with the state of the linearized dynamical system (2.11) at time $t + \Delta t$, which means that we ask for the following equality to hold:

$$X_t e^{K_t \Delta t} = X_t + \frac{b(X_t)}{b_x(X_t)} \left(e^{b_x(X_t)\Delta t} - 1 \right).$$

From the above, we obtain the constant K_t very easily:

$$K_t = \frac{1}{\Delta t} \log \left(1 + \frac{b(X_t)}{X_t b_x(X_t)} \left(e^{b_x(X_t)\Delta t} - 1 \right) \right).$$

Notice that K_t depends on t only through the state of the process $X_t = x$. Hence we denote this constant by K_x to make the notation more consistent throughout the book. Given that the stochastic integral is a Gaussian random variable, it is clear that the transition density for $X_{t+\Delta t}$ given X_t is Gaussian as well. In particular, we have that $X_{t+\Delta t}|X_t = x \sim N(E_x, V_x)$, where

$$E_x = x + \frac{b(x)}{b_x(x)} \left(e^{b_x(x)\Delta t} - 1 \right), \tag{2.12}$$

$$V_x = \sigma^2 \frac{e^{2K_x \Delta t} - 1}{2K_x}, \tag{2.13}$$

with

$$K_x = \frac{1}{\Delta t} \log \left(1 + \frac{b(x)}{x b_x(x)} \left(e^{b_x(x)\Delta t} - 1 \right) \right).$$

So it is possible to use the method of drawing from the conditional law to simulate the increments of the process as in Section 2.10, simulating `X[i+1]` according to $N(E_x, V_x)$, where $x =$ `X[i]`. It is easy to implement this simulation scheme, and it is presented in Listing 2.8. The function `sde.sim.ozaki` will be called by `sde.sim` when `method` is equal to "`ozaki`." This function assumes that the diffusion coefficient is a constant and that the drift function depends only on the state variable `x`. These assumptions are checked inside the `sde.sim` interface as follows.

[4] Hence the name "local linearization method."

```
if (method == "ozaki"){
      vd <- all.vars(drift)
      vs <- all.vars(sigma)
      if(length(vd)!=1 || length(vs)>0)
        stop("drift must depend on `x' and volatility must be constant")
      if((length(vd) == 1) && (vd != "x"))
        stop("drift must depend on `x'")
     X <- sde.sim.ozaki(X0, t0, Dt, N, d1, d1.x, s1)
}
```

Please note that a constant drift is not admissible since K_x and hence V_x are not well defined.

```
"sde.sim.ozaki" <-
function(X0, t0, Dt, N, d1, d1.x, s1){
   X <- numeric(N+1)
   B <- function(x) d1(1,x)
   Bx <- function(x) d1.x(1,x)
   S <- s1(1,1)
   X[1] <- X0
   for(i in 2:(N+1)){
    x <- X[i-1]
    Kx <- log(1+B(x)*(exp(Bx(x)*Dt)-1)/(x*Bx(x)))/Dt
    Ex <- x + B(x)/Bx(x)*(exp(Bx(x)*Dt)-1)
    Vx <- S^2 * (exp(2*Kx*Dt) -1)/(2*Kx)
    X[i] <- rnorm(1, mean=Ex, sd=sqrt(Vx))
   }
 X
}
```

Listing 2.8. R code for the Ozaki simulation scheme.

Of course, the Ozaki method coincides with the Euler method if the drift is linear. The reader can try the following lines of code.

```
> # ex 2.12.R
> set.seed(123)
> X <- sde.sim(drift=expression(-3*x), method="ozaki")
> set.seed(123)
> Y <- sde.sim(drift=expression(-3*x))
> plot(X)
> lines(as.numeric(time(Y)), Y, col="red")
```

2.11.2 The Shoji-Ozaki method

An extension of the previous method to the more general case in which the drift is allowed to depend on the time variable also and the diffusion coefficient can vary is the Shoji-Ozaki method (see [204], [205], and [206]). Consider the stochastic differential equation

$$\mathrm{d}X_t = b(t, X_t)\mathrm{d}t + \sigma(X_t)\mathrm{d}W_t\,,$$

where b is two times continuously differentiable in x and continuously differentiable in t and σ is continuously differentiable in x. We already know that it is always possible to transform this equation into one with a constant diffusion coefficient using the Lamperti transform of Section 1.11.4. So one can start by considering the nonhomogeneous stochastic differential equation

$$dX_t = b(t, X_t)dt + \sigma dW_t\,,$$

which is different from (2.10) in that the drift function also depends on variable t. Now the local linearization method is developed by studying the behavior of b locally. We skip all the details, which can be found in the original works [207] and [206], but the main point is that the equation above is approximated locally on $[s, s + \Delta s)$ with

$$dX_t = (L_s X_t + tM_s + N_s)dt + \sigma dW_t, \qquad t \geq s,$$

where

$$L_s = b_x(s, X_s), \qquad M_s = \frac{\sigma^2}{2} b_{xx}(s, X_s) + b_t(s, X_s)\,,$$
$$N_s = b(s, X_s) - X_s b_x(s, X_s) - sM_s\,.$$

The next step is to consider the transformed process $Y_t = e^{-L_s t}X_t$, which has the explicit solution

$$Y_t = Y_s + \int_s^t (M_s u + N_s)e^{-L_s u}du + \sigma \int_s^t e^{-L_s u}dW_u,$$

from which the discretization of X_t can be easily obtained and reads as

$$X_{s+\Delta s} = A(X_s)X_s + B(X_s)Z,$$

where

$$A(X_s) = 1 + \frac{b(s, X_s)}{X_s L_s}\left(e^{L_s \Delta s} - 1\right) + \frac{M_s}{X_s L_s^2}\left(e^{L_s \Delta s} - 1 - L_s \Delta s\right), \tag{2.14}$$

$$B(X_s) = \sigma \sqrt{\frac{e^{2L_s \Delta s} - 1}{2L_s}}, \tag{2.15}$$

and $Z \sim N(0, 1)$. From the above, it follows that

$$X_{s+\Delta s} | X_s = x \sim N(A(x)x, B^2(x))\,.$$

This method is also quite easy to implement and is just a modification of the previous method. For the sake of simplicity, we will call this the "`shoji`" method to distinguish it in the R code. The only difference with respect to all previous methods is that we also need to specify the partial derivative of the drift coefficient with respect to variable t. This will be an argument of `sde.sim` called `drift.t` and eventually calculated using R symbolic differentiation. We skip the corresponding code and present just Listing 2.9, which implements the simulation part.

```
sde.sim.shoji <- function(X0, t0, Dt, N, d1, d1.x, d1.xx, d1.t, s1){
  X <- numeric(N+1)
  S <- s1(1,1)
  X[1] <- X0
```

```
    for(i in 2:(N+1)){
      x <- X[i-1]
      Lx <- d1.x(Dt,x)
      Mx <- S^2 * d1.xx(Dt,x)/2 + d1.t(Dt,x)
      Ex <- (x + d1(Dt,x)*(exp(Lx*Dt)-1)/Lx +
            Mx*(exp(Lx*Dt) -1 -Lx*Dt)/Lx^2)
      Vx <- S^2*(exp(2*Lx*Dt)-1)/(2*Lx)
      X[i] <- rnorm(1, mean=Ex, sd=sqrt(Vx))
    }
  X
}
```

Listing 2.9. R code for the Shoji-Ozaki simulation scheme.

Please note that in this case as well, a drift function not depending on x is not admissible since $A(X_s)$ is not well-defined. One more thing to note is that the Ozaki, Shoji-Ozaki, and Euler methods draw increments from a Gaussian law with mean E_x and variance V_x that in the case of the Euler scheme are $E_x = x + b(x,t)\mathrm{d}t$ and $V_x = V = \sigma^2\mathrm{d}t$. For the Euler method, the variance V_x is independent from the previous state of the process $X_t = x$, and this property is inherited from the independence of the increments of the Brownian motion. On the contrary, V_x for the Ozaki and Shoji-Ozaki methods depend on the previous state of the process and differ from the value of the constants K_x and L_x, respectively. Even in the linear case, the Shoji-Ozaki method performs differently from the Euler and Ozaki methods. The difference is in the fact that the Shoji-Ozaki method also takes into account the stochastic behavior of the discretization because of the Itô formula. Of course, in the linear homogeneous case the Euler, Shoji-Ozaki, and Ozaki methods coincide. One added value in using the Shoji-Ozaki method over the Ozaki and Euler methods is that it is more stable if the time Δ is large. In fact, not surprisingly, the Euler scheme tends to explode in non-linear cases when Δ is large enough. The following example shows some empirical evidence of this fact. We simulate the solution of $\mathrm{d}X_t = (5 - 11X_t + 6X_t^2 - X_t^3)\mathrm{d}t + \mathrm{d}W_t$, $X_0 = 5$ for $\Delta = 0.1$ and $\Delta = 0.25$. For small values of Δ, all three methods gave similar results, but this is not the case for $\Delta = 0.25$, as can be seen in Figure 2.3, produced with the following code.

```
> # ex 2.13.R
> bX <- expression((5 - 11 * x + 6 * x^2 - x^3))
> x0 <- 5
> DT <- 0.1
> par(mfrow=c(2,3))
> set.seed(123)
> X <- sde.sim(drift=bX, delta=DT,X0=x0)
> plot(X,main="Euler")
> set.seed(123)
> Y <- sde.sim(drift=bX, method="ozaki",delta=DT,X0=x0)
> plot(Y,main="Ozaki")
> set.seed(123)
> Z <- sde.sim(drift=bX, method="shoji",delta=DT,X0=x0)
> plot(Z,main="Shoji-Ozaki")
>
> DT <- 0.25
> set.seed(123)
> X <- sde.sim(drift=bX, delta=DT,X0=x0)
> plot(X, main="Euler")
```

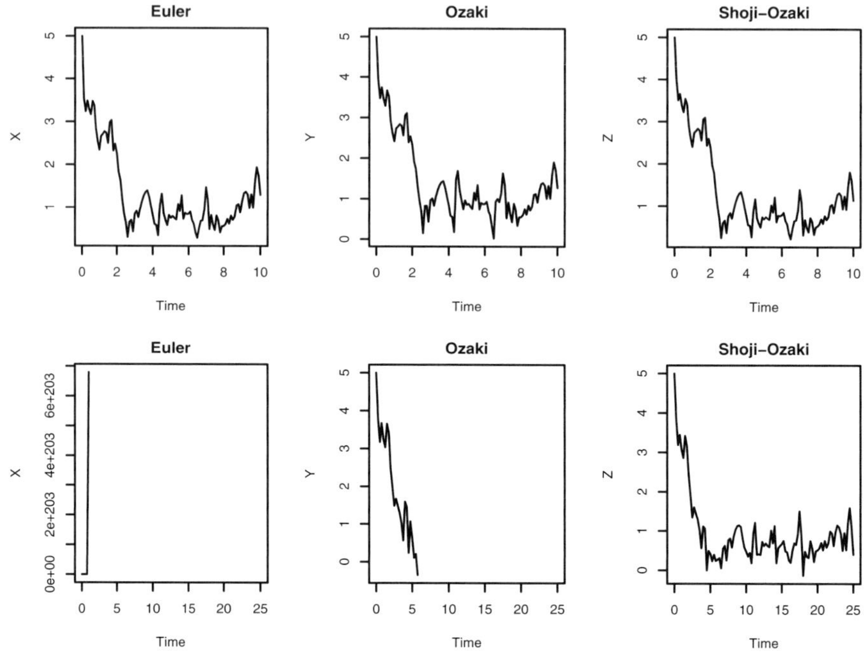

Fig. 2.3. Different performance of the Euler, Ozaki, and Shoji-Ozaki methods for different values of Δ (top: 0.1; bottom: 0.25). The Euler scheme explodes for high values of Δ.

```
> set.seed(123)
> Y <- sde.sim(drift=bX, method="ozaki",delta=DT,X0=x0)
> plot(Y,main="Ozaki")
> set.seed(123)
> Z <- sde.sim(drift=bX, method="shoji",delta=DT,X0=x0)
> plot(Z,main="Shoji-Ozaki")
```

Further properties of the method

As for the properties of this method, the authors show that the Shoji-Ozaki discretization performs well in terms of one-step-ahead error in mean absolute and mean square values. In particular, the mean absolute one-step-ahead error is of order $O(\Delta t^2)$ and the mean square one-step-ahead error is of order $O(\Delta t^3)$ as $\Delta t \to 0$. These errors are measured in terms of the distance between the true trajectory and the approximated trajectory. In particular, the mean square error attains the optimum rate in the sense of Rümelin [198].

Nonconstant diffusion coefficient

If the original stochastic differential equation X_t does not have a constant diffusion coefficient, it is always possible to apply the Lamperti transform of

Section 1.11.4 to obtain a new process Y_t that has a unitary diffusion coefficient. So one can simulate the path of $Y_t = F(X_t)$ and then use the inverse transform F^{-1} to get X_t's path. So, for example, following [207], consider the stochastic differential equation

$$\mathrm{d}X_t = (\alpha_0 + \alpha_1 X_t + \alpha_2 X_t^2 + \alpha_3 X_t^3)\mathrm{d}t + \sigma X_t^\gamma \mathrm{d}W_t\,,$$

where $(\alpha_0 = 6, \alpha_1 = -11, \alpha_2 = 6, \alpha_3 = -1, \gamma = 0.5, \sigma = 1)$. The transformation

$$F(x) = \frac{1}{\sigma}\int_0^x \frac{1}{u^\gamma}\mathrm{d}u = \frac{1}{\sigma}\frac{x^{1-\gamma}}{1-\gamma}$$

and its inverse

$$F^{-1}(y) = (\sigma y(1-\gamma))^{\frac{1}{1-\gamma}}$$

can be used to apply the Shoji-Ozaki method to the stochastic differential equation above. The drift function of process Y_t is then

$$b_y(t,x) = \frac{b(t,x)}{\sigma x^\gamma} - \frac{\gamma\sigma}{2}x^{\gamma-1},$$

which has to be calculated in $F^{-1}(y)$. For the particular choice of σ and γ, we have that $F^{-1}(y) = (y/2)^2$. Hence the process Y_t satisfies

$$\mathrm{d}Y_t = \frac{23 - 11Y_t^2 + \frac{3}{2}Y_t^4 - \frac{1}{2^4}Y_t^6}{2Y_t}\mathrm{d}t + \mathrm{d}W_t, \quad Y_0 = 2\sqrt{X_0}\,.$$

Once we have the trajectory of Y_t, we can get a trajectory of X_t by means of the transformation $X_t = (Y_t/2)^2$. The following lines of code show how to proceed.

```
> # ex 2.14.R
> bY <- expression( (23-11*x^2+1.5*x^4-(x^6)/(2^4))/(2*x) )
> bX <- expression( (6-11*x+6*x^2-x^3) )
> sX <- expression( sqrt(x) )
>
> set.seed(123)
> X <- sde.sim(drift=bX, sigma=sX)
> plot(X)
> set.seed(123)
> Y <- sde.sim(drift=bY, X0 = 2, method="shoji")
> plot((Y/2)^2)
```

2.12 Exact sampling

Very recently there appeared a new proposal for an exact sampling algorithm that, when feasible, is also easy to implement (see [27] and [29]). This method is a rejection sampling algorithm (see, e.g., [227], [68]) for diffusion processes. The rejection sampling algorithm for finite-dimensional random variables is as follows. Suppose f and g are two densities with respect to some measure in $\mathbb{R}^d$ and such that $f(x) \le \epsilon g(x)$ for some $\epsilon > 0$. If one wants to simulate pseudorandom numbers from f and knows how to simulate from g, then one can use the following algorithm:

1. Sample y from g, $y \sim g$.
2. Sample u from the uniform law, $u \sim U(0,1)$.
3. If $u < \epsilon f(y)/g(y)$, retain y; otherwise iterate from 1.

Then y will be f-distributed. This algorithm needs some modifications if it is to be applied to continuous-time processes such as diffusions because in principle there is the need to generate a whole continuous path of the diffusion[5] before accepting/rejecting it and not just a simple number. Luckily, such an exact sampling algorithm relies on the fact that the rejection rule can be made equivalent to the realization of some event related to point processes and hence simpler to handle. We skip here all the details, but these and other considerations constitute the core of the algorithm proposed in [27]. The algorithm is given for a diffusion with unit diffusion coefficient

$$\mathrm{d}X_t = b(X_t)\mathrm{d}t + \mathrm{d}W_t, \qquad 0 \le t \le T, \quad X_0 = x\,. \tag{2.16}$$

Again, if this is not the case, the Lamperti transform in Section 1.11.4 can be used. From now on, we discuss the algorithm for the exact simulation of the random variable X_Δ for some $\Delta > 0$ and initial value $X_0 = x$. We assume that $b(\cdot)$ satisfies the usual conditions for the existence of the stochastic differential equation (2.16) and also the following assumption.

Assumption 2.1

(i) The derivative b_x of b exists.
(ii) There exist k_1 and k_2 such that $k_1 \le \frac{1}{2}b^2(x) + \frac{1}{2}b_x(x) \le k_2$ for any $x \in \mathbb{R}$.

Let us denote $\phi(x) = \frac{1}{2}b^2(x) + \frac{1}{2}b_x(x) - k_1$. A further requirement is that $0 \le \phi(x) \le M$ for any $x \in \mathbb{R}$, with $M = (k_2 - k_1)$, which implies also that $\Delta \le 1/M$ for identifiability. Now set

$$A(z) = \int_0^z b(u)\mathrm{d}u$$

and

$$h(z) = \exp\left\{A(z) - \frac{(z-x)^2}{2\Delta}\right\}, \quad K = \int_{-\infty}^{\infty} h(u)\mathrm{d}u\,.$$

The function $\tilde{h}(x) = h(x)/K$ is a density function on $\mathbb{R}$ whenever $K < \infty$. The algorithm requires the ability to generate[6] pseudo random numbers from the density $\tilde{h}$. We present here a version of the exact algorithm (EA) in the simplified form as described in [56].

1. Simulate $Y_\Delta = y$ according to distribution $\tilde{h}$.

---

[5] The ratio $f(y)/g(y)$ of the algorithm should be a likelihood ratio in the case of a diffusion, as given by the Girsanov theorem.

[6] In this case, it is possible to draw from h using a reject sampling algorithm with Gaussian proposals.

2. Simulate $\tau = k$ from the Poisson distribution with intensity $\lambda = \Delta M$.
3. Draw $(T_i, V_i) = (t_i, v_i)$ according to $U[(0, \Delta) \times (0, M)]$, $i = 1, \dots, k$.
4. Generate a Brownian bridge starting at x at time 0 and ending in y at time Δ at time instants t_i; i.e., generate $Y_{t_i} = y_i$, $i = 1, \dots, k$.
5. Compute the indicator function

$$I = \prod_{i=1}^{k} \mathbf{1}_{\{\phi(y_i) \le v_i\}}.$$

6. If $I = 1$, the trajectory $(x, Y_{t_1} = y_1, \dots, Y_{t_k} = y_k, Y_\Delta = y)$ is accepted and Y_Δ is an exact draw of X_Δ. Otherwise, restart from step 1.

Two approaches are possible if one wants to simulate a process up to an arbitrary time T: either set $\Delta = T$ or set $\Delta = T/N$ and iterate the algorithm N times. As in [56], we suggest keeping only the last value of X_Δ and simulating the next one $X_{2\Delta}$ (conditionally on $X_\Delta = y$) up to the final time T. The advantage of this approach is that we get a path of the process on a regular grid, which makes this path compatible with the other schemes presented in this book. The EA algorithm is implemented in Listing 2.10, and the relevant changes to the `sde.sim` function are given below.

```
sde.sim <- function (t0 = 0, T = 1, X0 = 1, N = 100, delta, drift, sigma,
    drift.x, sigma.x, drift.xx, sigma.xx, drift.t, method = c("euler",
        "milstein", "KPS", "milstein2", "cdist","ozaki","shoji","EA"),
    alpha = 0.5, eta = 0.5, pred.corr = T, rcdist = NULL, theta = NULL,
    model = c("CIR", "VAS", "OU", "BS"),
    k1, k2, phi, max.psi = 1000, rh, A){

# (...)

    if(method == "ozaki" || method == "shoji" || method == "EA")
        needs.dx <- TRUE

# (...)

     if (method == "EA")
      X <- sde.sim.ea(X0, t0, Dt, N, d1, d1.x, k1, k2, phi, max.psi, rh, A)

# (...)
}
```

Remarks on the method

The hypothesis of boundedness of ϕ can be too restrictive, and some relaxation is possible as described in [27] with some modifications of the algorithm (known as EA2 and EA3 schemes, in contrast with the EA1 algorithm of Listing 2.10).

It is important to mention that the probability of the event $I = 1$ in the algorithm above, which is the probability of accepting a simulated path, exponentially decreases to zero as $\Delta \to 0$ and is at least e^{-1}, which justifies using this rejection algorithm from the point of view of efficiency.

One more important remark is that approximation schemes usually affect the statistical procedures in Monte Carlo experiments. On the contrary, the EA method does not influence the estimators (see [27]).

```
"sde.sim.ea" <-
function(X0, t0, Dt, N, d1, d1.x, k1, k2, phi, psi, max.psi, rh, A){

   psi <- function(x) 0.5*d1(1,x)^2 + 0.5*d1.x(1,x)

   if(missing(k1)){
    cat("k1 missing, trying numerical minimization...")
    k1 <- optimize(psi, c(0, max.psi))$obj
    cat(sprintf("(k1=%5.3f)\n",k1))
   }
   if(missing(k2)){
    cat("k2 missing, trying numerical maximization...")
    k2 <- optimize(psi, c(0, max.psi),max=TRUE)$obj
    cat(sprintf("(k2=%5.3f)\n",k2))
  }

  if(missing(phi))
    phi <- function(x) 0.5*d1(1,x) + 0.5*d1.x(1,x) - k1
  else
   phi <- function(x) eval(phi)

  M <- k2-k1
  if(M==0)
   stop("`k1' = `k2' probably due to numerical maximization")

  if(Dt>1/M)
    stop(sprintf("discretization step greater than 1/(k2_k1)"))

   if(missing(A))
     A <- function(x) integrate(d1, 0, x)

  if(missing(rh)){
   rh <- function(){
    h <- function(x) exp(A(x) - x^2/(2*Dt))
    f <- function(x) h(x)/dnorm(x,sqrt(Dt))
    maxF <- optimize(f,c(-3*Dt, 3*Dt),max=TRUE)$obj
    while(1){
     y <- rnorm(1)
     if( runif(1) < f(y)/maxF )
      return(y)
   }
  }
 }

  x0 <- X0
  X <- numeric(N)
  X[1] <- X0
  rej <- 0
  j <- 1
  while(j <= N){
   y <- x0+rh()
   k <- rpois(1,M*Dt)
   if(k>0){
    t <- runif(k)*Dt
    v <- runif(k)*M
    idx <- order(t)
    t <- c(0, t[idx], Dt)
    v <- v[idx]

    DT <- t[2:(k+2)] - t[1:(k+1)]
    W <- c(0,cumsum(sqrt(DT) * rnorm(k+1)))
    Y <- x0 + W -(W[k+2] -y+x0)*t/Dt
```

```
    if( prod(phi(Y[2:(k+1)]) <= v) == 1){
     j <- j+1
     x0 <- Y[k+2]
     X[j] <-  Y[k+2]
    } else {
     rej <- rej +1
   }
  }
 }
 cat(sprintf("rejection rate: %5.3f\n",rej/(N+rej)))
 X
}
```

Listing 2.10. R code for the EA1 simulation algorithm.

Periodic drift example (SINE process)

Consider the following example from [27]. We have a process satisfying the stochastic differential equation

$$\mathrm{d}\xi_t = \sin(\xi_t)\mathrm{d}t + \mathrm{d}W_t, \quad \xi_0 = 0\,. \tag{2.17}$$

The drift $b(x) = \sin(x)$ satisfies the usual hypotheses, and is differentiable and such that

$$-\frac{1}{2} = k_1 \leq \frac{1}{2}b(u)^2 + \frac{1}{2}b_x(u) = \frac{1}{2}\sin(u)^2 + \frac{1}{2}\cos(u) \leq k_2 = \frac{5}{8};$$

hence $M = k_2 - k_1 = 9/8$ and $A(u) = 1 - \cos(u)$. The following code simulates an exact path of the SINE process.

```
> # ex 2.15.R
> set.seed(123)
> d <- expression(sin(x))
> d.x <- expression(cos(x))
> A <- function(x) 1-cos(x)
> sde.sim(method="EA", delta=1/20, X0=0, N=500, drift=d,
+      drift.x = d.x, A=A) -> X
k1 missing, trying numerical minimization...(k1=-0.500)
k2 missing, trying numerical maximization...(k2=0.625)
rejection rate: 0.215
> plot(X, main="Periodic drift")
```

A more complicated example: the hyperbolic process

Another example taken from [56] is the following. Consider the modified Cox-Ingersoll-Ross process (see Section 1.13.6) solution to

$$\mathrm{d}X_t = -\theta_1 X_t \mathrm{d}t + \theta_2\sqrt{1 + X_t^2}\mathrm{d}W_t$$

with $\theta_1 + \theta_2^2/2 > 0$. Using the Lamperti transform, we get the new process $Y_t = F(X_t)$, which satisfies the stochastic differential equation

$$\mathrm{d}Y_t = -(\theta_1/\theta_2 + \theta_2/2)\tanh(\theta_2 Y_t)\mathrm{d}t + \mathrm{d}W_t$$

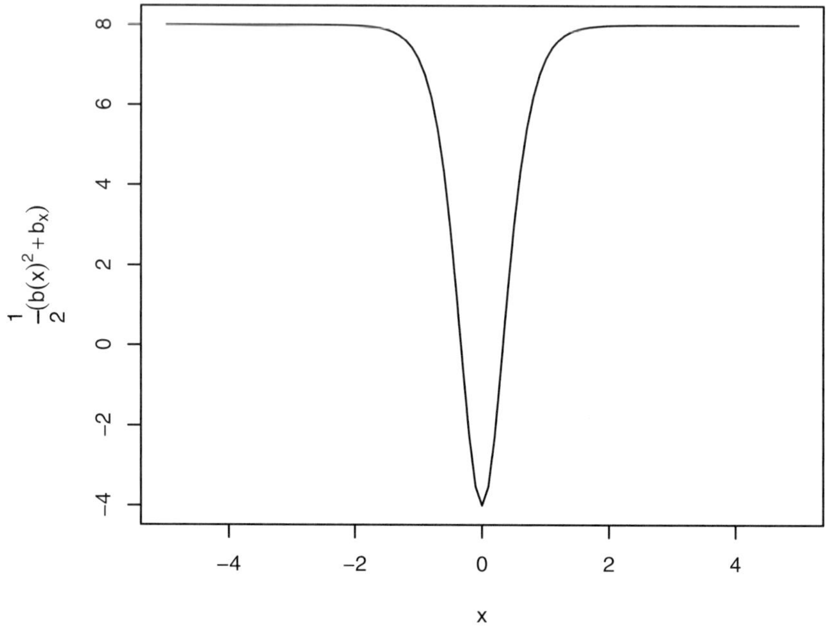

Fig. 2.4. Graph of the curve $\frac{1}{2}(b(x)^2 + b_x(x))$.

with $Y_0 = F(X_0)$ (see Section 1.13.6 for details). Choose $\theta_1 = 6$ and $\theta_2 = 2$. In this case,

$$\frac{1}{2}(b^2(x) + b_x(x)) = \frac{-8 + 16(\sinh(2x))^2}{2(\cosh(2x))^2},$$

and it is easy to show that

$$-4 = k_1 \le \frac{1}{2}(b(x)^2 + b_x(x)) \le k_2 = 8$$

(see also Figure 2.4) and

$$A(x) = \int_0^x -4\tanh(2u)\mathrm{d}u = -2\log(\cosh(2x)).$$

Hence $M = k_2 - k_1 = 12$, $0 \le \phi(x) \le 1/M$, and the constant $K = \int_{\mathbb{R}} e^{A(x) - \frac{x^2}{2T}}$ can be found numerically.

Once we have simulated a path of Y_t using the exact algorithm, we can obtain a trajectory of X_t using the inverse of F; i.e., $X_t = F^{-1}(Y_t) = \sinh(\theta_2 Y_t)$. The next R code performs the simulation, and both the original and the transformed paths of the process are shown in Figure 2.5.

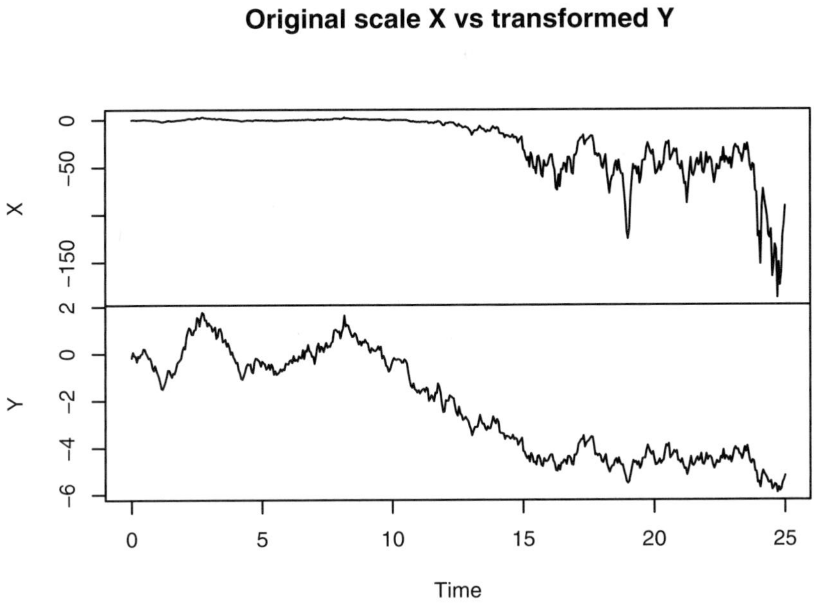

Fig. 2.5. Simulation of the process Y_t solution of (1.54) and its transformed version $X_t = \sinh(2Y_t)$ using the EA1 algorithm.

```
> # ex 2.16.R
> set.seed(123)
> d <- expression(-4*tanh(2*x))
> d.x <- expression(-(4 * (2/cosh(2 * x)^2)))
> A <- function(x) -(0.5+6/4)*log(cosh(2*x))
> X0 <- rt(1, df=4)/2
> F <- function(x) log(x + sqrt(1+x^2))/2
> Y0 <- F(X0)
> sde.sim(method="EA", delta=1/20, X0=Y0, N=500, drift=d,
+       drift.x=d.x, A=A, k1=-4,k2=8) -> Y
rejection rate: 0.474
> X <- sinh(Y)
> ts(cbind(X,Y),start=0,delta=1/20) -> XY
> plot(XY,main="Original scale X vs transformed Y")
```

The Cox-Ingersoll-Ross and Ornstein-Uhlenbeck processes and EA algorithm

Consider the Cox-Ingersoll-Ross process

$$\mathrm{d}X_t = (\theta_1 - \theta_2 X_t)\mathrm{d}t + \theta_3\sqrt{X_t}\mathrm{d}W_t, \qquad X_0 = 10\,,$$

for which the Lamperti transform

$$F(x) = \int_0^x \frac{1}{\theta_3\sqrt{u}}\mathrm{d}u = \frac{2\sqrt{x}}{\theta_3}$$

gives $Y_t = F(X_t)$, satisfying

$$\mathrm{d}Y_t = \frac{4\theta_1 - \theta_3^2}{2\theta_3^2 Y_t}\mathrm{d}t - \frac{\theta_2}{2}Y_t\mathrm{d}t + \mathrm{d}W_t\,.$$

The transformed drift $b(x) = \frac{4\theta_1-\theta_3^2}{2\theta_3^2 x} - \frac{\theta_2}{2}x$ is not bounded in zero from above or below and hence the EA algorithm is not applicable to this process. The same situation occurs for the Ornstein-Uhlenbeck process of equation (1.39).

2.13 Simulation of diffusion bridges

As we will see in the next chapter, exact likelihood inference for discretely observed diffusion processes is not always possible because the likelihood is not available in many cases. Section 3.3.2 describes the *simulated likelihood method*, which consists in estimating the transition density between two consecutive observations using the Monte Carlo approach. In this situation, the ability to simulate paths between two observations is essential. To this end an MCMC-algorithm was proposed in [196], and the exact method of Section 2.12 can also be applied. A new simple method that applies to ergodic diffusion processes has recently been introduced in [37]. This method relies on the *time-reversibility* property of the ergodic diffusion process and essentially consists in the simulation of two paths of a diffusion process, one moving forward in time and another one moving backward in time. If the two trajectories intersect, then the combined path is a realization of the bridge. Let (l, r) with $-\infty \leq l \leq r \leq +\infty$ be the state space of the diffusion process X solution to

$$\mathrm{d}X_t = b(X_t)\mathrm{d}t + \sigma(X_t)\mathrm{d}W_t$$

and take a and b as two points in the state space of X. A solution of the previous equation in the interval $[t_1, t_2]$ such that $X_{t_1} = a$ and $X_{t_2} = b$ is called a (t_1, x_1, t_2, x_2)-diffusion bridge. Time reversibility of an ergodic diffusion is assured by a mild set of conditions (see, e.g., [135]).

Let $m(x)$ and $s(x)$ be, respectively, the speed measure (1.22) and the scale measure (1.21) of the diffusion X. Under Assumption 1.5, we know that the diffusion X is also ergodic with invariant density proportional to the speed measure up to a normalizing constant. Our interest is in the simulation of a $(0, a, 1, b)$-diffusion bridge. Let W^1 and W^2 be two independent Wiener processes and define X^1 and X^2 as solutions to

$$\mathrm{d}X_t^i = b(X_t^i)\mathrm{d}t + \sigma(X_t^i)\mathrm{d}W_t^i$$

with $X_0^1 = a$ and $X_0^2 = b$.

Fact 2.1 (Theorem 1 in [37]) *Let* $\tau = \inf\{0 \leq t \leq 1 | X_t^1 = X_{1-t}^2\}$, *where the* inf *over the empty set is taken to be* ∞. *Define*

$$Z_t = \begin{cases} X_t^1 & \text{if } 0 \le t \le \tau, \\ X_{1-t}^2 & \text{if } \tau < t \le 1. \end{cases}$$

Then, the distribution of $\{Z_t\}_{0\le t\le 1}$ *conditional on the event* $\{\tau \le 1\}$ *is equal to the conditional distribution of* $\{X_t\}_{0\le t\le 1}$ *given* $X_0 = a$ *and* $X_1 = b$*; i.e.,* Z *is a* $(0, a, 1, b)$*-diffusion bridge.*

2.13.1 The algorithm

We now assume that our interest is in the simulation of a $(0, a, \Delta, b)$-diffusion bridge; i.e., a bridge on the generic interval $[0, \Delta]$. The algorithm consists in simulating two independent diffusion processes X^1 and X^2 using one of the previous methods (e.g., the Euler or Milstein scheme) on the time interval $[0, \Delta]$ with discretization step $\delta = \Delta/N$ and applying a rejection sampling procedure. Let $Y_{i\delta}^1$ and $Y_{i\delta}^2$, $i = 0, 1, \ldots, N$, be independent simulations of X^1 and X^2. If either $Y_{i\delta}^1 \ge Y_{(N-i)\delta}^2$ and $Y_{(i+1)\delta}^1 \le Y_{(N-(i+1))\delta}^2$ or $Y_{i\delta}^1 \le Y_{(N-i)\delta}^2$ and $Y_{(i+1)\delta}^1 \ge Y_{(N-(i+1))\delta}^2$, a crossing has been realized. Hence, let $\nu = \min\{i \in (1, \ldots, N) | Y_{i\delta}^1 \le Y_{(N-i)\delta}^2\}$ if $Y_0^1 \ge Y_\Delta^2$ and $\nu = \min\{i \in (1, \ldots, N) | Y_{i\delta}^1 \ge Y_{(N-i)\delta}^2\}$ if $Y_0^1 \le Y_\Delta^2$, and define

$$B_{i\delta} = \begin{cases} Y_{i\delta}^1 & \text{for } i = 0, 1, \ldots, \nu - 1, \\ Y_{(N-i)\delta}^2 & \text{for } i = \nu, \ldots, N. \end{cases}$$

Then B is a simulation of a $(0, a, \Delta, b)$-diffusion bridge. If no crossing happened, start again by simulating $Y_{i\delta}^1$ and $Y_{i\delta}^2$ and iterate until a crossing of the two trajectories is realized.

The nice feature of this method is that this algorithm produces trajectories with the same order (weak or strong) of approximation as the method used to simulate Y^1 and Y^2.

As for the exact algorithm, it is also interesting to evaluate the rejection rate of the method. The rejection probability (i.e., the probability of no crossings) depends on the drift and diffusion coefficients as well as the points a and b and the length of the time interval Δ. Simulation experiments (see [37]) show that if a and b are not too distant and Δ is relatively small, which usually occurs in inference for discretely observed diffusion processes, the rejection rate is acceptable. Usually, when the exact algorithm is feasible, it is more efficient. The nice property of the present approach is that the algorithm is relatively simple and works for the class of generic time-homogeneous ergodic diffusion processes.[7]

The following code illustrates how to implement the algorithm above. Although a `DBridge` function exists in the `sde` package which we discuss later,

[7] We recall again that version 3 of the exact sampling algorithm exists, which does not require bounds on the coefficients but is mathematically more involved [28].

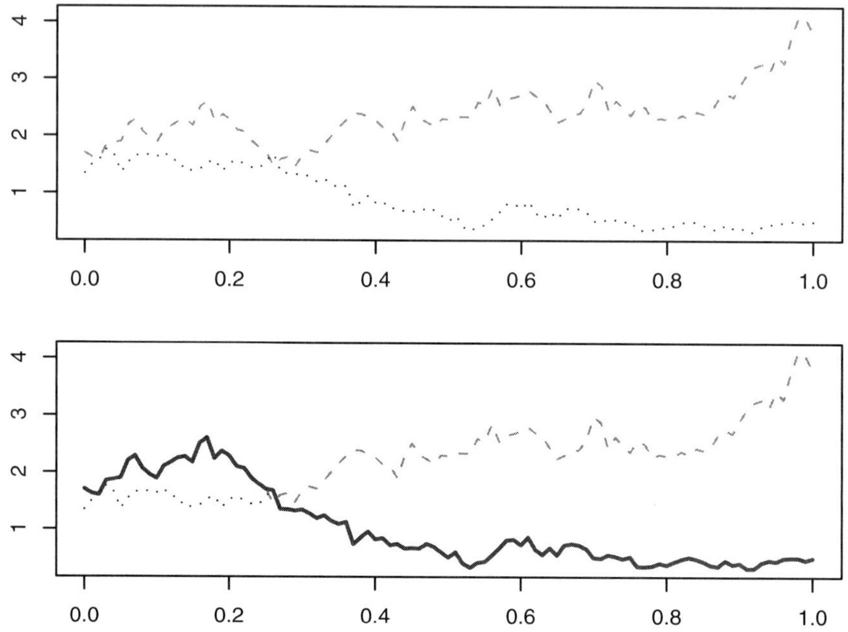

Fig. 2.6. A simulated path of a diffusion bridge (continuous line). The bridge is obtained by merging two paths of diffusion processes (dotted and dashed lines) at the first crossing.

we present the algorithm in an intuitive version in what follows. The following code creates two trajectories, Y1 and Y2, the first starting at $a = 1.7$ and the second starting at $b = 0.5$.

```
> # ex2.17.R
> drift <- expression((3-x))
> sigma <- expression(1.2*sqrt(x))
> a <- 1.7
> b <- 0.5
> set.seed(123)
> Y1 <- sde.sim(X0=a, drift=drift, sigma=sigma, T=1, delta=0.01)
> Y2 <- sde.sim(X0=b, drift=drift, sigma=sigma, T=1, delta=0.01)
> Y3 <- ts(rev(Y2), start=start(Y2), end=end(Y2),deltat=deltat(Y2))
```

The second trajectory is then time-reversed into Y3.

```
> id1 <- Inf
> if(Y1[1]>=Y3[1]){
+   if(!all(Y1>Y3))
+     min(which(Y1 <= Y3))-1 -> id1
+ } else {
+   if(!all(Y1<Y3))
+   min(which(Y1 >= Y3))-1 -> id1
+ }
> if(id1==0 || id1==length(Y1)) id1 <- Inf
```

The code then calculates the index id1 of the first crossing time of the two trajectories and merges Y1 and Y3 appropriately if this time (the time of crossing) is finite.

```
> par(mfrow=c(2,1))
> plot(Y1, ylim=c(min(Y1,Y2), max(Y1,Y2)),col="green",lty=2)
> lines(Y3,col="blue",lty=3)
>
> if(id1==Inf ){
+   cat("no crossing")
+ } else {
+   plot(Y1, ylim=c(min(Y1,Y2), max(Y1,Y2)),col="green",lty=2)
+ lines(Y3,col="blue",lty=3)
+ B <- ts(c(Y1[1:id1], Y3[-(1:id1)]), start=start(Y1),end=end(Y1),
+       frequency=frequency(Y1))
+ lines(B,col="red",lwd=2)
+ }
```

Figure 2.6 shows the trajectory of Y1 and Y3 (top) and the simulated path of the diffusion bridge (bottom). The following code uses the DBridge function as an interface to the algorithm above. The function has the following interface similar to BBridge for the simulation of the Brownian bridge:

```
DBridge(x = 0, y = 0, t0 = 0, T = 1, delta, drift, sigma, ...)
```

The variable arguments ... are passed directly to the sde.sim function, which is called internally. This allows selection of any simulation scheme for the diffusion, the default being the default of the sde.sim function. The code for DBridge simulates a (t_0, x, T, y)-diffusion bridge. The next code provides an example of how it is used and the output is given in Figure 2.7.

```
> # ex2.17.R (cont.)
> d <- expression((3-x))
> s <- expression(1.2*sqrt(x))
> par(mar=c(3,3,1,1))
> par(mfrow=c(2,1))
> set.seed(123)
> X <- DBridge(x=1.7,y=0.5, delta=0.01, drift=d, sigma=s)
> plot(X)
> X <- DBridge(x=1,y=5, delta=0.01, drift=d, sigma=s)

no crossing, trying again...
> plot(X)
```

2.14 Numerical considerations about the Euler scheme

Consider for example the geometric Brownian motion X_t in (1.5). If $X_0 > 0$, this process is always positive, being an exponential functional of the Brownian motion. The application of the Euler approximation can lead to unexpected results. Indeed, the Euler scheme for X_t reads as

$$Y_{i+1} = Y_i(1 + \theta_1 \cdot \Delta t + \theta_2 \sqrt{\Delta t} Z),$$

and if Δt is too small, it can happen in one or more simulations that a pseudo random number Z is drawn from the Gaussian distribution such that

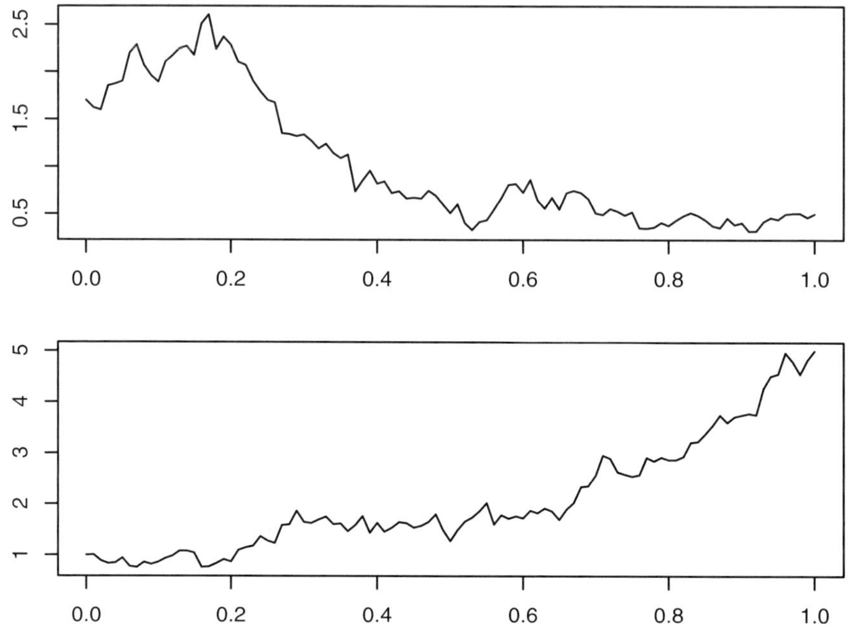

Fig. 2.7. Two simulated paths of diffusion bridges (t_1, x_1, t_2, x_2) using the `DBridge` function for the stochastic differential equation $\mathrm{d}X_t = (3 - X_t)\mathrm{d}t + 1.2\sqrt{X_t}\mathrm{d}W_t$: $(0, 1.7, 1, 0.5)$ (top) and $(0, 1, 1, 5)$ (bottom).

$$Z < -\frac{1 + \theta_1 \Delta t}{\theta_2 \sqrt{\Delta t}}$$

and therefore Y_{i+1} takes negative values. In this case, this is not the same phenomenon of absorption as in the CEV process of Section 2.5 but just a matter of the approximation method used. In fact, the Euler scheme is guaranteed to converge to the mathematical description of the geometric Brownian motion, but simulation by simulation we cannot expect this result to happen every time. An empirical proof of the fact that this is not an absorption phenomenon is that false "absorption" does not occur on the transformed process $\log X_t$.

2.15 Variance reduction techniques

We now adapt the general concepts in Section 1.4 to the framework of stochastic differential equations. In this framework, interest is in the evaluation of the expected value of some functional ψ of the trajectory of the process solution of some stochastic differential equation. Let us denote by Z this expected value,

$$Z = \psi(\{X_t, 0 \le t \le T\}) = \psi(X),$$

with X the solution to the stochastic differential equation

$$dX_t = b(X_t)dt + \sigma(X_t)dW_t, \quad X_0 = x. \tag{2.18}$$

The Monte Carlo estimator of $\mathbb{E}Z$ is built as follows. Let $Y_i^j = Y^j(t_i)$, $j = 1, \dots, N$, $i = 1, \dots, n$, be the value of the approximating process Y^j at time t_i for the jth simulated path of the process X. Then, the estimator has the form

$$\hat{\mu}_N = \frac{1}{N} \sum_{j=1}^{N} \psi(Y^j).$$

For example, if we are interested in the expected value of X_T (i.e., $\mathbb{E}Z = \mathbb{E}\psi(X) = \mathbb{E}X_T$), the Monte Carlo estimator reads as

$$\hat{\mu}_N = \frac{1}{N} \sum_{j=1}^{N} Y_n^j.$$

From the general Monte Carlo results, we know that $\hat{\mu}_N$ is unbiased and has a variance equal to $\text{Var}\psi(Y_n)/N$, and we also know that the length of the confidence intervals shrinks to 0 at speed $N^{-\frac{1}{2}}$. Once again, if the process itself has large a variance, the confidence interval might be too big to be used to assess the quality of the estimator and hence the need for variance reduction techniques.

2.15.1 Control variables

From Section 1.4.2, we know that in order to reduce the variance of $\mathbb{E}Z$, one possibility is to apply the control variable technique. In particular, we need to rewrite $\mathbb{E}Z$ as $\mathbb{E}(Z - Y) + \mathbb{E}(Y)$, for which $\mathbb{E}Y$ can be calculated explicitly. For obvious reasons, this is easy to implement when $\mathbb{E}Y = 0$. When dealing with stochastic differential equations, a good hint is clearly to build such a random variable Y on top of Brownian motion. Indeed, we know that if $(H(t), 0 \le t \le T)$ is an Itô integrable process, then

$$\mathbb{E}\left(\int_0^T H(s)dW(s)\right) = 0.$$

Thus, in principle, the role of Y might be taken by a properly chosen stochastic integral. In fact, there is a general result that allows us to rewrite any square-integrable random variable, adapted to the natural filtration of the Brownian motion, in terms of its expected value and a stochastic integral of some process H. This theorem, called the *predicted representation* theorem (see [130] or [193]), is rather general, but unfortunately the explicit formula for the process

H is quite difficult to find in general. In [167], one formula based on Malliavin calculus is provided, but it is hard to implement. In general, each case study might provide different ways to obtain control variables. A result descending from the Feynman-Kac theorem relates the construction of such a process to the solution of a partial differential equation. Theorem 5.4.2 in [156] assumes that $u(t,x)$ is a function of class $C^{1,2}$ with bounded derivatives in x and solution of

$$\begin{cases} \left(\frac{\partial u}{\partial t} + \mathcal{L}u\right)(t,x) = f(x) \\ u(T,x) = g(x), \end{cases}$$

where $\mathcal{L}$ is the infinitesimal generator (see (1.28)) of the diffusion process solution of (2.18). Setting

$$Z = g(X_T) - \int_0^T f(X_s)\mathrm{d}s$$

and

$$Y = \int_0^T \frac{\partial u}{\partial x}(s, X_s)\sigma(X_s)\mathrm{d}W_s,$$

then

$$\mathbb{E}Z = Z - Y.$$

This theorem is the key to finding the control variable that are interests us. But still the expression of Y involves partial derivatives of the function u, and hence in practice the approach is to find an approximation $\bar{u}$ of u that is simple to handle and put it in the expression of Y. There are a lot of heuristics behind the application of this method in concrete cases, and we show one from [134]. Suppose we want to calculate the average price of a call option,

$$\mathbb{E}Z = \mathbb{E}\left(e^{-rT}\left(\frac{1}{T}\int_0^T S_u \mathrm{d}u - K\right)_+\right), \tag{2.19}$$

where S is the geometric Brownian motion in (1.5). If $\sigma \simeq 0.5$, $r \simeq 1$, and $T \simeq 1$, then the integral

$$\frac{1}{T}\int_0^T S_u \mathrm{d}u$$

is "close" to

$$\exp\left\{\frac{1}{T}\int_0^T \log(S_u)\mathrm{d}u\right\}.$$

Hence, we can set

$$Y = e^{-rT}\left(e^Z - K\right)_+$$

and

$$Z = \frac{1}{T}\int_0^T \log(S(s))\mathrm{d}s$$

and use Y as the control variable. Moreover, Z is easy to simulate, as it is essentially a Gaussian random variable. This approach successfully reduces the variance of the Monte Carlo estimator. (Details on how (2.19) is related to the previous theorem can be found in Section 5.2.6 of [156].) Here we don't give R code for this case (although it is quite easy to implement). as this example is quite peculiar.

2.16 Summary of the function sde.sim

The package `sde` further generalizes the function `sde.sim`. In particular, it is possible to generate `M` independent trajectories of the same process with one single call of the function by just specifying a value for `M` (which is 1 by default). For `M>=2`, the function `sde.sim` returns an object of class `mts` "multi-dimensional time series." This is quite convenient in order to avoid loops in the case of Monte Carlo replications. The function `sde.sim` has a rather flexible interface, which matches the following definition:

```
sde.sim(t0 = 0, T = 1, X0 = 1, N = 100, delta, drift, sigma,
   drift.x, sigma.x, drift.xx, sigma.xx, drift.t,
   method = c("euler", "milstein", "KPS", "milstein2", "cdist",
   "ozaki","shoji","EA"), alpha = 0.5, eta = 0.5, pred.corr = T,
   rcdist = NULL, theta = NULL,
   model = c("CIR", "VAS", "OU", "BS"),
   k1, k2, phi, max.psi = 1000, rh, A, M=1)
```

A complete description of all of the parameters can be found on the manual page of the `sde` package in the Appendix B of this book. Here we mention that this interface allows us to simulate a stochastic differential equation by specifying the drift and the diffusion coefficient and a simulation scheme, or by specifying a model among them that admits well-known distributional results, by specifying a conditional distribution density. The following code shows an example of such flexibility.

```
> # ex2.18.R
> # Ornstein-Uhlenbeck process
> set.seed(123)
> d <- expression(-5 * x)
> s <- expression(3.5)
> sde.sim(X0=10,drift=d, sigma=s) -> X
> plot(X,main="Ornstein-Uhlenbeck")
>
> # Multiple trajectories of the O-U process
> set.seed(123)
> sde.sim(X0=10,drift=d, sigma=s, M=3) -> X
> plot(X,main="Multiple trajectories of O-U")
>
> # Cox-Ingersoll-Ross process
> # dXt = (6-3*Xt)*dt + 2*sqrt(Xt)*dWt
> set.seed(123)
> d <- expression( 6-3*x )
> s <- expression( 2*sqrt(x) )
> sde.sim(X0=10,drift=d, sigma=s) -> X
```

```
> plot(X,main="Cox-Ingersoll-Ross")
>
> # Cox-Ingersoll-Ross using the conditional distribution "rcCIR"
>
> set.seed(123)
> sde.sim(X0=10, theta=c(6, 3, 2), rcdist=rcCIR, method="cdist") -> X
> plot(X, main="Cox-Ingersoll-Ross")
>
> set.seed(123)
> sde.sim(X0=10, theta=c(6, 3, 2), model="CIR") -> X
> plot(X, main="Cox-Ingersoll-Ross")
```

2.17 Tips and tricks on simulation

We conclude briefly with some general remarks and things to remember before starting Monte Carlo analysis based on simulated paths of the processes seen so far. In general, it is recommended to apply the Lamperti transform[8] to eliminate the dependency of the diffusion coefficient from the state of the process during simulation. We have seen that many methods rely on this transformation without loss of generality (e.g., Ozaki, Shoji, Exact Algorithm). Also, the Euler and Milstein methods may benefit from this preliminary transformation.

If the conditional distribution of the process is known, which is rarely the case, then a simulation method based on this should be used. For example, the simulation of the Cox-Ingersoll-Ross process should be done in this way because simulations based on the discretization of the corresponding stochastic differential equation may lead to unwanted results such as negative values of the process.

In principle, if time is not a major constraint and the model satisfies the right conditions on the drift (which we saw are not satisfied by the Cox-Ingersoll-Ross process), then the exact algorithm must be used. It is worth mentioning that code more efficient than what we present here can be written so time efficiency of EA is not necessarily a concern. This will probably be done in the next version of the `sde` package, and hence this comment only applies to the current implementation.

We also mention that when there is no need to simulate the path of the process on a regular grid of points $t_i = i\Delta/T$, $i = 0, 1, \ldots, N$, $N\Delta = T$ like we did, then the EA algorithm is even faster. In fact, in our approach, we generate different points in between the time instants t_i and t_{i+1} but then keep just the last one and iterate this simulation N times. On the contrary, the algorithm can be used to simulate the path up to time T. In this case, the algorithm generates a random grid of points and simulated values of the process, and then Brownian bridges can be used between the points of the random grid. Of course, the way we use the EA algorithm avoids any dependency of the simulation scheme from the estimation part, as we will note in the following chapters.

---

[8] Of course, when the transform is well-defined and can be obtained in explicit analytic form and not by numerical integration.

When the interest is in the simulation of diffusion bridges, the algorithm presented in Section 2.13 is a good candidate.

The Ozaki and Shoji-Ozaki methods can be good ways of simulating a path when other methods do not apply (which is the case for unbounded nonlinear drift functions).

If the grid of points is relatively small, we have seen that most discretization methods perform equally well but the Euler method can still be unstable in some particular situations: see the counterexample on the geometric Brownian motion process in Section 2.5. Then a higher order of the approximation is always welcome.

Antithetic sampling and variance reduction techniques might be used when functionals of the processes are of interest. Unfortunately, the control variable approach is always an ad hoc art.

3

Parametric Estimation

In this chapter we consider parametric estimation problems for diffusion processes sampled at discrete times. We can imagine different schemes of observation:

- *Large sample scheme*: In this scheme, the time Δ between two consecutive observations is fixed and the number of observations n increases. In this case, the window of observation $[0, n\Delta = T]$ also increases with n. In this framework, which is considered the most natural, additional assumptions on the underlying continuous model are required such as stationarity and/or ergodicity.
- *High-frequency scheme*: In this case, $\Delta = \Delta_n$ goes to zero as n increases, and the window of observation $[0, n\Delta_n = T]$ is fixed. Neither stationarity nor ergodicity is needed.
- *Rapidly increasing design*: Δ_n shrinks to zero as n grows, but the window of observation $[0, n\Delta_n]$ also increases with n; i.e., $n\Delta_n \to \infty$. In this case, stationarity or ergodicity is needed. Further, the mesh Δ_n should go to zero at a prescribed rate $n\Delta_n^k \to 0$, $k \geq 2$. For high values of k, this is not a severe constraint because this means that Δ_n goes to zero but slowly.

Interesting reviews on this subject can be found, for example, in [213], [30], [212] and [128]. A vast collection of results, the majority of which we tried to cover here, can be found in the monograph [192]. Before going into the details of each approach, we need to discuss the underlying continuous model.

The underlying continuous model

Consider the one-dimensional, time-homogeneous stochastic differential equation

$$\mathrm{d}X_t = b(X_t, \theta)\mathrm{d}t + \sigma(X_t, \theta)\mathrm{d}W_t, \tag{3.1}$$

where $\theta \in \Theta \subset \mathbb{R}^p$ is the multidimensional parameter and θ_0 is the true parameter to be estimated. The functions $b : \mathbb{R} \times \Theta \to \mathbb{R}$ and $\sigma : \mathbb{R} \times \Theta \to$

S.M. Iacus, *Simulation and Inference for Stochastic Differential Equations*,
doi: 10.1007/978-0-387-75839-8_3,

$(0, \infty)$ are known and such that the solution of (3.1) exists.[1] The state space of the process is denoted by $I = (l, r)$, and $-\infty \le l < r \le +\infty$ is an open set and the same for all θ. Moreover, for any $\theta \in \Theta$ and any random variable ξ with support in I, equation (3.1) has a unique strong solution for $X_0 = \xi$. When ergodicity is required, additional assumptions to guarantee the existence of the *invariant distribution* $\pi_\theta(\cdot)$ should be imposed. In this case, the solution of (3.1) with $X_0 = \xi \sim \pi_\theta$ is strictly stationary and *ergodic*. We have seen different sets of sufficient conditions (more can be found in [130], [131], or [165]), and when $\pi_\theta(\cdot)$ exists it has the form

$$\pi_\theta(x) = \frac{1}{M(\theta)\sigma^2(x,\theta)s(x,\theta)}, \tag{3.2}$$

where

$$s(x,\theta) = \exp\left\{-2\int_{x_0}^{x} \frac{b(y,\theta)}{\sigma^2(y,\theta)}\mathrm{d}y\right\}$$

for some $x_0 \in I$ and $M(\theta)$ the normalizing constant. As seen in Section 1.13, the function s is called the scale measure and $m(x) = \pi_\theta(x) \cdot M(\theta)$ is called the speed measure. The distribution of X with $X_0 \sim \pi_\theta$ is denoted by P_θ, and under P_θ, $X_t \sim \pi_\theta$ for all t. We further denote by $p_\theta(t, \cdot|x)$ the conditional density (or transition density) of X_t given $X_0 = x$. As X is time-homogeneous, $p_\theta(t, \cdot|x)$ is just the density of X_{s+t} conditional on $X_s = x$ for all $t \ge 0$. In some cases, we will use the notation $p(t, \cdot|x, \theta)$. As already mentioned, the transition probabilities in most of the cases are not known in explicit analytic form. On the contrary, the invariant density is easier to obtain (up to the normalizing constant). We now reintroduce the infinitesimal generator (see also Section 1.10) of the diffusion X in the multidimensional parametric case. The operator $\mathcal{L}_\theta$ defined as

$$\mathcal{L}_\theta f(x,\theta) = b(x,\theta)f_x(x,\theta) + \frac{1}{2}\sigma^2(x,\theta)f_{xx}(x,\theta) \tag{3.3}$$

is called the infinitesimal generator of the diffusion. Here $f(\cdot)$ is a twice-continuous differentiable function $f : \mathbb{R} \times \Theta \to \mathbb{R}$, where $f_x(\cdot)$ and $f_{xx}(\cdot)$ are the first and second partial derivatives of $f(\cdot)$ with respect to argument x (see [197] for more details).

In the continuous case, it is quite straightforward to estimate the parameters efficiently. In particular, θ (at least the subset of parameters concerning the diffusion part of (3.1)) can be calculated rather than estimated from the quadratic variation of the process (see (1.27)) since, for all $t \ge 0$,

$$<X, X>_t = \lim_{n\to\infty} \sum_{k=1}^{2^n} \left(X_{t\wedge k/2^n} - X_{t\wedge(k-1)/2^n}\right)^2 = \int_0^t \sigma^2(X_s,\theta)\mathrm{d}s$$

[1] Any other equivalent set of assumptions, as the ones presented in Chapter 1, are equally good.

as $n \to \infty$ in probability under P_θ. The rest of the parameters present only in the drift part of (3.1) can be estimated using the maximum likelihood approach. Indeed, once the diffusion coefficient is known, which we just say is always true in principle (i.e., $\sigma(x,\theta) = \sigma(x)$) the likelihood function of X is given by

$$L_T(\theta) = \exp\left(\int_0^T \frac{b(X_s,\theta)}{\sigma^2(X_s)} \mathrm{d}X_s - \frac{1}{2}\int_0^T \frac{b^2(X_s,\theta)}{\sigma^2(X_s)} \mathrm{d}s\right). \tag{3.4}$$

Therefore θ can be estimated by maximizing $L_T(\theta)$. Please note that $L_T(\theta)$ is just the Radon-Nikodým derivative appearing in (1.36). A comprehensive set of results for this model (for both parametric and nonparametric statistical problems) can be found in [149] (and in [148] for small diffusion asymptotics).

The discrete-time observations

We assume that the process is observed at discrete times $t_i = i\Delta_i$, $i = 0, 1, \ldots, n$, and $T = n\Delta_n$. In some cases, the sampling rate has to be constant $\Delta_i = \Delta$ or such that $\max_i \Delta_i < \Delta$ for some fixed Δ; in other cases, Δ_n varies and it is assumed that $n\Delta_n^k \to 0$ for some power $k \geq 2$. The asymptotics is considered as $n \to \infty$, which is equivalent to $T \to \infty$. In the following, we will denote $X_{t_i} = X_{i\Delta_i}$ just by X_i to simplify the writing. We further denote by $\mathcal{F}_n = \sigma\{X_{t_i}, i \leq n\}$ the σ-field generated by the first n observations with $\mathcal{F}_0$ the trivial σ-field. The discrete counterpart of $L_T(\theta)$ we are interested in, conditional on X_0, is given by

$$L_n(\theta) = \prod_{i=1}^n p_\theta\left(\Delta, X_i | X_{i-1}\right) p_\theta(X_0), \tag{3.5}$$

which can be derived using the Markov property of X (see, e.g., [14]). We denote by $\ell_n(\theta) = \log L_n(\theta)$ the log-likelihood function

$$\begin{aligned} \ell_n(\theta) = \log L_n(\theta) &= \sum_{i=1}^n \ell_i(\theta) + \log(p_\theta(X_0)) \\ &= \sum_{i=1}^n \log p_\theta\left(\Delta, X_i | X_{i-1}\right) + \log(p_\theta(X_0)). \end{aligned} \tag{3.6}$$

Usually $p_\theta(X_0)$ is not known unless the process is assumed to be in a stationary regime but, even in this case, it is not always easy to determine $p_\theta(X_0)$. If the number of observations increases with time, one can assume that the relative weight of $p_\theta(X_0)$ in the whole likelihood $L_n(\theta)$ decreases, so we will assume that $p_\theta(X_0) = 1$ from now on without mentioning it any further. In the following, we will use a dot " $\dot{}$ " or multiple dots " $\ddot{}$ " for single or multiple times differentiation with respect to the vector θ and $\partial_{\theta_i} f$ for $\frac{\partial}{\partial \theta_i} f$ and $\partial_{\theta_i}^k f$ for

$\frac{\partial^k}{\partial \theta_i^k} f$ to keep formulas compact but still understandable. When the transition density is differentiable, we can define the score (vector) function

$$\dot{\ell}_n(\theta) = \sum_{i=1}^{n} \dot{\ell}_i(\theta) = \sum_{i=1}^{n} \begin{pmatrix} \partial_{\theta_1} \ell_i(\theta) \\ \vdots \\ \partial_{\theta_p} \ell_i(\theta) \end{pmatrix} \tag{3.7}$$

and the Fisher information matrix for θ,

$$i_n(\theta) = \sum_{i=1}^{n} \mathbb{E}_\theta\{\dot{\ell}_i(\theta)\dot{\ell}_i(\theta)^T\}, \tag{3.8}$$

where "T" denotes the transposition operator. Since $p_\theta(t, \cdot|x)$ is usually not known explicitly, so are $L_n(\theta)$ and all the derived quantities. There are different ways to deal with this problem and we will show some options in what follows. Still, there are quite important models for which the transition density is known in explicit form. Hence we start with exact likelihood inference for these models.

3.1 Exact likelihood inference

As mentioned, maximum likelihood estimation on the true likelihood of the process is a rare case. In [35], [189], [190], and [62], sufficient conditions for the consistency and asymptotic normality of these maximum likelihood estimators are given. In particular, in [62] consistency and asymptotic normality are proved, irrespective of the size of Δ, which seems a good property in applications. The reader might want to read the original references or consider Section 3.3 in [192] as a starting point. We are not going to give the complete set of conditions here (which hold also for multidimensional diffusion processes). Still, we present a smaller set of hypotheses that are the basic set used by many methods. These assumptions mimic the ones presented in Chapter 1 but are adapted to the parametric framework.

Assumption 3.1 (Linear growth assumption) *There exists a constant K independent of θ such that, for all x,*

$$|b(x,\theta)| + |\sigma(x,\theta)| \le K(1 + |x|) .$$

Assumption 3.2 (Global Lipschitz assumption) *There exists a constant K independent of θ such that*

$$|b(x,\theta) - b(y,\theta)| + |\sigma(x,\theta) - \sigma(y,\theta)| \le K|x - y| .$$

Assumption 3.3 (Positiveness of diffusion coefficient)

$$\inf_x \sigma^2(x,\theta) > 0 .$$

Assumption 3.4 (Bounded moments) *For all $k > 0$, all moments of order k of the diffusion process exist and are such that*

$$\sup_t \mathbb{E}|X_t|^k < \infty \,.$$

Assumption 3.5 (Smoothness of the coefficients) *The two coefficients b and σ and their derivatives in θ (eventually up to order 3) are smooth in x and of polynomial growth order in x uniformly on θ.*

The last assumption is usually specified in a different manner for different methods, as we will show. This set of assumptions is also completed by technical conditions to ensure the proper rate of convergence and the existence of Fisher information of the experiment. When the parameters in the drift and diffusion coefficients are separated, the rate of convergence is always faster for the parameter in the diffusion coefficient, and a two-step approach to find estimators is possible as in [231]. Usually the rate of convergence for the diffusion part is $\sqrt{n}$ and, under the condition $n\Delta_n^3 \to 0$, the estimator for the parameters in the drift has rate of convergence $\sqrt{n\Delta_n}$ (in some cases, $n\Delta_n^2 \to 0$ is required).

We now present some code for direct calculation of the estimators or numerical maximization using the `mle` function from the package `stats4` (which is included in all recent distributions of R), which is just an interface to one of the optimizers available in R. This interface is worth using because its output is more digestible to statisticians. For example, the estimated variance of the estimators is provided by inverting the Hessian matrix at the optimum, and approximated confidence intervals for the estimates can be obtained. The `mle` optimizes the negative log-likelihood and needs named arguments. It is also quite flexible in that it allows one to obtain maximum likelihood estimators of a subset of the parameters conditioned on the fact that the complimentary set of parameters has been fixed.

Please note that in the code that follows we specify some initial values for θ using the option `start` of the `mle` function. This is of course necessary any time an optimizer is used. We will return to this aspect later. Another interesting fact is that `mle` allows for box-constrained optimization. In the examples below we will use the option `method = ''L-BFGS-B''` to specify the box-constrained optimization method and `lower` to specify a vector of lower bounds in which the algorithm searches for a solution. The parameter `lower` also accepts `-Inf` (which is the default for a non constrained search), and the analogue parameter `upper` accepts `+Inf` (again the default; hence we don't need to specify it).

3.1.1 The Ornstein-Uhlenbeck or Vasicek model

We already introduced the Vasicek or Ornstein-Uhlenbeck model in Section 1.13.1. This process solves the stochastic differential equation

$$dX_t = (\theta_1 - \theta_2 X_t)dt + \theta_3 dW_t, \qquad X_0 = x_0\,, \tag{3.9}$$

with $\theta_1, \theta_2 \in \mathbb{R}$, $\theta_3 \in \mathbb{R}_+$ and it is ergodic for $\theta_2 > 0$. We have also shown its exact conditional and stationary densities. In particular, the conditional density $p_\theta(t, \cdot|x)$ is the density of a Gaussian law with mean and variance respectively

$$m(t,x) = \mathbb{E}_\theta(X_t|X_0 = x) = \frac{\theta_1}{\theta_2} + \left(x_0 - \frac{\theta_1}{\theta_2}\right) e^{-\theta_2 t} \tag{3.10}$$

and

$$v(t,x) = \mathrm{Var}_\theta(X_t|X_0 = x) = \frac{\theta_3^2\left(1 - e^{-2\theta_2 t}\right)}{2\theta_2}\,. \tag{3.11}$$

The following code, easy to implement, takes care of maximum likelihood estimation of the Vasicek model. The function `OU.lik` needs as input the three parameters and assumes that sample observations of the process are available in the current R workspace in the `ts` object `X`. The function then calls the `dcOU` function, which evaluates the conditional density of the process. In principle, `dcOU`, which is a vectorized function, also accepts different Δ's, but in practice our Δ will be considered fixed because we are using a `ts` object.[2]

```
dcOU <- function(x, t, x0, theta, log = FALSE){
  Ex <- theta[1]/theta[2]+(x0-theta[1]/theta[2])*exp(-theta[2]*t)
  Vx <- theta[3]^2*(1-exp(-2*theta[2]*t))/(2*theta[2])
  dnorm(x, mean=Ex, sd = sqrt(Vx), log=log)
}
OU.lik <- function(theta1, theta2, theta3){
  n <- length(X)
  dt <- deltat(X)
  -sum(dcOU(X[2:n], dt, X[1:(n-1)], c(theta1,theta2,theta3), log=TRUE))
}
```

The `dcOU` is part of the `sde` package and is one of the `[dpqr]cOU` functions related to the conditional density of the Ornstein-Uhlenbeck process. We simulate a solution of $dX_t = (3 - X_t)dt + 2dW_t$ and try to estimate the parameters. For this example, the vector of parameters is $\theta = c(\theta_1, \theta_2, \theta_3) = (3, 1, 2)$.

```
> # ex3.01.R
> require(stats4)
> require(sde)
> set.seed(123)
> X <- sde.sim(model="OU", theta=c(3,1,2), N=1000, delta=1)
> mle(OU.lik, start=list(theta1=1, theta2=0.5, theta3=1),
+       method="L-BFGS-B", lower=c(-Inf,0,0)) -> fit
> summary(fit)
Maximum likelihood estimation

Call:
mle(minuslogl = OU.lik, start = list(theta1 = 1, theta2 = 0.5,
    theta3 = 1), method = "L-BFGS-B", lower = c(-Inf, 0, 0))

Coefficients:
```

[2] It could be that in the future the package `sde` will evolve in this direction, adopting the `zoo` class of objects.

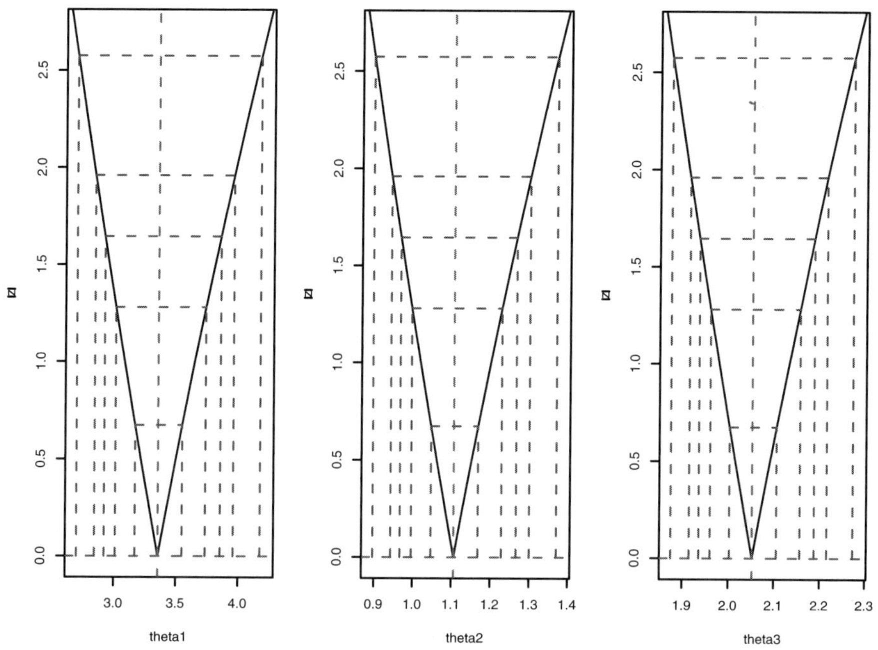

Fig. 3.1. Profile likelihood for the three parameters of the Vasicek process.

```
       Estimate Std. Error
theta1 3.355322 0.28159504
theta2 1.106107 0.09010627
theta3 2.052815 0.07624441

-2 log L: 3366.389
```

At this point, it seems reasonable to extract and plot the profile likelihood for each of the parameters (see Figure 3.1) and get the confidence intervals and the variance-covariance matrix as follows.

```
> # ex3.01.R (cont.)
> prof <- profile(fit)
> par(mfrow=c(1,3))
> plot(prof)
> par(mfrow=c(1,1))
> vcov(fit)
           theta1      theta2      theta3
theta1 0.07929576 0.024620718 0.016634557
theta2 0.02462072 0.008119141 0.005485549
theta3 0.01663456 0.005485549 0.005813209
> confint(fit)
Profiling...
           2.5 %   97.5 %
theta1 2.8449242 3.961076
theta2 0.9433407 1.300659
theta3 1.9147160 2.216142
```

What happens if the asymptotics is not realized?

As we have seen, the likelihood inference has been quite successful, but let us look at the next example.

```
> # ex3.01.R (cont.)
> set.seed(123)
> X <- sde.sim(model="OU", theta=c(3,1,2), N=1000, delta=1e-3)
> mle(OU.lik, start=list(theta1=1, theta2=0.5, theta3=1),
+        method="L-BFGS-B", lower=c(-Inf,0,0)) -> fit2
> summary(fit2)
Maximum likelihood estimation

Call:
mle(minuslogl = OU.lik, start = list(theta1 = 1, theta2 = 0.5,
    theta3 = 1), method = "L-BFGS-B", lower = c(-Inf, 0, 0))

Coefficients:
        Estimate  Std. Error
theta1 13.303899 6.78851336
theta2  5.411345 3.08469142
theta3  1.984735 0.04448513

-2 log L: -2704.316
```

In this case, the likelihood inference fails (apart from the diffusion coefficient). The reason for this is that the assumption $n\Delta_n \to \infty$ does not (reasonably) hold for the second case for which `N * delta` is equal to 1. Please note that there is no difference in the estimation part but only on the data used to make the inference. This is where many methods may fail to work in practice because the data do not conform to the assumptions. We will return to this point later in the book.

Some particular model specifications

Two particular cases are interesting for the Ornstein-Uhlenbeck model. If $\theta_1 = 0$, then the stochastic differential equation becomes $\mathrm{d}X_t = -\theta_2 X_t \mathrm{d}t + \theta_3 \mathrm{d}W_t$ and the only parameters of interest are θ_2 and θ_3. In this case, if the sampling rate Δ is fixed, the maximum likelihood estimator of θ_2 is available in explicit form and takes the form

$$\hat{\theta}_{2,n} = -\frac{1}{\Delta} \log \left(\frac{\sum\limits_{i=1}^{n} X_{i-1} X_i}{\sum\limits_{i=1}^{n} X_{i-1}^2} \right), \tag{3.12}$$

which is defined only if $\sum_{i=1}^{n} X_{i-1} X_i > 0$. This estimator is consistent and asymptotically Gaussian (see, e.g., [180]). The maximum likelihood estimator of θ_3^2 is given by

$$\hat{\theta}_{3,n}^2 = \frac{2\hat{\theta}_{2,n}}{n(1 - e^{-2\Delta\hat{\theta}_{2,n}})} \sum_{i=1} \left(X_i - X_{i-1} e^{-\Delta\hat{\theta}_{2,n}} \right)^2, \tag{3.13}$$

where $\hat{\theta}_{2,n}$ is from (3.12). Now we compare the explicit maximum likelihood estimators with the ones obtained numerically. We choose $\theta_2 = 3$ and $\theta_3 = 2$. We will fix the parameter $\theta_1 = 0$ in the likelihood when we optimize using `mle`. The maximum likelihood estimators can be found numerically using the following code.

```
> # ex3.02.R
> set.seed(123)
> X <- sde.sim(model="OU", theta=c(0,3,2), N=1000, delta=1)
> mle(OU.lik, start=list(theta2=1.5, theta3=1), fixed=list(theta1=0),
+   method="L-BFGS-B", lower=c(0,0)) -> fit
> summary(fit)
Maximum likelihood estimation

Coefficients:
       Estimate Std. Error
theta2 3.889088  1.5438246
theta3 2.254119  0.4487823

-2 log L: 2411.654
>
```

Explicit estimators give

```
> # ex3.02.R (cont.)
> n <- length(X)
> tmp.sum <- sum(X[1:(n-1)]*X[2:n])
> dt <- deltat(X)
> theta2.hat <- ifelse(tmp.sum>0, -log(tmp.sum/sum(X[1:(n-1)]^2))/dt ,NA)
> theta2.hat
[1] 3.888998
> theta3sq.hat <- 2*theta2.hat/((n-1)*(1-exp(-2*dt*theta2.hat))) *
+      sum((X[2:n]-X[1:(n-1)]*exp(-dt*theta2.hat))^2)
> sqrt(theta3sq.hat)
[1] 2.254092
```

The results are consistent but in this case, `mle` is worth using because, as said before, one can obtain the variance and covariace matrix and confidence intervals numerically. Next is the model with a unitary diffusion coefficient.

```
> # ex3.02.R (cont.)
> set.seed(123)
> X <- sde.sim(model="OU", theta=c(0,3,1), N=1000, delta=1)
> mle(OU.lik, start=list(theta2=1.5),
+        method="L-BFGS-B", lower=c(0), fixed=c(theta1=0, theta3=1)) -> fit
> summary(fit)
Maximum likelihood estimation

Coefficients:
       Estimate Std. Error
theta2 3.077711  0.1360534

-2 log L: 1026.040
```

3.1.2 The Black and Scholes or geometric Brownian motion model

The Black and Scholes, or geometric Brownian motion model solves the stochastic differential equation

$$\mathrm{d}X_t = \theta_1 X_t \mathrm{d}t + \theta_2 X_t \mathrm{d}W_t, \quad X_0 = x_0, \quad \theta_1 \in \mathbb{R}, \theta_2 \in \mathbb{R}_+ .$$

The conditional density function $p_\theta(t,\cdot|x)$ is log-normal with mean and variance

$$m(t,x) = xe^{\theta_1 t}, \qquad v(t,x) = x^2 e^{2\theta_1 t}\left(e^{\theta_2^2 t} - 1\right).$$

Therefore

$$p_\theta(t,y|x) = \frac{1}{\theta_2 y\sqrt{2\pi t}} \exp\left\{-\frac{(\log y - (\log x + (\theta_1 - \frac{1}{2}\theta_2^2)t))^2}{2\theta_2^2 t}\right\}.$$

R has the `[dqpr]lnorm` functions to evaluate the density, the quantiles, and the cumulative distribution or generate pseudo random numbers from the log-normal distribution. So in this case the code is quite easy to implement (see formula (1.43) for the details). The package `sde` contains the set of functions `[dqpr]lBS`.

```
dcBS <- function(x, t, x0, theta, log = TRUE){
   ml <- log(x0) + (theta[1]-theta[2]^2/2)*t
   sl <- sqrt(t)*theta[2]
   lik <- dlnorm(x, meanlog = ml, sdlog = sl, log=TRUE)
  if(!log)
   lik <- exp(lik)
  lik
}
BS.lik <- function(theta,sigma) {
 n <- length(X)
 dt <- deltat(X)
 -sum(dcBS(x=X[2:n], t=dt, x0=X[1:(n-1)], theta=c(theta,sigma),
  log=TRUE))
}
```

We can easily optimize that:

```
> # ex3.03.R
> set.seed(123)
> X <- sde.sim(model="BS", theta=c(.5,.2), delta=0.01)
> mle(BS.lik, start=list(theta1=1,  theta2=1),
+     method="L-BFGS-B", lower=c(0.01,0.01)) -> fit
> coef(fit)
  theta1    theta2
0.6773086 0.1816525
> length(X)*deltat(X)
[1] 1.01
>
> set.seed(123)
> X <- sde.sim(model="BS", theta=c(.5,.2), N=1000, delta=0.01)
> mle(BS.lik, start=list(theta1=1,  theta2=1),
+     method="L-BFGS-B", lower=c(0.01,0.01)) -> fit
> coef(fit)
  theta1    theta2
0.5319061 0.1982440
> length(X)*deltat(X)
[1] 10.01
>
> set.seed(123)
> X <- sde.sim(model="BS", theta=c(.5,.2), N=5000, delta=0.01)
> mle(BS.lik, start=list(theta1=1,  theta2=1),
+     method="L-BFGS-B", lower=c(0.01,0.01)) -> fit
> coef(fit)
  theta1    theta2
0.4986410 0.1988985
> length(X)*deltat(X)
[1] 50.01
```

In the example above, we have shown empirically that the convergence of the MLE to the true value actually happens as N increases.

3.1.3 The Cox-Ingersoll-Ross model

Another interesting process is the Cox-Ingersoll-Ross model [59] solution to the stochastic differential equation

$$\mathrm{d}X_t = (\theta_1 - \theta_2 X_t)\mathrm{d}t + \theta_3\sqrt{X_t}\mathrm{d}W_t, \quad X_0 = x_0 > 0\,, \tag{3.14}$$

where $\theta_1, \theta_2, \theta_3 \in \mathbb{R}_+$. If $2\theta_1 > \theta_3^2$, the process is strictly positive; otherwise it is only nonnegative. Under this configuration of the parameters, the conditional density $p_\theta(t, \cdot|x)$ follows a non-central χ^2 distribution,

$$p_\theta(t, y|x) = ce^{-u-v}\left(\frac{u}{v}\right)^{q/2} I_q(2\sqrt{uv}), \qquad x, y \in \mathbb{R}_+\,,$$

where

$$c = \frac{2\theta_2}{\theta_3^2(1 - e^{-\theta_2 t})}\,, \qquad q = \frac{2\theta_1}{\theta_3^2} - 1\,,$$
$$u = cxe^{-\theta_2 t}, \qquad v = cy\,.$$

Here $I_q(\cdot)$ is the modified Bessel function of the first kind of order q (see, e.g., [1]),

$$I_q(x) = \sum_{k=0}^{\infty}\left(\frac{x}{2}\right)^{2k+q}\frac{1}{k!\Gamma(k+q+1)}, \qquad x \in \mathbb{R},$$

where $\Gamma(\cdot)$ is the Gamma function, $\Gamma(z) = \int_0^\infty x^{z-1}e^{-x}\mathrm{d}x$, $z \in \mathbb{R}_+$.

We have already seen in (1.48) that the conditional density $p_\theta(t, y|x_0)$ can also be written in terms of Bessel functions. R provides different versions of Bessel functions, and what we need here is the besselI function. To avoid the explosion of the Bessel function $I_q(\cdot)$, it is worth using the exponentially rescaled version, which R provides. besselI(x,q) can return either $I_q(x)$ or $e^{-x}I_q(x)$, but unfortunately, for large values of x and q, the current implementation in R overflows. The same applies to the non-central chi-squared distribution under some circumstances even though its code is a bit more robust. This might be a problem when the conditional density is used to evaluate the likelihood on true data because we have mentioned that for a realistic set of values of the triplet (θ, β, σ) and sampling rate Δ, we usually obtain that the argument of the Bessel function gets larger (see Section 1.13.3). Hence a remedy must be found in such cases. The following code calculates the exponentially rescaled Bessel function using asymptotic expansion.[3] (see [1])

```
expBes <- function(x,nu){
 mu <-  4*nu^2
 A1 <-  1
```

[3] Thanks to Diethelm Wuertz for suggesting this approach.

```
 A2 <- A1 * (mu-   1) / (1 * (8*x))
 A3 <- A2 * (mu-   9) / (2 * (8*x))
 A4 <- A3 * (mu- 25) / (3 * (8*x))
 A5 <- A4 * (mu- 49) / (4 * (8*x))
 A6 <- A5 * (mu- 81) / (5 * (8*x))
 A7 <- A6 * (mu-121) / (6 * (8*x))
 1/sqrt(2*pi*x) * (A1 - A2 + A3 - A4 + A5 - A6 + A7)
}
```

The function `expBes` will be used internally in the `sde` package instead of `besselI` for the evaluation of the conditional density on true data (the likelihood of the process) for the reasons explained. This version of the exponentially rescaled Bessel function is even about ten times faster than the implementation that uses the noncentral chi-squared distribution. Again for reasons of potential numerical overflow, we show two versions of the code that calculates the log-likelihood of a CIR process. If we proceed as in the previous examples, we need $\ell_i(\theta)$ in the form

$$\ell_i(\theta) = \log c - (u+v) + \frac{q}{2}\log\left(\frac{u}{v}\right) + \log I_q(2\sqrt{uv}).$$

If we use the exponentially rescaled Bessel function, what we have is

$$\log\left(e^{-x} I_q(x)\right) = \log I_q(x) - x,$$

and we are going to use this trick in the following code for the `CIR.lik` function.

```
dcCIR <- function(x, t, x0, theta, log = FALSE){
   c <- 2*theta[2]/((1-exp(-theta[2]*t))*theta[3]^2)
   ncp <- 2*c*x0*exp(-theta[2]*t)
   df <- 4*theta[1]/theta[3]^2
   u <- c*x0*exp(-theta[2]*t)
   v <- c*x
   q <- 2*theta[1]/theta[3]^2 -1
   lik <- (log(c) - (u+v) + q/2 * log(v/u) + log(expBes( 2*sqrt(u*v), q))
    +  2*sqrt(u*v))
  if(!log)
   lik <- exp(lik)
  lik
}

CIR.lik <- function(theta1,theta2,theta3) {
 n <- length(X)
 dt <- deltat(X)
 -sum(dcCIR(x=X[2:n], t=dt, x0=X[1:(n-1)], theta=c(theta1,theta2,theta3),
   log=TRUE))
}
```

Next comes the inefficient version of the same likelihood based on the chi-squared distribution.

```
# inefficient version based on noncentral chi^2 density
dcCIR2 <-function (x, t, x0, theta, log = FALSE)
{
   c <- 2*theta[2]/((1-exp(-theta[2]*t))*theta[3]^2)
   ncp <- 2*c*x0*exp(-theta[2]*t)
   df <- 4*theta[1]/theta[3]^2
   lik <- (dchisq(2 * x * c, df = df, ncp = ncp, log = TRUE)
    + log(2*c))
```

```
  if(!log)
    lik <- exp(lik)
  lik
}
CIR.lik2 <- function(theta1,theta2,theta3) {
 n <- length(X)
 dt <- deltat(X)
 -sum(dcCIR2(x=X[2:n], t=dt, x0=X[1:(n-1)], theta=c(theta1,theta2,theta3),
   log=TRUE))
}
```

The following example shows the difference in speed of the two methods, provided that they work with almost the same numerical accuracy, although the Bessel version might be more stable under some circumstances. The reader might want to refer to the references [168], [182], [169] and [72] for extended discussions on the topic.

```
> # ex3.04.R
> set.seed(123)
> X <- sde.sim(X0=.1, model="CIR", theta=c(.2, 0.06, 0.15),
+          N=2500, delta=0.1)
> # ex3.04.R
> system.time(L1 <-CIR.lik(.1,.1,.1))
[1] 0.006 0.006 0.013 0.000 0.000
> print(L1, digits=12)
[1] -2122.73613963
> system.time(L2 <- CIR.lik2(.1,.1,.1))
[1] 0.160 0.003 0.192 0.000 0.000
> print(L2, digits=12)
[1] -2122.73612612
```

There might be cases in which one of the two ways of calculating the conditional distribution does not work because the approximation tends to explode. The following is an example in which the chi-square approximation fails, but the contrary might be true as well for other sets of parameters.

```
> # ex3.04.R (cont.)
> mle(CIR.lik, start=list(theta1=.1,  theta2=.1,theta3=.3),
+   method="L-BFGS-B",lower=c(0.001,0.001,0.001), upper=c(1,1,1)) -> fit
> fit
Coefficients:
   theta1      theta2      theta3
0.19490172 0.06096784 0.14848362
> mle(CIR.lik2, start=list(theta1=.1,  theta2=.1,theta3=.3),
+   method="L-BFGS-B",lower=c(0.001,0.001,0.001), upper=c(1,1,1)) -> fit
Error in optim(start, f, method = method, hessian = TRUE, ...) :
   L-BFGS-B needs finite of 'fn'
```

The Cox-Ingersoll-Ross conditional density can also be approximated using the *Poisson mixing-Gamma* characterization of a density (see, e.g., [68]), which is an infinite sum of Gamma functions weighted by Poisson weights (see, e.g., formula (9) in [233]). This is just the law of Feller's Poisson-driven Gamma process introduced in [84] with a fractional parameter. In [233], the reader can find a simulation study for small samples concerning the use of this approximation. Although this approximation appears to work well in the presence of high volatility, the likelihood might easily diverge in other cases. Other asymptotic expansions related to Bessel functions can be found in [178] and [171].

3.2 Pseudo-likelihood methods

Another way of obtaining estimators is to use some approximation scheme. These approximation schemes do not approximate the transition density directly but the path of the process in such a way that the discretized version of the process has a likelihood that is usable.

3.2.1 Euler method

Consider a process solution of the general stochastic differential equation

$$\mathrm{d}X_t = b(X_t,\theta)\mathrm{d}t + \sigma(X_t,\theta)\mathrm{d}W_t\,.$$

If the coefficients of the stochastic differential equation above are constant over small intervals $[t, t+\Delta t)$, then the Euler scheme produces the discretization

$$X_{t+\Delta t} - X_t = b(X_t,\theta)\Delta t + \sigma(X_t,\theta)(W_{t+\Delta t} - W_t),$$

and the increments $X_{t+\Delta t} - X_t$ are then independent Gaussian random variables with mean $b(X_t,\theta)\Delta t$ and variance $\sigma^2(X_t,\theta)\Delta t$. Therefore the transition density of the process can be written as

$$p_\theta(t,y|x) = \frac{1}{\sqrt{2\pi t\sigma^2(x,\theta)}}\exp\left\{-\frac{1}{2}\frac{(y-x-b(x,\theta)t)^2}{t\sigma^2(x,\theta)}\right\}. \tag{3.15}$$

This approximation is good if Δt is very small; otherwise some bias is introduced. Moreover, if the parameters in the vector θ are different for the drift and the diffusion parts, some reasonable results can be obtained. So we assume now that $\sigma(x,\theta) = \sigma > 0$ is constant and that all the other parameters are in the drift coefficient $b(x,\theta)$; i.e., σ is not one of the parameters in θ. We further suppose that the discretization step is constant and that Assumptions 3.3 and 3.4 hold. Moreover, we need to control the growth of the drift coefficient as follows,

Assumption 3.6 (Polynomial growth condition) *There exist $L > 0$ and $m > 0$ (independent of θ) such that $|b(x,\theta)| \le L(1+|x|^m)$, $\theta \in \Theta$.*

Then the log-likelihood of the discretized process is

$$\ell_n(\theta) = -\frac{1}{2}\left\{\sum_{i=1}^n \frac{(X_i - X_{i-1} - b(X_{i-1},\theta)\Delta)^2}{\sigma^2\Delta} + n\log(2\pi\sigma^2\Delta)\right\}.$$

The equation above is also called the *locally Gaussian approximation.* Given that σ^2 is constant, the maximization of the log-likelihood is equivalent to the maximization of the function

$$\sum_{i=1}^n (X_i - X_{i-1})b(X_{i-1},\theta) - \frac{\Delta}{2}\sum_{i=1}^n b^2(X_{i-1},\theta)\,. \tag{3.16}$$

Under the additional assumption $n\Delta_n^3 \to 0$, the maximum likelihood estimator built on the pseudo-likelihood above is consistent and asymptotically normal [231]. In [231], the result is proved to hold for a diffusion coefficient that can also depend on the space variable and for a multidimensional diffusion process. In that case, a preliminary estimator of the diffusion coefficient is needed to estimate θ further. When θ has been estimated, an update of the estimator of the diffusion coefficient can be obtained. Further, [231] and [86] (for the one-dimensional case) showed that a consistent estimator of σ^2 is

$$\hat{\sigma}^2 = \frac{1}{n\Delta}\sum_{i=1}^{n}(X_i - X_{i-1})^2. \tag{3.17}$$

This algorithm is quite easy to implement. We start by writing the code for the function (3.16) to be maximized.

```
dcEuler <- function(x, t, x0, theta, drift){
dd <- drift(x0, theta)
 (x-x0)*dd - 0.5*t*dd^2
}
Euler.lik <- function(theta,beta){
  n <- length(X)
  dt <- deltat(X)
   -sum(dcEuler(X[2:n], dt, X[1:(n-1)], c(theta,beta), mydrift))
}
```

Then we simulate a path for the Vasicek process with parameters ($\theta = 5, \beta = 3, \sigma = 2$), we construct a drift function to pass as b in (3.16), and we maximize Euler's pseudo-likelihood with respect to θ and β and calculate the estimator of σ^2.

```
> # ex3.05.R
> set.seed(123)
> X <- sde.sim(model="OU", theta =c(5,3,2), N=2500)
> mle(Euler.lik, start=list(theta1=1.5,  theta2=1),
+     method="L-BFGS-B", lower=c(0,0)) -> fit
>
> summary(fit)
Maximum likelihood estimation
Coefficients:
       Estimate Std. Error
theta1 16.47255   3.768991
theta2  8.59521   1.953179

-2 log L: -19.59694
> sqrt(mean((X[2:length(X)] - X[1:(length(X)-1)])^2)/deltat(X))
[1] 1.977273
>
> X <- sde.sim(model="OU", theta =c(5,3,2), delta=0.1, N=2500)
> mle(Euler.lik, start=list(theta1=1.5,  theta2=1),
+     method="L-BFGS-B", lower=c(0,0)) -> fit
>
> summary(fit)
Maximum likelihood estimation
Coefficients:
       Estimate Std. Error
theta1 4.440512 0.14397788
theta2 2.704831 0.07886108

-2 log L: -1176.404
```

```
> sqrt(mean((X[2:length(X)] - X[1:(length(X)-1)])^2)/deltat(X))
[1] 1.865060
```

As can be seen, the impact of the Δ is not negligible in the Euler approximation. We can now compare these estimates with the exact maximum likelihood estimators optimizing the `OU.lik` true likelihood of Section 3.1.1

```
> # ex3.05.R (cont.)
> mle(OU.lik, start=list(theta1=1.5, theta2=1, theta3=1),
+    method="L-BFGS-B", lower=c(0,0,0)) -> fit2
> summary(fit2)
Maximum likelihood estimation
Coefficients:
       Estimate Std. Error
theta1 5.177466 0.33307169
theta2 3.153729 0.18748140
theta3 2.013842 0.03312004

-2 log L: 4091.324
```

from which it seems that the estimates are comparable. The higher variability of the true maximum likelihood estimators is due to the fact that we jointly estimate three parameters. The package `sde` implements the complete Euler approximated likelihood as in Listing 3.1, which allows specification of both drift and diffusion coefficients as functions of the initial point x_0, time `t0`, and parameter vector θ.

```
dcEuler <- function(x, t, x0, t0, theta, d, s, log=FALSE){
 dnorm(x, mean = x0 - d(t0, x0, theta)*t, sd= sqrt(t)*s(t0,x0,theta),
    log=log)
}
```

Listing 3.1. Euler conditional likelihood.

The effect of Δ on the Euler approximation

It is worth noting that this kind of approach must, in general, be used only as a *last resort* method if Δ does not shrink to zero. If Δ is not small, the estimates include bias even for the very simple Ornstein-Uhlenbeck model. Indeed, consider the Vasicek process. For this model, both Euler's pseudo transition density and the true transition density are Gaussian. From equations (3.10) and (3.11), we have that

$$m(\Delta, x) = xe^{-\theta_2\Delta} + \frac{\theta_1}{\theta_2}\left(1 - e^{-\theta_2\Delta}\right), \qquad v(\Delta, x) = \frac{\theta_3^2\left(1 - e^{-2\theta_2\Delta}\right)}{2\theta_2},$$

and for the Euler approximation we have

$$m^{Euler}(\Delta, x) = x(1 - \theta_2\Delta) + \theta_1\Delta, \qquad v^{Euler}(\Delta, x) = \theta_3^2\Delta,$$

but m^{Euler} and v^{Euler} coincide with the true values only if $\Delta \to 0$. Another example of inappropriate use of the Euler scheme is taken from [160]. Consider the geometric Brownian motion

$$\mathrm{d}X_t = \theta_1 X_t \mathrm{d}t + \theta_2 X_t \mathrm{d}W_t\,.$$

We already know from Section 1.13.2 that this process has log-normal increments, but the Euler scheme assumes zero-mean Gaussian increments with variance $\theta_2^2\Delta$. From the Euler discretization, it emerges that maximum likelihood estimators of θ_1 and θ_2 are

$$\hat\theta_{1,n} = \frac{1}{n\Delta}\sum_{i=1}^{n}\left(\frac{X_i}{X_{i-1}} - 1\right), \quad \hat\theta_{2,n}^2 = \frac{1}{n\Delta}\sum_{i=1}^{n}\left(\frac{X_i}{X_{i-1}} - 1 - \hat\theta_{1,n}^2\Delta\right)^2.$$

For fixed and not negligible Δ, the following asymptotics are true as $n \to \infty$:

$$\hat\theta_{1,n} \xrightarrow{P} \frac{1}{\Delta}\left(e^{\theta_1\Delta} - 1\right) \neq \theta_1, \quad \hat\theta_{2,n} \xrightarrow{P} \frac{1}{\Delta}e^{2\theta_1\Delta}\left(e^{\theta_2^2\Delta} - 1\right) \neq \theta_2.$$

So the estimators are not event consistent. One natural fix to this problem is to modify the sampling scheme a bit so that Δ_n also shrinks to 0, for example, with $n\Delta_n = T$ fixed. Phillips [184] termed the limit of this sampling scheme the "continuous data recording" to distinguish it from inference on true "continuous-time observations." In this case, the author proved consistency for the estimator $\hat\theta_{1,n}$ and non consistency of $\hat\theta_{2,n}$ (see also [183]). In the case above, other consistent estimators for θ_1 exist when Δ do not shrink; for example (see again [160]),

$$\tilde\theta_{1,n} = \frac{1}{\Delta}\log(1 + \hat\theta_{1,n}), \quad \tilde\theta_{1,n}' = \frac{1}{n\Delta}\log(X_n/X_0) + \frac{1}{2}\hat\theta_{2,n}^2.$$

The considerations above are, for example, motivations for other refinements in the sampling scheme like the one presented in Section 3.3.1.

3.2.2 Elerian method

Elerian [76] proposed to use the transition density derived from the Milstein scheme (see Section 2.2)

$$\begin{aligned} X_{t+\mathrm{d}t} &= X_t + b(t, X_t)\mathrm{d}t + \sigma(t, X_t)(W_{t+\mathrm{d}t} - W_t) \\ &\quad + \frac{1}{2}\sigma(t, X_t)\sigma_x(t, X_t)((W_{t+\mathrm{d}t} - W_t)^2 - \mathrm{d}t)\,. \end{aligned}$$

When the process has constant volatility or at least $\sigma_x \simeq 0$, the transition density proposed by Elerian reduces to the transition density of the Euler scheme in Section 3.2.1. Elerian transition density looks like the one presented in the following result.

Fact 3.1 (Theorem 2.1 in [76]) *Suppose that the usual conditions for the existence of a unique solution of the stochastic differential equation* $\mathrm{d}X_t = b(t, X_t)\mathrm{d}t + \sigma(t, X_t)\mathrm{d}W_t$, $X_0 = x_0$, *are satisfied. Introduce the notation*

$$Af(t,x) = \sigma(t,x)\frac{\partial}{\partial x}f(t,x)$$

and

$$b^*(t,x) = b(t,x) - \frac{1}{2}\sigma(t,x)\sigma_x(t,x)\,.$$

Suppose that the following set of additional conditions on the coefficients b and σ hold true for all t, s in $[0,T]$ and all x in $\mathbb{R}$:

$$|b^*(t,x)| + |Ab(t,x)| \le K(1+|x|), \quad |\sigma(t,x)| + |A\sigma(t,x)| \le K(1+|x|),$$

$$|A^2\sigma(t,x)| \le K(1+|x|),$$

and

$$|g(t,x) - g(s,x)| \le K(1+|x|)|s-t|^{\frac{1}{2}}, \tag{3.18}$$

where condition (3.18) on the generic function $g(\cdot,\cdot)$ is assumed to hold for $b^*(\cdot,\cdot)$, $\sigma(\cdot,\cdot)$, and $A\sigma(\cdot,\cdot)$. In all the inequalities above, K is a (generic) constant not dependent on the discretization step. Then, the transition density of the Milstein scheme can be written as

$$p^{Elerian}(t,y|x) = \frac{z^{-\frac{1}{2}}\cosh(\sqrt{Cz})}{|A|\sqrt{2\pi}}e^{-\frac{C+z}{2}},$$

where

$$A = \frac{\sigma(x)\sigma_x(x)t}{2}, \qquad B = -\frac{\sigma(x)}{2\sigma_x(x)} + x + b(x)t - A\,,$$

$$z = \frac{y-B}{A}, \qquad C = \frac{1}{\sigma_x^2(x)t}\,.$$

The approximation is valid if $\sigma_x \neq 0$ and $z > 0$.

For the implementation, it is usually better to use the exponential version of $\cosh(x) = (e^x + e^{-x})/2$.

```
dcElerian <- function(x, t, x0, t0, theta, d, s, sx, log=FALSE){
 A <- s(t0, x0, theta)*sx(t0, x0, theta)*t/2
 B <- -s(t0, x0,theta)/(2*sx(t0, x0,theta)) + x0 + d(t0, x0, theta)*t - A
 z <- (x-B)/A
 C <- 1/((s(t0, x0,theta)^2)*t)
 lik <- (exp(-(C+z)/2) *(exp(sqrt(C*z)) +
  exp(-sqrt(C*z)))/(2 * sqrt(z) * abs(A) * sqrt(2*pi)))
 if(log)
  lik <- log(lik)
 lik
}
```

Listing 3.2. Elerian conditional likelihood.

One thing worth mentioning is that this transition density is not Gaussian or symmetric. Listing 3.2 implements this approximate conditional likelihood. `dcElerian` needs as input three functions `d`, `s`, and `sx` (respectively $b(t,x;\theta)$, $\sigma(t,x;\theta)$, and $\sigma_x(t,x;\theta)$), and all must be functions of the initial point `x` and the parameter vector `theta`. Here and in the following, we explicitly require `d`, `s`, etc., to be true R functions and not expressions and we further require to specify all the derivatives. This is done to increase the speed of execution of the code. In fact, if we use R expressions and further calculate all the necessary partial derivatives as in `sde.sim`, the code will slow down because this conditional density has to be called several times to obtain the likelihood and then again for different values of the parameter `theta` during the optimization step.

3.2.3 Local linearization methods

We saw in Section 2.11 that another approach to approximate the solution of a stochastic differential equation is to use a local linearization method. We discussed the Ozaki method for homogeneous stochastic differential equations and the Shoji-Ozaki method for the non-homogeneous case. The functions `dcOzaki` and `dcShoji` return the value of the transition density $p_\theta(t_0+t, x|x_0)$. Even if the Ozaki method is for a homogeneous diffusion process with constant volatility,

$$\mathrm{d}X_t = b(X_t)\mathrm{d}t + \sigma \mathrm{d}W_t,$$

the drift and diffusion coefficients must be specified as functions of `t`, `x` and `theta`. We just recall that the transition density for the Ozaki method is Gaussian with mean E_x and variance V_x as in equations (2.12) and (2.13) respectively.

```
dcOzaki <- function(x, t, x0, t0, theta, d, dx, s, log=FALSE){
  Lx <- dx(t0,x0,theta)
  Kx <- log(1+d(t0,x0,theta)*(exp(Lx*t)-1)/(x0*Lx))/t
  Ex <- x0 + d(t0,x0,theta)/Lx*(exp(Lx*t)-1)
  Vx <- s(t0,x0,theta)^2 * (exp(2*Kx*t) -1)/(2*Kx)
  dnorm(x, mean=Ex, sd=sqrt(Vx),log=log)
}
```

Listing 3.3. Ozaki conditional likelihood.

For the Shoji-Ozaki method, the transition density is Gaussian with mean $E_x = A(x)$ and variance $V_x = B(x)$, where $A(x)$ and $B(x)$ are defined in (2.14) and (2.15), respectively. So both conditional likelihoods can be easily implemented as in Listings 3.3 and 3.4.

```
dcShoji <- function(x, t, x0, t0, theta, d, dx, dxx, dt, s, log=FALSE){
    Lx <- dx(t0,x0,theta)
    Mx <- s(t0,x0,theta)^2 * dxx(t0,x0,theta)/2 + dt(t0,x0,theta)
    Ex <- (x0 + d(t0,x0,theta)*(exp(Lx*t)-1)/Lx +
          Mx*(exp(Lx*t) -1 -Lx*t)/Lx^2)
    Vx <- s(t0,x0,theta)^2*(exp(2*Lx*t)-1)/(2*Lx)
    dnorm(x, mean=Ex, sd=sqrt(Vx),log=log)
}
```

Listing 3.4. Shoji-Ozaki conditional likelihood.

3.2.4 Comparison of pseudo-likelihoods

We have already mentioned that the Euler scheme and hence the Euler conditional likelihood have good performance when the time interval Δ is small and the process is well-behaved. It is difficult to make a generic comparison of these approximated likelihood methods without relying on some particular model. So, at present we are not trying to summarize the result but just show an application of these methods to one sample path. The reader might want to read the original papers of the authors who proposed the methods to understand which case better fits his need. As a motivating example we, consider the following Cox-Ingersoll-Ross process

$$\mathrm{d}X_t = (0.5 - 0.2X_t)\mathrm{d}t + \sqrt{0.05X_t}\mathrm{d}W_t\,.$$

In the notation of Section 1.13.3, the true parameter is the vector $(\theta_1, \theta_2, \theta_3) = (0.5, 0.2, \sqrt{0.05})$. We perform approximate maximum likelihood estimation using some of the methods introduced so far. We also compare the profile likelihoods for θ_2 given the true values of θ_1 and θ_3 in order to show the speed of convergence of the approximated likelihood to the true likelihood, which is known for this model. For this model, we need to make explicit the quantities

$$b(t,x) = \theta_1 - \theta_2 x, \quad b_x(t,x) = -\theta_2, \quad b_{xx}(t,x) = b_t(t,x) = 0,$$

$$\sigma(x) = \theta_3\sqrt{x}, \quad \sigma_x(x) = \frac{\theta_3}{2\sqrt{x}}, \quad \sigma_{xx}(t,x) = -\frac{\theta_3}{4x^{\frac{3}{2}}}\,.$$

To make use of the Ozaki and Shoji-Ozaki methods, we need to transform a process into one with constant volatility using the transform

$$F(X_t) = \frac{1}{\theta_3}\int_0^{X_t} \frac{1}{\sqrt{u}}\mathrm{d}u = \frac{2\sqrt{X_t}}{\theta_3}, \quad F^{-1}(y) = \left(\frac{\theta_3 y}{2}\right)^2.$$

The transformed process $Y_t = F(X_t)$ satisfies a stochastic differential equation with unitary diffusion coefficient and drift function b_Y defined as follows

$$b_Y(t,x) = \frac{\theta_1 - \theta_2 x}{\theta_3\sqrt{x}} - \frac{\theta_3}{4\sqrt{x}}.$$

We rewrite b_Y at point $F^{-1}(y)$ as

$$b_Y(t, F^{-1}(y)) = \frac{4\theta_1 - \theta_3^2}{2\theta_3^2 y} - \frac{\theta_2}{2}y;$$

hence Y_t is the solution of $\mathrm{d}Y_t = b_Y(t, F^{-1}(Y_t))\mathrm{d}t + \mathrm{d}W_t$ with initial condition $Y_0 = F(X_0)$. We also need the first and second partial derivatives of b_Y with respect to y,

$$\frac{\partial}{\partial y}b_Y = -\frac{\theta_2}{2} - \frac{4\,\theta_1 - \theta_3^2}{2y^2\theta_3^2}, \qquad \frac{\partial^2}{\partial y^2}b_Y = \frac{4\,\theta_1 - \theta_3^2}{y^3\theta_3^2}\,.$$

The next code prepares the approximated likelihood functions, and in order to compare them with the true likelihood of the process, we need to adjust the Ozaki and Shoji-Ozaki likelihoods by the Jacobian of the transform; i.e., the transition density of $X_t|X_s$ can be obtained from that of $Y_t = 2\sqrt{X_t}/\theta_3$, multiplying p_y by the Jacobian $1/(\theta_3\sqrt{X})$ so

$$\log p_X(t, x|x_0; \theta) = \log p_Y(t, F(x)|F(x_0); \theta) - \frac{1}{2}\log(\theta_3\sqrt{x}).$$

```
# ex3.06.R
require(sde)

d <- function(t,x,theta) theta[1]-theta[2] * x
dx <- function(t,x,theta) -theta[2]
dxx <- function(t,x,theta) 0
dt <- function(t,x,theta) 0
s <- function(t,x,theta)  theta[3]*sqrt(x)
sx <- function(t,x,theta) theta[3]/(2*sqrt(x))
sxx <- function(t,x,theta) -theta[3]/(4*x^1.5)

d2 <- function(t,x,theta){
 (4*theta[1]-theta[3]^2)/(2*x*theta[3]^2) -theta[2]*x/2 }
d2x <- function(t,x,theta){
 -theta[2]/2 - (4*theta[1]-theta[3]^2)/(2*x^2*theta[3]^2)}
d2xx <- function(t,x,theta) (4*theta[1]-theta[3]^2)/(x^3*theta[3]^2)
d2t <- function(t,x,theta) 0
s2 <- function(t,x,theta)  1
s2x <- function(t,x,theta) 0
s2xx <- function(t,x,theta) 0

Euler.LIK <- function(theta) {
   sum(dcEuler(X[2:n], t[2:n], X[1:(n-1)], t[1:(n-1)],
     c(0.5, theta, sqrt(0.05)), d,s, TRUE),na.rm=TRUE)
}

Elerian.LIK <- function(theta) {
   sum(dcElerian(W[2:n], t[2:n], W[1:(n-1)], t[1:(n-1)],
    c(0.5, theta, sqrt(0.05)), d2, s2, s2x, TRUE)
     -0.5*log(X[2:n]*0.05),na.rm=TRUE)
}

Ozaki.LIK <- function(theta) {
   sum(dcOzaki(W[2:n], t[2:n], W[1:(n-1)], t[1:(n-1)],
     c(0.5, theta, sqrt(0.05)), d2, d2x, s2, TRUE)
      -0.5*log(X[2:n]*0.05),na.rm=TRUE)
}

Shoji.LIK <- function(theta) {
   sum(dcShoji(W[2:n], t[2:n], W[1:(n-1)], t[1:(n-1)],
    c(0.5, theta, sqrt(0.05)), d2, d2x,d2xx, d2t, s2, TRUE)
     -0.5*log(X[2:n]*0.05),na.rm=TRUE)
}

True.LIK <- function(theta) {
   sum(dcCIR(X[2:n], deltat(X), X[1:(n-1)],
    c(0.5, theta, sqrt(0.05)), TRUE),na.rm=TRUE)
}

pTrue <- function(x) True.LIK(x)
pEuler <- function(x) Euler.LIK(x)
pElerian <- function(x) Elerian.LIK(x)
pOzaki <- function(x) Ozaki.LIK(x)
pShoji <- function(x) Shoji.LIK(x)
```

Notice that in the code above, to make the Elerian method work in this example, we used the transformed process. This implies that the Euler and Elerian methods do not differ too much. The user might want to test the bad performance in this example when the Elerian method is applied to the original process. In general, this applies to several other methods and Durham and Gallant [74] suggest that the Lamperti transform (1.34) always be applied to the original process. This rule seems to be worth using in both simulation and estimation applications. What follows now is the approximation experiment. We simulate one long trajectory of the Cox-Ingersoll-Ross process with parameters $(\theta_1, \theta_2, \theta_3) = (0.5, 0.2, \sqrt{0.05})$. More precisely, we set `N` equal to 500000 and $\Delta = 0.001$ and then resample the trajectory for different values of Δ to show the convergence of approximations to the true likelihood. To resample the trajectory, we use the function `window`. We compare the true likelihood against the Euler, Elerian, Ozaki and Shoji-Ozaki (just "Shoji" in the plot and table) methods for fixed values of $\theta_1 = 0.5$ and $\theta_3 = \sqrt{0.05}$ as a function of θ_2. We also optimize these profile likelihoods as functions of θ_2 because what really matters in practice is the behavior of the approximating likelihoods in a neighborhood of the true value of the parameter. To optimize the likelihoods, we use the function `optimize` directly instead of `mle`.

```
# ex3.06.R (cont)
set.seed(123)
X1 <- sde.sim(model="CIR", theta=c(0.5, 0.2, sqrt(0.05)),
 X0=2,delta=.001, N=500000)
xx <- seq(0.001,0.4, length=50)

par(mfrow=c(2,2))
est <- NULL
for(dt in c(4, 2,1,.5)){
 X <- window(X1, deltat=dt)
 W <- 2*sqrt(X)/sqrt(0.05)
 t <- as.numeric(time(X))
 n <- length(X)
 cat(sprintf("number of observations: %d, Delta=%3.2f\n",n,dt))
 dEuler <- sapply(xx, pEuler)
 dTrue <- sapply(xx, pTrue)
 dElerian <- sapply(xx, pElerian)
 dOzaki <- sapply(xx, pOzaki)
 dShoji <- sapply(xx, pShoji)

 mx <- max(c(dTrue,dEuler,dShoji,dOzaki,dElerian,na.rm=TRUE))
 mn <- min(c(dTrue,dEuler,dShoji,dOzaki,dElerian,na.rm=TRUE))

 matplot(xx,cbind(dTrue,dEuler,dOzaki,dShoji,dElerian),type="l",
  ylim=c(mn,mx),xlab="",ylab="approx",
  main=sprintf("N=%d, Delta=%3.2f",n,dt),lty=1:5,col=1:5)
 legend(.15,0.6*(mx+mn), lty=1:5,col=1:5,
  legend=c("True", "Euler", "Ozaki", "Shoji","Elerian"))

 tmp <- c(n, dt, optimize(pTrue, c(0.01,.4),max=T)$max,
  optimize(pEuler, c(0,.4),max=T)$max,
  optimize(pElerian, c(0,1),max=T)$max,
  optimize(pOzaki, c(0,.4),max=T)$max,
  optimize(pShoji, c(0,.4),max=T)$max)
 est <- rbind(est, tmp)
}
dimnames(est)[[2]] <- c("N","Delta","True","Euler",
 "Elerian", "Ozaki", "Shoji")
```

```
dimnames(est)[[1]] <- 1:4
print(est)
par(mfrow=c(1,1))
```

The result of the approximation as a function of Δ can be inspected graphically in Figure 3.2, generated using the `matplot` command. Table 3.1 presents different estimated values of θ_2 for different values of Δ and for all the methods considered.

Table 3.1. Different estimates of the parameter θ_2 for a simulated path of the Cox-Ingersoll-Ross model of parameters $(\theta_1, \theta_2, \theta_3) = (0.5, 0.2, \sqrt{0.05})$ and maximum likelihood estimates on the true likelihood (True) and several approximation methods for different values of the discretization step Δ.

N	Δ	True	Euler	Elerian	Ozaki	Shoji
126	4.0	0.2074	0.1944	0.1929	0.1954	0.1945
251	2.0	0.1931	0.1926	0.1916	0.1930	0.1925
501	1.0	0.1925	0.1926	0.1922	0.1929	0.1926
1001	0.5	0.1922	0.1922	0.1920	0.1924	0.1922

3.3 Approximated likelihood methods

In this section, we present methods that differ from the previous in that they do not try to approximate the paths of a diffusion but instead provide direct approximation of the likelihood.

3.3.1 Kessler method

Kessler [137] proposed to use a higher-order Itô-Taylor expansion to approximate the mean and variance of the conditional density. We cannot go into details, but roughly speaking the Itô-Taylor expansion is the expansion

$$\mathbb{E}_\theta(\phi(X_i)|\mathcal{F}_{i-1}) = \phi(X_{i-1}) + \Delta_n \mathcal{L}_\theta \phi(X_{i-1}) + \frac{1}{2}\Delta_n^2 \mathcal{L}_\theta^2 \phi(X_{i-1}) + \cdots , \quad (3.19)$$

where $\mathcal{L}_\theta$ is the infinitesimal generator of the diffusion and $\phi(x)$ the appropriate function (see, e.g., [107]). This approximation is valid for $n \to \infty$, $\Delta_n \to 0$, $n\Delta_n \to \infty$ and $n\Delta_n^2 \to 0$; i.e., under the so-called "rapidly increasing experimental design" (see, e.g., [190]). Other additional hypotheses apart from the one needed for ergodicity are that $\inf_{x,\theta} \sigma^2(x, \theta) > 0$, the process X_t has finite moments of all order, and the diffusion and drift coefficients and their derivatives (up to order 2) in x are three times continuously differentiable in θ and of polynomial growth. These assumptions, corresponding to Assumptions 3.1 to 3.5, can be found in detail in Section 2 of [137], and the main result

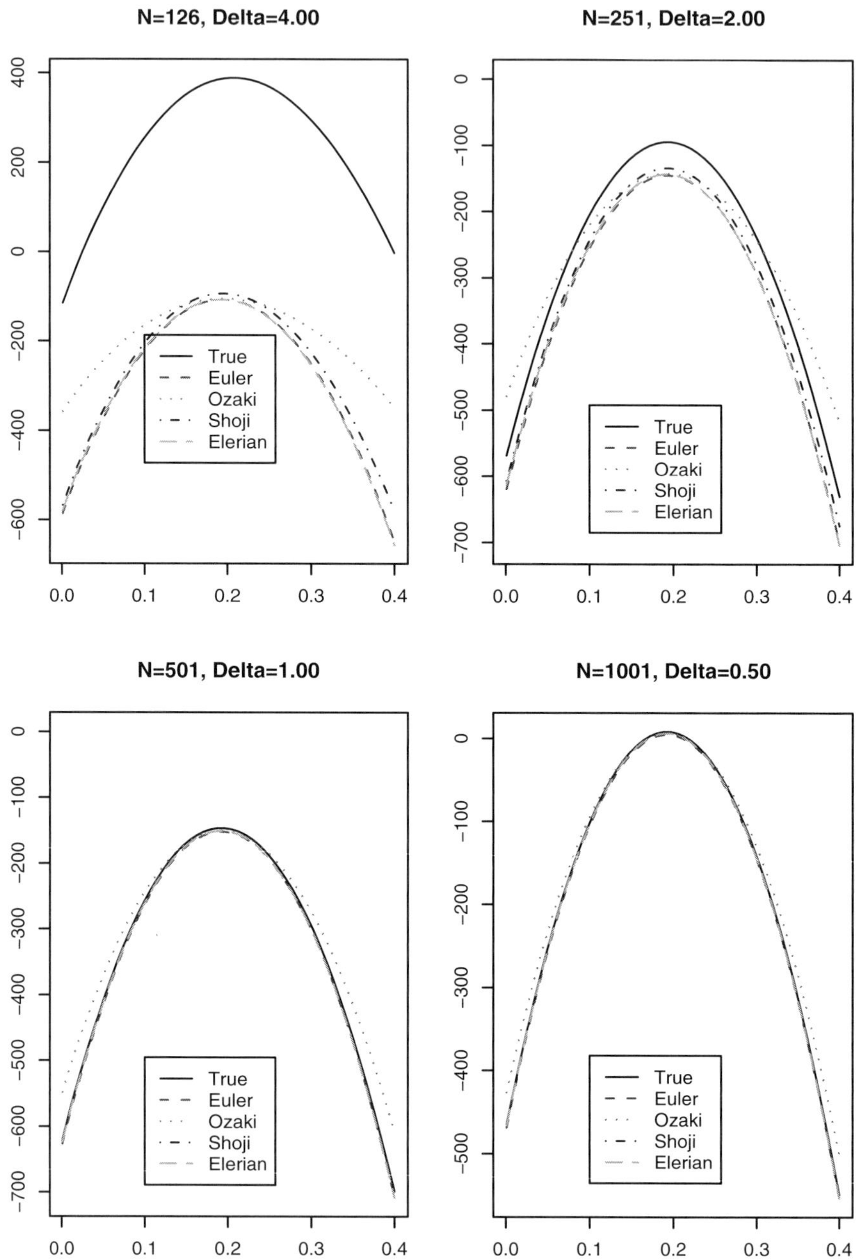

Fig. 3.2. Different approximations of the true likelihood (black, solid line) for different values of the discretization step Δ. For small values of Δ, most methods converge to the true likelihood, and this example is no exception.

on approximation is contained in Lemmas 1 and 2 of its Section 3. The final approximation results in a conditional Gaussian density with parameters E_x and V_x as follows:

$$\begin{aligned}
E_x =& x + b(t,x)\mathrm{d}t + \left(b(t,x)b_x(t,x) + \frac{1}{2}\sigma^2(t,x)b_{xx}(t,x)\right)\frac{(\mathrm{d}t)^2}{2},\\
V_x =& x^2 + \left(2b(t,x)x + \sigma^2(t,x)\right)\mathrm{d}t + \Big\{2b(t,x)(b_x(x)x + b(t,x)\\
&+ \sigma(t,x)\sigma_x(t,x)) + \sigma^2(t,x)(b_{xx}(t,x)x + 2b_x(t,x) + \sigma_x^2(t,x)\\
&+ \sigma(t,x)\sigma_{xx}(t,x))\Big\}\frac{(\mathrm{d}t)^2}{2} - E_x^2.
\end{aligned}$$

For some trajectories, V_x might be negative for some values of θ or x, and hence we return `NA` in our code of Listing 3.5.

```
dcKessler <- function(x, t, x0, t0, theta, d, dx, dxx,  s, sx, sxx,
         log=FALSE){
 Ex <- (x0 + d(t0,x0,theta)*t + (d(t0,x0,theta)*dx(t0,x0,theta)+
    (s(t0,x0,theta)^2 * dxx(t0,x0,theta))/2)*(t^2)/2)
 Vx <- (x0^2 +(2*d(t0,x0,theta)*x0 + s(t0,x0,theta)^2)*t +
 (2*d(t0,x0,theta)*(dx(t0,x0,theta)*x0+d(t0,x0,theta)+
 s(t0,x0,theta)*sx(t0,x0,theta))+  s(t0,x0,theta)^2*(dxx(t0,x0,theta)*x0
  + 2*dx(t0,x0,theta) +  sx(t0,x0,theta)^2 +
  s(t0,x0,theta)*sxx(t0,x0,theta)))*(t^2)/2 - Ex)
 if(Vx < 0) return(NA)
 dnorm(x, mean = Ex, sd=sqrt(Vx),log=log)
}
```

Listing 3.5. Kessler conditional likelihood.

In the framework consider by Kessler, the following result holds for the maximum likelihood estimator.

Fact 3.2 (Theorem 1 in [137]) *Suppose hypotheses A1 to A4 and A5[2] in [137] are satisfied. If* $\theta = (\theta_1, \theta_2) \in \Theta$, Θ *a compact subset of* $\mathbb{R}^2$, $b(x,\theta) = b(x,\theta_1)$, *and* $\sigma(x,\theta) = \sigma(x,\theta_2)$, *then the maximum likelihood estimators obtained on the conditional likelihood above are consistent and asymptotically normal; i.e.,*

$$\begin{pmatrix} \sqrt{n\Delta_n}(\hat\theta_{1,n} - \theta_{1,0}) \\ \sqrt{n}(\hat\theta_{2,n} - \theta_{2,0}) \end{pmatrix} \stackrel{d}{\to} N(0, K_0)$$

with

$$K_0 = \begin{bmatrix} \left(\int\left(\frac{\partial_{\theta_1} b(x,\theta_{1,0})}{\sigma(x,\theta_{2,0})}\right)^2 \pi_0(\mathrm{d}x)\right)^{-1} & 0 \\ 0 & 2\left(\int\left(\frac{\partial_{\theta_2}\sigma(x,\theta_{2,0})}{\sigma^2(x,\theta_{2,0})}\right)^2 \pi_0(\mathrm{d}x)\right)^{-1} \end{bmatrix},$$

where $\theta_{1,0}$ *and* $\theta_{2,0}$ *are the true values of the parameter and* $\pi_0(\cdot)$ *is the density with respect to the Lebesgue measure of the invariant density of the diffusion process.*

To ease the notation, we denoted by ∂_θ the partial derivatives of the coefficients b and σ with respect to the parameters of interest. The nice feature of this method is that it is possible to obtain approximated confidence intervals for the estimates of the parameters thanks to the distributional results above. Of course, the K_0 above depends on the invariant density $\pi_0(\cdot)$ and the evaluation of the integrals. For some processes, this can be explicitly calculated; otherwise numerical integration is required using the fact that the invariant density $\pi_0(\cdot)$ is an explicit function of the drift and diffusion coefficients, as formula (3.2) shows. We leave this "ad hoc" analysis to the reader, but for rough numerical approximated confidence intervals one can rely on the usual `mle` output given that asymptotic normality is provided by the result above.

3.3.2 Simulated likelihood method

The simulated likelihood method was proposed independently by Pedersen [180] and Santa-Clara [200]. The idea is relatively simple but smart. Let $p_\theta(\Delta, y|x)$ be the true transition density of $X_{t+\Delta}$ at point y given $X_t = x$. When the time step Δ is too large, we have seen that the Euler approximation usually gives a poor estimate of $p_\theta(\Delta, y|x)$. The idea is then to consider a smaller $\delta << \Delta$, for example $\delta = \Delta/N$ for N large enough, and then use the Chapman-Kolmogorov equation as follows:

$$p_\theta(\Delta, y|x) = \int p_\theta(\delta, y|z) p_\theta(\Delta - \delta, z|x) \mathrm{d}z = \mathbb{E}_z\{p_\theta(\delta, y|z)|\Delta - \delta\},$$

which means that $p_\theta(\Delta, y|x)$ is seen as the expected value over all possible transitions of the process from time $t + (\Delta - \delta)$ to $t + \Delta$, taking into account that the process was in x at time t. To realize this program, it is necessary to simulate trajectories of the process X from t to $t + (\Delta - \delta)$ using the Euler scheme, which is assumed to be valid because δ has been chosen small enough. If the Euler scheme is involved, then $p_\theta(\Delta - \delta, z|x)$ is nothing but the Euler transition density, which is Gaussian. So it is necessary to simulate M trajectories of the process in order to estimate $p_\theta(\Delta, y|x)$ by the Monte Carlo method. Let $\phi_\theta(\delta, y|z)$ be the Euler transition density of equation (3.15). To calculate an estimate of $p_\theta(\Delta, y|x)$, we need to simulate M trajectories starting at $X_0 = x$ using Euler method with time step δ up to time $\Delta - \delta$. We denote the last value of the mth trajectory as z_m, $m = 1, \ldots, M$. Finally, we obtain the Monte Carlo estimate as

$$\hat{p}_\theta^{(N,M)}(\Delta, X_{t+\Delta}|x) = \frac{1}{M} \sum_{i=1}^{M} \phi_\theta(\delta, X_{t+\Delta}|z_i).$$

In principle, by increasing the number of simulated trajectories M, one can obtain any degree of accuracy in the estimation provided that the process is regular enough and δ is also small enough so that all the conditions for the Euler approximation are satisfied. In [180], it was proved that

$$\hat{p}_\theta^{(N,M)}(\Delta, X_{t+\Delta}|x) \to p_\theta(\Delta, X_{t+\Delta}|x) \quad \text{for } N \to \infty, M \to \infty.$$

In [112], it is possible to find some comments on the performance of this method as a function of N and M. Usually, a large number of simulations M are needed, and N may vary from 5 to 10 for reasonable estimates. A typical variance reduction technique can be used here to reduce the computational burden. Indeed, in generating the M trajectories, once the Gaussian pseudo random numbers are generated, the same sequence can be used after a sign change to generate a symmetric trajectory (we discussed this approach in Section 1.4.3). Calculating the approximation of the whole likelihood is a really intensive computational task because for each couple of observations M trajectories of length N are needed. If this approximation has to be used as a function of the parameters θ in order to get maximum likelihood estimates of the parameters, this task may require a huge amount of time. A discussion about an efficient approach to simulated maximum likelihood estimation can be found in [74] and in many recent papers on the subject. In Listing 3.6, we present the very basic implementation in R via the function `dcSim`. The `sde` package contains a function with the same name written in C in order to give reasonable computational times.

```
dcSim <- function(x0,x, t, d, s, theta, M=10000, N=5){
 delta <- t/N
 N <- N-1
 x1 <- numeric(N)
 x2 <- numeric(N)
 w <- numeric(M)
 for(j in seq(1,M,by=2)){
  z <- rnorm(N-1)
  x1[1] <- x0
  x2[1] <- x0
  for(i in 2:N){
   x1[i] <- x1[i-1] + d(0,x1[i-1],theta)*delta +
              s(0, x1[i-1],theta)*sqrt(delta)*z[i-1]
   x2[i] <- x2[i-1] + d(0,x2[i-1],theta)*delta -
              s(0, x2[i-1],theta)*sqrt(delta)*z[i-1]
  }
  w[j] <- x1[N]
  w[j+1] <- x2[N]
 }
 mean(dcEuler(x,delta,w,0,theta,drift,sigma), na.rm=TRUE)
}
```

Listing 3.6. Simulated conditional likelihood.

The next example and Figure 3.3 show that the approximation of the likelihood depends heavily on the number of subintervals N. The next code calculates the conditional density by simulation for the Cox-Ingersoll-Ross model $\mathrm{d}X_t = \theta_1(\theta_2 - x)\mathrm{d}t + \theta_3\sqrt{X_t}\mathrm{d}W_t$ with $\theta = (2, 0.02, 0.15)$. We know the target conditional density, which we generate using the function `dcCIR`.[4] This density is highly skewed, so it is a good target to check against.

```
> # ex3.07.R
```

[4] Recall the different parametrization of the CIR model in the function `dcCIR`.

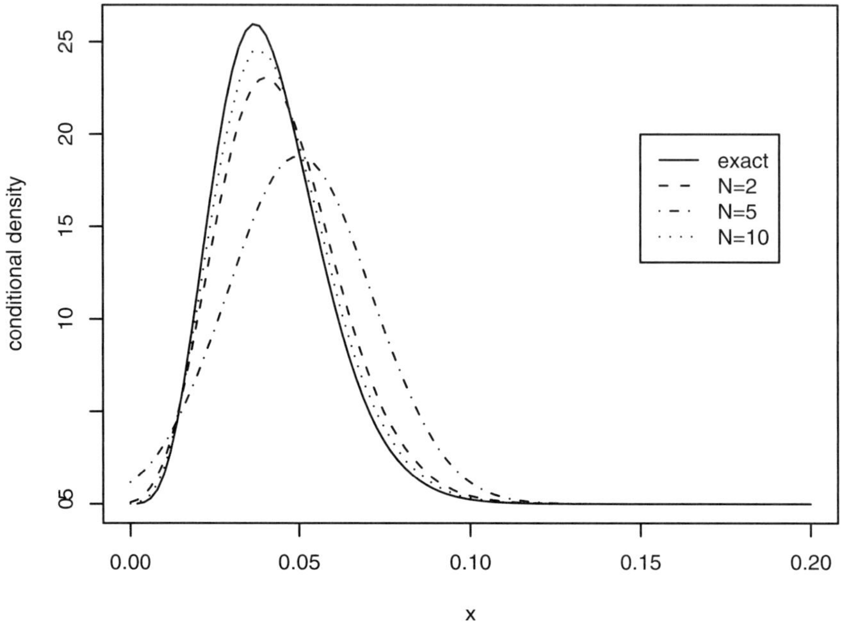

Fig. 3.3. Simulated conditional density $p_\theta(\Delta = 0.5, y|x = 0.8)$ for the Cox-Ingersoll-Ross model. As the number of subintervals N increases from 2 to 10, the approximation gets better ($M = 50000$). The true conditional density is marked as a solid line and the approximation for $N = 10$ with dots.

```
> d1 <- function(t,x,theta) theta[1]*(theta[2]-x)
> s1 <- function(t,x,theta) theta[3]*sqrt(x)
> from <- 0.08
> x <- seq(0,0.2, length=100)
> sle10 <- NULL
> sle2 <- NULL
> sle5 <- NULL
> true <- NULL
> set.seed(123)
> for(to in x){
+  sle2 <- c(sle2, dcSim(from, to, 0.5, d1, s1, theta=c(2,0.02,0.15),
+                 M=50000,N=2))
+  sle5 <- c(sle5, dcSim(from, to, 0.5, d1, s1, theta=c(2,0.02,0.15),
+                 M=50000,N=5))
+  sle10 <- c(sle10, dcSim(from, to, 0.5, d1, s1, theta=c(2,0.02,0.15),
+                 M=50000,N=10))
+  true <- c(true, dcCIR(to, 0.5, from, c(2*0.02,2,0.15)))
+ }
> par(mar=c(5,5,1,1))
> plot(x, true, type="l", ylab="conditional density")
> lines(x, sle2, lty=4)
> lines(x, sle5, lty=2)
> lines(x, sle10, lty=3)
> legend(0.15,20, legend=c("exact","N=2", "N=5", "N=10"), lty=c(1,2,4,3))
```

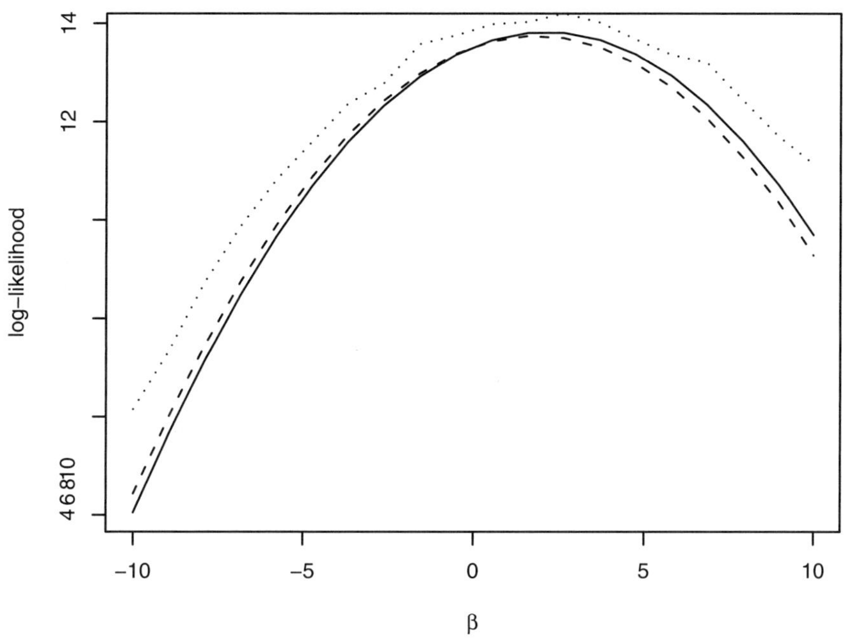

Fig. 3.4. Exact, Euler (dashed line), and simulated (dotted line) likelihoods of simulated data from an Ornstein-Uhlenbeck process. In this case, $M = 10000$ and $N = 5$.

To compute the whole log-likelihood, the package `sde` provides a function called `SIMloglik`. This function has a simple interface that requires the data, the vector or parameters, the drift and diffusion coefficients, and then the parameters M and N as in the `dcSim` function. We compare the exact likelihood, the Euler likelihood, and the simulated likelihood in the next code. Figure 3.4 shows the actual performance.

```
> # ex3.08.R
> set.seed(123)
> d <- expression(-1*x)
> s <- expression(2)
> sde.sim(drift=d, sigma=s,N=50,delta=0.01) -> X
> S <- function(t, x, theta) sqrt(theta[2])
> B <- function(t, x, theta) -theta[1]*x
>
> true.loglik <- function(theta) {
+  DELTA <- deltat(X)
+  lik <- 0
+  for(i in 2:length(X))
+   lik <- lik + dnorm(X[i], mean=X[i-1]*exp(-theta[1]*DELTA),
+    sd = sqrt((1-exp(-2*theta[1]*DELTA))*theta[2]/(2*theta[1])),TRUE)
+  lik
+ }
>
> xx <- seq(-10,10,length=20)
> sapply(xx, function(x) true.loglik(c(x,4))) -> py
```

```
> sapply(xx, function(x) EULERloglik(X,c(x,4),B,S)) -> pz
> sapply(xx, function(x) SIMloglik(X,c(x,4),B,S,M=10000,N=5)) -> pw
> par(mar=c(5,5,1,1))
> plot(xx,py, type="l", xlab=expression(beta), ylab="log-likelihood")
> lines(xx, pz, lty=2)
> lines(xx, pw, lty=3)
```

As mentioned, this method is computationally demanding and, if possible, other approaches are preferable to perform likelihood inference for discrete diffusions. To conclude, we mention that a relevant variant of this method is the importance sampler, which is well-described in [74]. Further, links between MCMC analysis and simulated likelihood methods can be found in [78]. Finally, another importance sampler based on the exact algorithm of Section 2.12 can be found in [27] and [29]. We do not discuss these approaches here.

3.3.3 Hermite polynomials expansion of the likelihood

Another way of approximating the transition density of a diffusion is to expand it in a way similar to the i.i.d. case. The analogy is the following (see [6]). Consider a standardized sum of random variables to which the central limit theorem (CLT) applies. For large sample sizes n, by the CLT one can use the limiting distribution $N(0,1)$, but if the sample size is not large, one should expand the limiting distribution (for example, by Edgeworth expansion) to increase the accuracy of the approximation. The idea proposed by Aït-Sahalia (see [8] and [6]) is to use such an approach to approximate the unknown transition density of a diffusion and treat the asymptotics as $\Delta \to 0$ in parallel to the one as $n \to \infty$ in the i.i.d. case.

The standard method in the i.i.d. case was originally introduced by Cramér [60]. Let $p(z)$ be the target density to expand around the standard normal $N(0,1)$. With this method, the expansion converges rapidly when the target density is almost normal or, more precisely, when its tails are sufficiently thin in the sense that $\exp(z^2/2)p'(z)^2$ must be integrable. In some cases, as in many diffusion models in finance, the Gaussian assumption is not an option. For example, if the process X is a geometric Brownian motion (see Section 1.7), the right tail of the corresponding log-normal density $p(z)$ is too large to converge, as it is of order $z^{-1}\exp(-\log^2(z))$ as $z \to +\infty$ (see, e.g., [6]). Furthermore, for any $N(0,\nu)$ with $\nu > 2$, the expansion diverges. In his papers [8] and [6], Aït-Sahalia proposed to proceed differently: instead of attempting a direct expansion on $p(z)$, the density is first transformed into something that is close to the normal and then this transformed density is expanded using Hermite polynomials. To this end, the original process X is first transformed into another process, whose transition density is as expected. We now present here the Aït-Sahalia method. A simple exposition can be found in [188], but the reader should refer to the original papers [8] and [6] for more detailed analysis. Given that Hermite polynomials are needed in this setup, we introduce them here.

Hermite polynomials

Hermite polynomials $H_j(z)$ are defined as

$$H_j(z) = e^{\frac{z^2}{2}} \frac{\mathrm{d}^j}{\mathrm{d}z^j} e^{-\frac{z^2}{2}}, \qquad j \geq 0. \tag{3.20}$$

This is not the "usual" definition of Hermite polynomials. Indeed, they are usually defined (see, e.g., [1], Chapter 22) as

$$H_j^{prob}(z) = (-1)^n e^{\frac{z^2}{2}} \frac{\mathrm{d}^j}{\mathrm{d}z^j} e^{-\frac{z^2}{2}}, \qquad j \geq 0,$$

in probability or as

$$H_j^{phys}(z) = (-1)^n e^{z^2} \frac{\mathrm{d}^j}{\mathrm{d}z^j} e^{-z^2}, \qquad j \geq 0,$$

in physics, and $H_j^{phys}(z) = \sqrt{2^j} H_j^{prob}(\sqrt{2}z)$. Of course, $H_j^{prob}(z)$ is preferred in probability because it can be expressed in terms of the density function $\phi(z) = e^{-z^2/2}/\sqrt{2\pi}$ of the standard Gaussian law. Definition (3.20) is more convenient for the development we are about to introduce because it simplifies the notation.

To make the method quite general and useful in financial applications, Aït-Sahalia introduces different sets of conditions on the coefficients b and σ of the stochastic differential equations. In particular, growth conditions at infinity are replaced by assumptions on the sign of the drift near the boundaries. Furthermore, the class of processes is not restricted to stationary diffusions only. Let D_X be the domain of the diffusion X. The following two cases are considered: $D_X = (-\infty, +\infty)$ and $D_X = (0, +\infty)$. This second case is relevant in financial market models of interest rates or asset prices, where usually b and/or σ violate the linear growth condition near the boundaries or σ is defined in such a way that $\lim_{x\to 0^+} \sigma(x, \theta) = 0$.

Assumption 3.7 (Smoothness of the coefficients) *The functions $b(x, \theta)$ and $\sigma(x, \theta)$ are infinitely differentiable in x and three times continuously differentiable in θ for all $x \in D_X$ and $\theta \in \Theta$.*

Assumption 3.8 (Nondegeneracy) *Two cases may happen:*

1. *If $D_X = (-\infty, +\infty)$, then $\exists\, c : \sigma(x, \theta) > c > 0$ for all $x \in D_X$ and $\theta \in \Theta$.*
2. *If $D_X = (0, +\infty)$, then $\sigma(x, \theta)$ is nondegenerate in D_X; i.e., for each $\xi > 0$, $\exists\, c_\xi : \sigma(x, \theta) \geq c_\xi > 0$ for all $x \in [\xi, +\infty]$ and $\theta \in \Theta$. If in addition $\lim_{x\to 0^+} \sigma(x, \theta) = 0$, then $\exists\, \xi_0 > 0, \omega > 0, \rho \geq 0 : \sigma(x, \theta) \geq \omega x^\rho$ for all $0 < x \leq \xi_0$ and $\theta \in \Theta$.*

The first step consists in transforming the original process X into a new one $Y = F(X)$, where Y satisfies

$$\mathrm{d}Y_t = \mu_Y(Y_t, \theta)\mathrm{d}t + \mathrm{d}W_t. \tag{3.21}$$

We have already encountered the Lamperti transform

$$F(\gamma) = \int^{\gamma} \frac{du}{\sigma(u,\theta)},$$

in Section 2.12 (recall that the constant of integration is irrelevant). Moreover, under Assumption 3.8, $\sigma > 0$ on D_X, and then F is an increasing function and invertible for all $\theta \in \Theta$. By the Itô formula, we obtain

$$\mu_Y(y,\theta) = \frac{b(F^{-1}(y),\theta)}{\sigma(F^{-1}(y),\theta)} - \frac{1}{2}\sigma_x(F^{-1}(y),\theta)\,. \tag{3.22}$$

If $D_X = (\underline{x}, \bar{x})$, then F maps it to $D_Y = (\underline{y}, \bar{y})$, where $\underline{y} = \lim_{x\to\underline{x}^+} F(x)$ and $\bar{y} = \lim_{x\to\bar{x}^-} F(x)$. To keep the notation simple, the parameter space Θ is chosen in such a way that D_Y is independent of $\theta \in \Theta$.

The next hypothesis is crucial to control the behavior of the transformed (and original) diffusion near the boundaries. It is rather technical in its aspect and hence before stating it we explain its intuitive meaning. This assumption is formulated in terms of the coefficient μ_Y of Y but clearly it also forces conditions on the original coefficients b and σ of X. Assumption 3.9 restricts the growth of μ_Y to be at most linear near the boundary of D_Y when it has the wrong sign (i.e., positive sign near $\bar{y}$ and negative sign near $\underline{y}$). If μ_Y has the right sign near the boundaries, mean reversion occurs and the diffusion is pulled back away from the boundaries. Under this hypothesis, the diffusion Y cannot explode to $+\infty$ in a finite time, but in some cases the boundary 0 is attainable. We denote by $T_Y = \inf\{t \geq 0 : T_t \notin D_Y\}$ the first exit time of Y from D_Y.

Assumption 3.9 (Boundary behavior) *For all $\theta \in \Theta$, μ_Y and its derivatives with respect to y and θ have at most polynomial growth near the boundaries and $\lim_{y\to\underline{y}^+ \text{ or } \bar{y}^-} \lambda_Y(y,\theta) < +\infty$ with $\lambda_Y(y,\theta) = -(\mu_Y^2(y,\theta) + \partial\mu_Y(y,\theta)/\partial y)/2$.*

1. *Left boundary: If $\underline{y} = 0$, then $\exists\, \epsilon_0, \kappa, \alpha : \forall 0 < y \leq \epsilon_0$ and $\theta \in \Theta, \mu_Y(y,\theta) \geq \kappa y^{-\alpha}$, where either $\alpha > 1$ and $\kappa > 0$ or $\alpha = 1$ and $\kappa \geq 1$. If $\underline{y} = -\infty$, then $\exists\, E_0 > 0, K > 0 : \forall y \leq -E_0$ and $\theta \in \Theta, \mu_Y(y,\theta) \geq Ky$.*
2. *Right boundary: If $\bar{y} = +\infty$, then $\exists\, E_0 > 0, K > 0 : \forall y \geq E_0$ and $\theta \in \Theta, \mu_Y(y,\theta) \leq Ky$. If $\bar{y} = 0$, then $\exists\, \epsilon_0, \kappa, \alpha : \forall 0 > y \geq -\epsilon_0$ and $\theta \in \Theta, \mu_Y(y,\theta) \leq -\kappa|y|^{-\alpha}$, where either $\alpha > 1$ and $\kappa > 0$ or $\alpha = 1$ and $\kappa \geq 1/2$.*

The next result from [6] summarizes the intuition above.

Fact 3.3 (Proposition 1 in [6]) *Under Assumptions 3.7, 3.8, and 3.9, the stochastic differential equation (3.21) admits a weak solution $\{Y_t : t \geq 0\}$, unique in probability law, for every distribution of its initial value Y_0. The boundaries of D_Y are unattainable in the sense that $P(T_Y = +\infty) = 1$.*

Finally, we have the following.

Fact 3.4 (Proposition 2 in [6]) *Under Assumptions 3.7, 3.8, and 3.9, Y admits a transition density $p_Y(\Delta, y|y_0, \theta)$ that is continuously differentiable in $\Delta > 0$, infinitely differentiable in $y \in D_Y$ and $y_0 \in D_Y$, and three times continuously differentiable in $\theta \in \Theta$.*

Proposition 2 in [6] also ensures the good behavior of p_Y and its first derivative with respect to y so that the forthcoming expansion of p_Y converges. Moreover, as a corollary, the same result holds for the process X. Still, one further step is necessary to make p_Y more suitable for the Hermite expansion: for a given $\Delta > 0$, $\theta \in \Theta$, and $y \in \mathbb{R}$, we introduce the pseudo-normalized increment of Y as

$$Z = \Delta^{-1/2}(Y - y_0)\,.$$

Let $p_Y(\Delta, y|y_0, \theta)$ denote the transition density of $Y_{t+\Delta}|Y_t = y_0$ and define the density function of Z as

$$p_Z(\Delta, z|y_0, \theta) = \Delta^{1/2} p_Y(\Delta, \Delta^{1/2}z + y_0|y_0, \theta)\,. \tag{3.23}$$

Then

$$p_Y(\Delta, y|y_0, \theta) = \Delta^{-1/2} p_Z(\Delta, \Delta^{-1/2}(y - y_0)|y_0, \theta) \tag{3.24}$$

and

$$p_X(\Delta, x|x_0, \theta) = \sigma(x, \theta)^{-1} p_Y(\Delta, F(x)|F(x_0), \theta)\,. \tag{3.25}$$

The subsequent step consists in the approximation of the transition density for the process Z_t. This approximation is based on the Hermite expansion of $p_Z(\Delta, z|y_0, \theta)$ for fixed Δ, y_0, and θ,

$$p_Z^{(J)}(\Delta, z|y_0, \theta) = \phi(z) \sum_{j=0}^{J} \eta_Z^{(j)}(\Delta, z|y_0, \theta) H_j(z)\,. \tag{3.26}$$

The coefficients $\eta_Z^{(j)}$ are defined as[5]

$$\eta_Z^{(j)}(\Delta|y_0, \theta) = \frac{1}{j!} \int_{-\infty}^{+\infty} H_j(z) p_Z(\Delta, z|y_0, \theta) \mathrm{d}z \tag{3.27}$$

and can be expressed in terms of the moments of the diffusion, as will be shown later on. As in (3.24) and (3.25), we can define $p_Y^{(J)}$ from $p_Z^{(J)}$ and $p_X^{(J)}$ from $p_Y^{(J)}$ as

$$p_Y^{(J)}(\Delta, y|y_0, \theta) = \Delta^{-1/2} p_Z^{(J)}(\Delta, \Delta^{-1/2}(y - y_0)|y_0, \theta) \tag{3.28}$$

and

$$p_X^{(J)}(\Delta, x|x_0, \theta) = \sigma(x, \theta)^{-1} p_Y^{(J)}(\Delta, F(x)|F(x_0), \theta)\,. \tag{3.29}$$

The next result proves that the expansion (3.29) converges to (3.25) uniformly as $J \to \infty$.

[5] Given that $H_j(z)/\sqrt{j!}$ are orthonormal in $L^2(\phi)$.

Fact 3.5 (Theorem 1 in [6]) *Under Assumptions 3.7, 3.8, and 3.9, there exists $\bar{\Delta} > 0$ such that, for every $\Delta \in (0, \bar{\Delta})$, $\theta \in \Theta$, and $(x, x_0) \in D_X^2$, as $J \to \infty$,*

$$p_X^{(J)}(\Delta, x|x_0, \theta) \longrightarrow p_X(\Delta, x|x_0, \theta). \tag{3.30}$$

In addition, the convergence is uniform in θ and x_0 over compact subsets of D_X. If $\sigma(x, \theta) > c > 0$ on D_X, then the convergence is uniform in x on all D_X. If $D_X = (0, +\infty)$ and $\lim_{x\to 0^+} \sigma(x, \theta) = 0$, then the convergence is uniform in x in each interval of the form $[\epsilon, +\infty)$, $\epsilon > 0$.

Two more points still remain open: Could we actually calculate the $\eta_Z^{(j)}$? And, given the approximated sequence of $p_X^{(J)}$'s, does the maximum likelihood estimator, say $\hat{\theta}^{(J)}$, possess good properties, and how is it related to the nonexplicitly calculable true maximum likelihood estimator of θ?

Explicit expression of the approximation

In the very beginning of this section, we mentioned that this expansion is based on the moments of the process. To be more precise, the coefficients $\eta_Z^{(J)}$ can be expressed in terms of the moments of the conditional distribution of the process by using (3.27) and (3.24); i.e.,

$$\begin{aligned}
\eta_Z^{(j)}(\Delta|y_0, \theta) &= \frac{1}{j!}\int_{-\infty}^{+\infty} H_j(z) p_Z(\Delta, z|y_0, \theta)\mathrm{d}z \\
&= \frac{1}{j!}\int_{-\infty}^{+\infty} H_j(\Delta^{-1/2}(y - y_0)) p_Y(\Delta, y|y_0, \theta)\mathrm{d}y \\
&= \frac{1}{j!}\mathbb{E}_\theta \left(H_j(\Delta^{-1/2}(Y_{t+\Delta} - y_0))|Y_t = y_0\right).
\end{aligned} \tag{3.31}$$

But, for a smooth function g, Aït-Sahalia proved the Taylor expansion

$$\mathbb{E}_\theta\left(g(Y_{t+\Delta})|Y_t = y_0\right) = \sum_{i=1}^{n} (\mathcal{L}_\theta^i g)(y_0)\frac{\Delta^i}{i!} + \text{remainder},$$

where $\mathcal{L}_\theta$ is the infinitesimal generator of the diffusion Y (see, e.g., (3.3)) and $(\mathcal{L}_\theta g)$ looks like

$$(\mathcal{L}_\theta g)(y_0) = \mu_Y(y_0, \theta) g'(y_0) + \frac{1}{2} g''(y_0)$$

because Y has a unit diffusion coefficient. The notation $(\mathcal{L}_\theta^i g)$ means that the generator is applied recursively i times (e.g., $\mathcal{L}_\theta^0 g = g$, $\mathcal{L}_\theta^1 g = \mu_Y g' + \frac{1}{2}g''$, $\mathcal{L}_\theta^2 g = \mu_Y \mu_Y' g' + \mu_Y^2 g'' + \mu_Y g^{(3)} + \frac{1}{2}\mu_Y'' g' + \mu_Y' g'' + \frac{1}{4}g^{(4)}$, etc.) and $g^{(k)}$ means the kth derivative of g. The remainder in the Taylor expansion above vanishes as n increases if g is regular (for example if g has at most exponential growth).

Aït-Sahalia's idea is to first fix J and then choose H_j as g and fix n in such a way that the Taylor expansion has terms up to the order of at most $\Delta J/2$. Usually (see [126] and [6]) a number of Hermite polynomials up to $J = 6$ is enough because this method converges quite fast. Numerical experiments in [126] provide evidence of a very accurate approximation of the likelihood also for $J = 3$, at least in the cases of the Vasicek, CIR, and Black-Scholes processes. Hence we will adopt the value of $J = 6$ in the following and in the implementation. This implies also that up to six derivatives of the drift function μ_Y must be calculated or passed to the R function. What follows is now the explicit expression of the Taylor approximations of the $\eta_Z^{(j)}$ we will use in the implementation. Apart from $\eta_Z^{(0)} = 1$, these are quite lengthy expression but still in closed form [126].

$$\begin{aligned}\eta_Z^{(1)} &= -\mu_Y \Delta^{1/2} - (2\mu_Y\mu_Y' + \mu_Y'')\Delta^{3/2}/4 \\ &\quad - (4\mu_Y(\mu_Y')^2 + 4\mu_Y^2\mu_Y'' + 6\mu_Y'\mu_Y'' + 4\mu_Y\mu_Y^{(3)} + \mu_Y^{(4)})\Delta^{5/2}/24,\end{aligned}$$

$$\begin{aligned}\eta_Z^{(2)} &= (\mu_Y^2 + \mu_Y')\Delta/2 + (6\mu_Y^2\mu_Y' + 4(\mu_Y')^2 + 7\mu_Y\mu_Y'' + 2\mu_Y^{(3)})\Delta^2/12 \\ &\quad + (28\mu_Y^2(\mu')^2 + 28\mu_Y^2\mu_Y^{(3)} + 16(\mu_Y')^3 + 16\mu_Y^3\mu_Y'' + 88\mu_Y\mu_Y'\mu_Y'' \\ &\quad + 21(\mu_Y'')^2 + 32\mu_Y'\mu_Y^{(3)} + 16\mu_Y\mu_Y^{(4)} + 3\mu_Y^{(5)})\Delta^3/96,\end{aligned}$$

$$\begin{aligned}\eta_Z^{(3)} &= -(\mu_Y^3 + 3\mu_Y\mu_Y' + \mu_Y'')\Delta^{3/2}/6 - (12\mu_Y^3\mu_Y' + 28\mu_Y(\mu_Y')^2 \\ &\quad + 22\mu_Y^2\mu_Y'' + 24\mu_Y'\mu_Y'' + 14\mu_Y\mu_Y^{(3)} + 3\mu_Y^{(4)})\Delta^{5/2}/48,\end{aligned}$$

$$\begin{aligned}\eta_Z^{(4)} &= (\mu_Y^4 + 6\mu_Y^2\mu_Y' + 3(\mu_Y')^2 + 4\mu_Y\mu_Y'' + \mu_Y^{(3)})\Delta^2/24 \\ &\quad + (20\mu_Y^4\mu_Y' + 50\mu_Y^3\mu_Y'' + 100\mu_Y^2(\mu_Y')^2 + 50\mu_Y^2\mu_Y^{(3)} + 23\mu_Y\mu_Y^{(4)} \\ &\quad + 180\mu_Y\mu_Y'\mu_Y'' + 40(\mu_Y')^3 + 34(\mu_Y'')^2 + 52\mu_Y'\mu_Y^{(3)} + 4\mu_Y^{(5)})\Delta^3/240,\end{aligned}$$

$$\begin{aligned}\eta_Z^{(5)} &= -(\mu_Y^5 + 10\mu_Y^3\mu_Y' + 15\mu_Y(\mu_Y')^2 + 10\mu_Y^2\mu_Y'' \\ &\quad + 10\mu_Y'\mu_Y'' + 5\mu_Y\mu_Y^{(3)} + \mu_Y^{(4)})\Delta^{5/2}/120,\end{aligned}$$

$$\begin{aligned}\eta_Z^{(6)} &= (\mu_Y^6 + 15\mu_Y^4\mu_Y' + 15(\mu_Y')^3 + 20\mu_Y^3\mu_Y'' + 15\mu_Y\mu_Y^{(3)} + 45\mu_Y^2(\mu_Y')^2 \\ &\quad + 10(\mu_Y'')^2 + 15\mu_Y^2\mu_Y^{(3)} + 60\mu_Y\mu_Y'\mu_Y'' + 6\mu_Y\mu_Y^{(4)} + \mu_Y^{(5)})\Delta^3/720.\end{aligned}$$

We need to specialize the Hermite polynomials to the case $j = 0, 1, \ldots, 6$ as well: $H_0(z) = 1$, $H_1(z) = -z$, $H_2(z) = -1 + z^2$, $H_3(z) = 3z - z^3$, $H_4(z) = 3 - 6z^2 + z^4$, $H_5(z) = -15z + 10z^3 - z^5$, $H_6(z) = -15 + 45z^2 - 15z^4 + z^6$.

Implementation of the Hermite polynomial expansion approximation

To implement this scheme in our setup we need to isolate all the quantities in a more convenient way. So, given our discrete observations $X_0, X_1, \ldots, X_i$, from the process X we are interested in writing down the approximation of $p_X^{(J)}$ in terms of Δ, X_i, X_{i-1}, and θ; i.e., $p_X(\Delta, X_i|X_{i-1}, \theta)$. To get a notation consistent with the previous sections, we set $\tilde{p}_\theta(\Delta, X_i, X_{i-1}) = p_X(\Delta, X_i|X_{i-1}, \theta)$. Making use of (3.29), we can write it as

$$\tilde{p}_\theta(\Delta, X_i, X_{i-1}) = \frac{\Delta^{-1/2}}{\sigma(X_i, \theta)} p_Z^{(J)}(\Delta, \Delta^{-1/2}(F(X_i) - F(X_{i-1}))|F(X_{i-1}), \theta),$$

and the whole approximated likelihood will look like the product of i of the approximation above. Recall further that $p_Z^{(J)}$ is based on the $\eta_Z^{(j)}$'s, which in turn require the calculation of the drift of the transformed process μ_Y and its derivatives up to order 6. Finally, μ_Y (see (3.22)) depends on the drift and diffusion coefficients of X but also on the inverse function of F. In principle, to implement this algorithm, we need at least the X_i's, b, and σ, and then F, F^{-1}, and the $\eta_Z^{(j)}$'s can be calculated using R code. In practice, such an algorithm will imply too much R code just to build all the quantities. Numerical derivatives of these quantities will also be too inefficient from the point of view of speed and accuracy, and thus we assume for simplicity that F, μ_Y, and all its derivatives $\mu^{(k)}$, $k = 1, \ldots, 6$ are passed in explicit form to the code as a list of functions. Further, although an implementation in pure R code is straightforward, we prefer to go directly to the C level to obtain a faster algorithm. The function `HPloglik` in Listing 3.7 accepts as input a `ts` object `X`, a vector of parameters `theta`, a list `M` containing μ_Y and $\mu^{(k)}$, $k = 1, \ldots, 6$, the transform function `F`, and the diffusion coefficient `s`. By default, it returns the log-likelihood. To obtain the one-step conditional density, it is sufficient to pass it (e.g., the vector `X[1,2]`). All the functions are assumed to have three arguments `t`, `x`, and `theta`, in this order, where `t` and `x` are real numbers and `theta` a vector. The next section shows an application of this function.

```
HPloglik <- function (X, theta, M, F, s, log = TRUE)
```

Listing 3.7. Aït-Sahalia's Hermite polynomial approximation of the likelihood.

Seeing the quality of the approximation

We now show a concrete application of this method. We consider the Ornstein-Uhlenbeck process $\mathrm{d}X_t = -\beta X_t \mathrm{d}t + \sigma \mathrm{d}W_t$, where $\theta = (\beta, \sigma^2)$ and $D_X = (-\infty, +\infty)$. For this process we know that the transition density from X_t to $X_{t+\Delta}$ is the Gaussian law with mean $X_t e^{-2\beta\Delta}$ and variance $(1 - e^{-2\beta\Delta})\sigma^2/2\beta$. For this simple case $F(x) = x/\sigma$ and $\mu_Y(y, \theta) = -\beta y$. The next code sets up the parameters for the `HPloglik` function and generates a trajectory of the process.

```
> # ex3.09.R
>
> set.seed(123)
> d <- expression(-1*x)
> s <- expression(2)
> sde.sim(drift=d, sigma=s) -> X
sigma.x not provided, attempting symbolic derivation.
>
> M0 <- function(t, x, theta) -theta[1]*x
> M1 <- function(t, x, theta) -theta[1]
> M2 <- function(t, x, theta) 0
> M3 <- function(t, x, theta) 0
> M4 <- function(t, x, theta) 0
> M5 <- function(t, x, theta) 0
> M6 <- function(t, x, theta) 0
> mu <- list(M0, M1, M2, M3, M4, M5, M6)
>
> F <- function(t, x, theta) x/sqrt(theta[2])
> S <- function(t, x, theta) sqrt(theta[2])
> B <- function(t, x, theta) -theta[1]*x
>
> true.loglik <- function(theta) {
+  DELTA <- deltat(X)
+  lik <- 0
+  for(i in 2:length(X))
+   lik <- lik + dnorm(X[i], mean=X[i-1]*exp(-theta[1]*DELTA),
+    sd = sqrt((1-exp(-2*theta[1]*DELTA))*theta[2]/(2*theta[1])),TRUE)
+  lik
+ }
```

Then the true, the Euler, and the approximated log-likelihood functions for X are evaluated for $\beta \in (-3, +3)$ and $\sigma^2 = 4$. The graph of the true (dotted line), approximated (solid line), and Euler likelihoods (dashed line) of β (solid line) are plotted against one another in Figure 3.5.

```
> # ex3.09.R (cont)
> xx <- seq(-3,3,length=100)
> sapply(xx, function(x) HPloglik(X,c(x,4),mu,F,S)) -> px
> sapply(xx, function(x) true.loglik(c(x,4))) -> py
> sapply(xx, function(x) EULERloglik(X,c(x,4),B,S)) -> pz
>
> plot(xx,px,type="l",xlab=expression(beta),ylab="log-likelihood") # approx
> lines(xx,py, lty=3) # true
> lines(xx,pz, lty=2) # Euler
```

How does one interpret the quality of the approximation from Figure 3.5? From the statistical perspective, what is more important is that the inference made on the approximation is the best possible. Figure 3.5 shows that, at least in this example, the approximated maximum likelihood estimator built on $\tilde{p}_\theta^{Euler}$ will be more distant from the true value $\beta = 1$ than the one calculated on the Hermite polynomial approximation $\tilde{p}_\theta$. The next section will clarifies the situation a bit .

Behavior of the approximated maximum likelihood estimator

The other question that remains unsolved is whether the maximum likelihood estimator $\hat{\theta}_n^{(J)}$ obtained on the approximated likelihood $\tilde{p}_\theta$ possesses good properties and, for example, if it converges to the true maximum likelihood

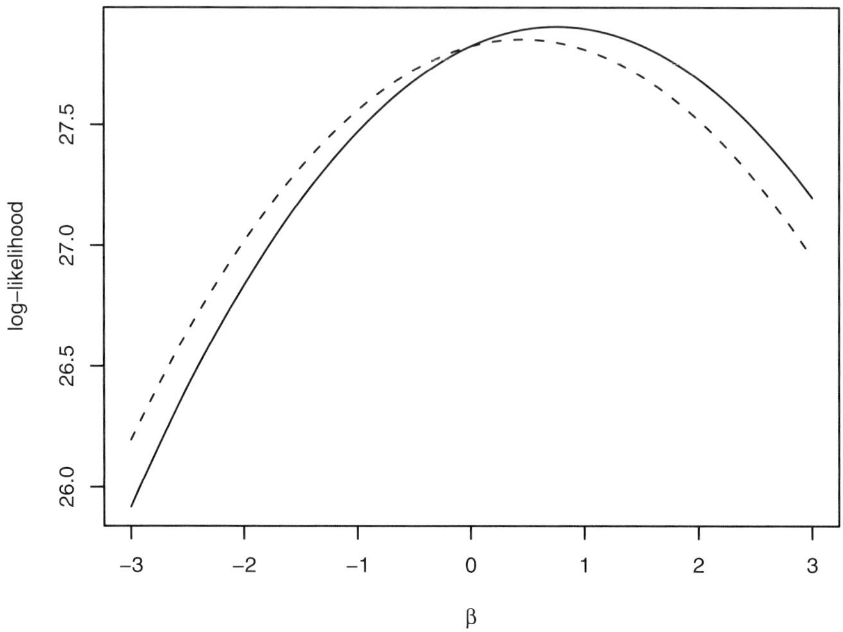

Fig. 3.5. True likelihood (dotted line) versus Hermite expansion approximation (solid line) and Euler approximation (dashed line). The Hermite expansion approximation graphically coincides with the true likelihood, so the dotted line is not visible.

estimator $\hat{\theta}_n$ calculated on the true (but usually unknown) likelihood of the process. In [6], it is shown that $\hat{\theta}_n^{(J)}$ converges to $\hat{\theta}_n$ as $J \to \infty$. Moreover, as the sample size increases ($n \to \infty$), it is always possible to find a J_n such that $\hat{\theta}_n^{(J_n)}$ converges to the true parameter, say θ_0. To obtain this result, first some identification conditions must be assumed in addition to Assumptions 3.7, 3.8, and 3.9. We denote $\ell_i(\theta) = \log p_\theta(\Delta, X_{i\Delta}|X_{(i-1)\Delta})$ and use one "$\dot{}$" or multiple dots to indicate differentiation one or multiple times with respect to θ. The score vector is then denoted by $\dot{\ell}_n(\theta) = \sum_{i=1}^n \dot{\ell}_i(\theta)$. From Fact 3.4 and its corollary, we know that up to three derivatives in θ exist, so we can define the following quantities:

$$\begin{aligned} i_n(\theta) &= \sum_{i=1}^n \mathbb{E}_\theta\{\dot{\ell}_i(\theta)\dot{\ell}_i(\theta)^T\}, & H_n(\theta) &= -\sum_{i=1}^n \ddot{\ell}_i(\theta)\,, \\ I_n(\theta) &= \operatorname{diag}\{i_n(\theta)\}, & T_n(\theta) &= -\sum_{i=1}^n \dddot{\ell}_i(\theta)\,. \end{aligned} \tag{3.32}$$

Assumption 3.10 (Identifiability) *The true value θ_0 belongs to Θ, $I_n(\theta)$ is invertible, and $I_n^{-1}(\theta) \longrightarrow 0$ and $n \to \infty$ almost surely and uniformly in*

$\theta \in \Theta$. Moreover, $R_n(\theta, \tilde\theta) = I_n^{-1/2} T_n(\tilde\theta) I_n^{-1/2}(\theta)$ is uniformly bounded in probability for all $\tilde\theta$ in a $I_n^{1/2}(\theta)$-neighborhood of θ.

If X is stationary and $[I_n(\theta)]_{kk} < \infty$ for all k (and uniformly in θ), then the convergence $I_n^{-1}(\theta) \longrightarrow 0$ is guaranteed. Proposition 3 in [6] gives results on the asymptotic distribution of the $\hat\theta_n$ and $\hat\theta_n^{(J)}$ under different situations. We refer the reader to the original paper for the details, but what is interesting to us is that this result specializes to the following one when X is a stationary diffusion:

$$n^{1/2}(\hat\theta_n - \theta_0) \stackrel{d}{\longrightarrow} N(0, i(\theta_0)^{-1}) \,. \tag{3.33}$$

In this case, the true maximum likelihood estimator is also efficient in the Fisher-Rao sense because $i(\theta_0)^{-1}$ is indeed the smallest possible asymptotic variance among that of all consistent and asymptotically normal estimators of θ_0. This property is shared by $\hat\theta_n^{(J)}$ due to the approximation results mentioned above.

MLE estimation for the Ornstein-Uhlenbeck process

Consider again the Ornstein-Uhlenbeck process satisfying the stochastic differential equation $\mathrm{d}X_t = -\beta X_t \mathrm{d}t + \sigma \mathrm{d}W_t$. In the stationary case (i.e., $\beta > 0$), result (3.33) specializes to

$$\sqrt{n}\left(\begin{pmatrix}\hat\beta_n \\ \hat\sigma_n^2\end{pmatrix} - \begin{pmatrix}\beta \\ \sigma^2\end{pmatrix}\right) \stackrel{d}{\longrightarrow} N\left(\begin{pmatrix}0 \\ 0\end{pmatrix}, V(\theta)\right),$$

where

$$V(\theta) = \begin{pmatrix} \frac{e^{2\beta\Delta}-1}{\Delta^2} & \frac{\sigma^2(e^{2\beta\Delta}-1-2\beta\Delta)}{\beta\Delta^2} \\ \frac{\sigma^2(e^{2\beta\Delta}-1-2\beta\Delta)}{\beta\Delta^2} & \frac{\sigma^4\left((e^{2\beta\Delta}-1)^2+2\beta^2\Delta^2(e^{2\beta\Delta}+1)+4\beta\Delta(e^{2\beta\Delta}-1)\right)}{\beta^2\Delta^2(e^{2\beta\Delta}-1)} \end{pmatrix}. \tag{3.34}$$

In the explosive case $\beta < 0$, the limit is no longer normal (this is in fact the LAMN case of Proposition 3 in [6]) and the convergence rate is different for $\hat\beta_n$ and $\hat\sigma_n^2$. The result is

$$\frac{e^{-(n+1)\beta\Delta}\Delta}{e^{-2\beta\Delta}-1}(\hat\beta_n - \beta) \stackrel{d}{\longrightarrow} \mathcal{C}\,, \qquad \sqrt{n}(\hat\sigma_n^2 - \sigma^2) \stackrel{d}{\longrightarrow} N(0, 2\sigma^4)\,,$$

where $\mathcal{C}$ is the standard Cauchy distribution; i.e., its density is $f(x) = 1/(\pi(1+x^2))$.

We first consider the joint estimation of β and σ^2 on the same trajectory of the process X of the previous section. We optimize the negative log-likelihood and we look at the estimates using the `mle` function.

```
> # ex3.09.R (cont)
> HP.negloglik <- function(BETA=3, SIGMA2=2)
+   -HPloglik(X,c(BETA,SIGMA2),mu,F,S)
> true.negloglik <- function(BETA=3, SIGMA2=2)
```

```
+    -true.loglik(c(BETA,SIGMA2))
> euler.negloglik <- function(BETA=3, SIGMA2=2)
+    -EULERloglik(X,c(BETA,SIGMA2),B,S)
>
> mle(true.negloglik,lower=c(0,0),method="L-BFGS-B") -> fit.true
> mle(HP.negloglik,lower=c(0,0),method="L-BFGS-B") -> fit.approx
> mle(euler.negloglik,lower=c(0,0),method="L-BFGS-B") -> fit.euler
>
> # we look at the estimates
> coef(fit.true)
     BETA    SIGMA2
0.4521549 3.3106872
> coef(fit.approx)
     BETA    SIGMA2
0.4521552 3.3106872
> coef(fit.euler)
     BETA    SIGMA2
0.4511489 3.2957707
```

For this sample, the joint estimation of β and σ^2 is not particularly good. The function `mle` returns an R object with much useful information. We can, for example, obtain the value of the negative log-likelihood at the estimates

```
> # ex3.09.R (cont)
> logLik(fit.true)
'log Lik.' 28.73278 (df=2)
> logLik(fit.approx)
'log Lik.' 28.73278 (df=2)
> logLik(fit.euler)
'log Lik.' 28.73278 (df=2)
```

and the approximate variance and covariance matrix obtained by inverting the Hessian matrix at the optimum.

```
> # ex3.09.R (cont)
> vcov(fit.true)
            BETA    SIGMA2
BETA   2.9874758 0.0987569
SIGMA2 0.0987569 0.2224775
> vcov(fit.approx)
             BETA     SIGMA2
BETA   2.98748109 0.09875703
SIGMA2 0.09875703 0.22247752
> vcov(fit.euler)
               BETA       SIGMA2
BETA   2.960589e+00 1.011108e-06
SIGMA2 1.011108e-06 2.172431e-01
```

Remember that the asymptotic variance (optimal) of the maximum likelihood estimator can be calculated from (3.34).

```
> # ex3.09.R (cont)
> beta <- 1
> sigma <- 2
> DELTA <- deltat(X)
> vbeta <- (exp(2*beta*DELTA)-1)/DELTA^2
> cv.bsigma <- sigma^2*(exp(2*beta*DELTA)-1-2*beta*DELTA)/(beta*DELTA^2)
> vsigma <- sigma^4 *((exp(2*beta*DELTA)-1)^2+
+    2*beta^2*DELTA^2*(exp(2*beta*DELTA)+1)+
+    4*beta*DELTA*(exp(2*beta*DELTA)-1))/(beta^2*DELTA^2*
+     (exp(2*beta*DELTA)-1))
> matrix(c(vbeta, cv.bsigma, cv.bsigma, vsigma),2,2)
           [,1]         [,2]
[1,] 202.013400     8.053601
[2,]   8.053601 12832.321070
```

From the Hessian matrix, we obtain (for example, for the true maximum likelihood estimator) the following.

```
> # ex3.09.R (cont)
> vcov(fit.true)*100 # the sample size
            BETA    SIGMA2
BETA   298.74758   9.87569
SIGMA2   9.87569  22.24775
```

Remember that `mle` is also useful if we want to find maximum likelihood estimators of a subset of the parameters. For example, suppose we know that $\sigma^2 = 4$. Then we can use `mle` to obtain just the estimate of β specifying the `fixed` option as follows.

```
> # ex3.09.R (cont)
> mle(true.negloglik,lower=c(0,0),fixed=list(SIGMA2=4),
+                method="L-BFGS-B") -> fit.true
> mle(HP.negloglik,lower=c(0,0),fixed=list(SIGMA2=4),
+                method="L-BFGS-B") -> fit.approx
> mle(euler.negloglik,lower=c(0,0),fixed=list(SIGMA2=4),
+                method="L-BFGS-B") -> fit.euler
> coef(fit.true)
    BETA   SIGMA2
0.757828 4.000000
> coef(fit.approx)
     BETA    SIGMA2
0.7578287 4.0000000
> coef(fit.euler)
     BETA    SIGMA2
0.4511337 4.0000000
>
> vcov(fit.true)
         BETA
BETA 3.552964
> vcov(fit.approx)
         BETA
BETA 3.552975
> vcov(fit.euler)
         BETA
BETA 3.593197
```

Further examples

We have seen a detailed analysis for the Ornstein-Uhlenbeck process. In this section, we present a few other models for which the expansion is available.

The complete Ornstein-Uhlenbeck model

This model is described by the stochastic differential equation

$$dX_t = \theta_1(\theta_2 - X_t)dt + \theta_3 dW_t.$$

As before, $F(x) = x/\theta_3$ and the transformed drift is $\mu_Y(y;\theta) = \theta_1\theta_2/\theta_3 - \theta_1 y$. Nothing else changes; hence only $\mu' = -\theta_1$ survives, all the other derivatives being zero.

The Black-Scholes-Merton model

This model is described by the stochastic differential equation

$$\mathrm{d}X_t = \theta_1 X_t \mathrm{d}t + \theta_2 X_t \mathrm{d}W_t\,.$$

For this model, $F(x) = (\log x)/\theta_2$ and $\mu_Y(y,\theta) = \theta_1/\theta_2 - \theta_2/2$. Therefore, all derivatives of μ_Y are zero.

The Cox-Ingersoll-Ross model

This model is described by the stochastic differential equation

$$\mathrm{d}X_t = \theta_1(\theta_2 - X_t)\mathrm{d}t + \theta_3\sqrt{X_t}\mathrm{d}W_t\,.$$

In this case, $F(x) = (2\sqrt{x})/\theta_3$ and $\mu_Y(y;\theta) = (4\theta_1\theta_2 - \theta_3^2)/(2\theta_3^2 y) - \theta_1 y/2$. So, we have the following derivatives for μ_Y:

$$\mu'(y,\theta) = -\frac{\theta_1}{2} - \frac{4\theta_1\theta_2 - \theta_3^2}{2y^2\theta_3^2}\,, \qquad \mu^{(2)}(y,\theta) = \frac{4\theta_1\theta_2 - \theta_3^2}{y^3\theta_3^2}\,,$$
$$\mu^{(3)}(y,\theta) = -\frac{3(4\theta_1\theta_2 - \theta_3^2)}{y^4\theta_3^2}\,, \qquad \mu^{(4)}(y,\theta) = \frac{12(4\theta_1\theta_2 - \theta_3^2)}{y^5\theta_3^2}\,,$$
$$\mu^{(5)}(y,\theta) = -\frac{60(4\theta_1\theta_2 - \theta_3^2)}{y^6\theta_3^2}\,, \qquad \mu^{(6)}(y,\theta) = \frac{360(4\theta_1\theta_2 - \theta_3^2)}{y^7\theta_3^2}\,.$$

CKLS model

The model described in Section 1.13.4 is the solution to the stochastic differential equation

$$\mathrm{d}X_t = \theta_1(\theta_2 - X_t)\mathrm{d}t + \theta_3 X_t^{\theta_4}\mathrm{d}W_t\,.$$

It is distributed on $(0,+\infty)$ if $\theta_1, \theta_2 > 0$ and $\theta_4 > \frac{1}{2}$. If $\theta_4 = 1/2$, this is the Cox-Ingersoll-Ross model. CKLS does not admit a closed-form transition density if $\theta_2 \neq 0$ (see [57]), so in this case the Hermite polynomial expansion is interesting. If $\theta_4 > 1$, then

$$F(x) = \frac{x^{1-\theta_4}}{\theta_3(\theta_4 - 1)}$$

and

$$\mu_Y(y;\theta) = \frac{\theta_4}{2(\theta_4 - 1)y} - \theta_1(\theta_4 - 1)y + \theta_1\theta_2\theta_3^{1/(\theta_4-1)}(\theta_4 - 1)^{\theta_4/(\theta_4-1)}y^{\theta_4/(\theta_4-1)}\,.$$

For $\frac{1}{2} < \theta_4 < 1$, the transformation is $F(x) = x^{1-\theta_4}/\{\theta_3(1-\theta_4)\}$ and $\mu_Y(y;\theta)$ is similar.

Aït-Sahalia model

This model (see Section 1.13.7) satisfies the nonlinear equation

$$\mathrm{d}X_t = (\theta_{-1}X_t^{-1} + \theta_0 + \theta_1 X_t + \theta_2 X_t^2)\mathrm{d}t + \theta_3 X_t^{\theta_4}\mathrm{d}W_t \,.$$

For $\theta_4 = \frac{3}{2}$, we have that $F(x) = 2/(\theta_3\sqrt{x})$ and (see [8])

$$\mu_Y(y;\theta) = \frac{\frac{3}{2} - 2\frac{\theta_2}{\theta_3^2}}{y} - \frac{\theta_1 y}{2} - \frac{\theta_0\theta_3^2 y^3}{8} - \frac{\theta_{-1}\theta_3^4 y^5}{32} \,.$$

Double-well potential

This process satisfies the stochastic differential equation

$$\mathrm{d}X_t = (X_t - X_t^3)\mathrm{d}t + \mathrm{d}W_t$$

and is interesting in that its transition density is bimodal, which means highly non-Gaussian. The expansion of this model is simple in that it has unit diffusion. Therefore, $F(x) = x$, μ_Y is just the drift, and all the derivatives are quite easy to obtain.

Ahn and Gao model

This model (see Section 1.13.10) is the transformation of the Cox-Ingersoll-Ross model

$$\mathrm{d}X_t = X_t(\theta_1 - (\theta_3^3 - \theta_1\theta_2)X_t)\mathrm{d}t + \theta_3 X_t^{\frac{3}{2}}\mathrm{d}W_t,$$

and the conditional distribution of this process can be obtained as

$$p_\theta(t, y|x_0) = \frac{1}{y^2} p_\theta^{CIR}(t, 1/y|1/x_0) \,,$$

where p_θ^{CIR} is the transition density of the Cox-Ingersoll-Ross model. A direct expansion of the transition density of the Ahn-Gao model is equally possible. In such a case, the transformed process has the same drift $\mu_Y(y;\theta)$ as the CIR model, but the transformation is now $F(x) = 2/(\theta_3\sqrt{x})$. We do not rewrite the quantities here.

A generic approach

We have seen that exact specification of the derivatives up to order six are needed in order to evaluate the Hermite expansion. We now write some generic code that the reader might find interesting that essentially leaves to R the task of obtaining the derivatives. This is not costless from the numerical viewpoint. As an example, we consider again the Cox-Ingersoll-Ross model for which we have already derived in the above the explicit derivatives of the drift μ_Y. We recall that we use the parametrization

$$\mathrm{d}X_t = \theta_1(\theta_2 - X_t)\mathrm{d}t + \theta_3\sqrt{X_t}\mathrm{d}W_t\,, \quad F(x) = (2\sqrt{x})/\theta_3\,,$$

and

$$\mu_Y(y;\theta) = (4\theta_1\theta_2 - \theta_3^2)/(2\theta_3^2 y) - \theta_1 y/2\,.$$

The following code explicitly defines the derivatives of μ_Y and calculates the Hermite polynomial expansion.

```
> # ex3.10.R
>
> F <- function(t, x, theta) 2*sqrt(x)/theta[3]
> S <- function(t, x, theta) theta[3]*sqrt(x)
>
> M0 <- function(t, x, theta)
+   (4*theta[1]*theta[2]-theta[3]^2)/(2*x*theta[3]^2) - 0.5*theta[1]*x
> M1 <- function(t, x, theta)
+   -0.5*theta[1]-(4*theta[1]*theta[2]-theta[3]^2)/(2*x^2*theta[3]^2)
> M2 <- function(t, x, theta)
+   (4*theta[1]*theta[2]-theta[3]^2)/(x^3*theta[3]^2)
> M3 <- function(t, x, theta)
+   -3*(4*theta[1]*theta[2]-theta[3]^2)/(x^4*theta[3]^2)
> M4 <- function(t, x, theta)
+   12*(4*theta[1]*theta[2]-theta[3]^2)/(x^5*theta[3]^2)
> M5 <- function(t, x, theta)
+   -60*(4*theta[1]*theta[2]-theta[3]^2)/(x^6*theta[3]^2)
> M6 <- function(t, x, theta)
+   360*(4*theta[1]*theta[2]-theta[3]^2)/(x^7*theta[3]^2)
> mu1 <- list(M0, M1, M2, M3, M4, M5, M6)
```

The next code prepares the vector `mu2`, which is created using the R function `D`. We start from the expression for μ_Y

```
m0 <- expression((4*theta1*theta2-theta3^2)/(2*x*theta3^2)-0.5*theta1*x)
```

because we need to apply symbolic derivatives and then need to create functions of the triplet `(t,x,theta)`, where `theta` is the vector of the parameters. We need to remap the parameters `theta1`, `theta2`, and `theta3` appearing in `m0` to the vector `theta[1:3]`. This remapping is done in the wrapper function `HPloglik2`.

```
> # ex3.10.R (cont)
> # we now ask R to calculate derivatives
> m0 <- expression((4*theta1*theta2-theta3^2)/(2*x*theta3^2)-0.5*theta1*x)
>
> params <- all.vars(m0)
> params <- params[-which(params=="x")]
> np <- length(params)
>
> # we construct derivatives by iteration
> for(i in 1:6){
+   esp <- get(sprintf("m%d",i-1))
+   assign(sprintf("m%d",i), D(esp, "x"))
+ }
>
> mu2 <- vector(7, mode="list")
> # `mu2' must be a list of functions in (t,x,theta)
> mu2[[1]] <- function(t,x,theta) eval(m0)
> mu2[[2]] <- function(t,x,theta) eval(m1)
> mu2[[3]] <- function(t,x,theta) eval(m2)
```

```
> mu2[[4]] <- function(t,x,theta) eval(m3)
> mu2[[5]] <- function(t,x,theta) eval(m4)
> mu2[[6]] <- function(t,x,theta) eval(m5)
> mu2[[7]] <- function(t,x,theta) eval(m6)
>
> # we need to remap theta[1:3] into (theta1,theta2,theta3)
> # hence we write a wrapper function that calls HPloglik
> # in the correct way
> HPloglik2 <- function(X, theta, mu, F, S){
+  sapply(1:np, function(x) assign(params[x], theta[x], .GlobalEnv))
+  HPloglik(X, theta, mu2, F, S)
+ }
```

We now show that the exact derivatives and the R symbolic derivatives gave the same approximation (as it should be), but providing explicit derivatives speeds up the code by 30 times. To this end, we first simulate a trajectory from the CIR model using the `sde.sim` function. Please note that `model` CIR in the `sde.sim` function assumes the parametrization $\mathrm{d}X_t = (\alpha - \kappa X_t)\mathrm{d}t + \sigma \mathrm{d}W_t$ which means that (α, κ, σ) in this parametrization corresponds to $(\theta_1 = \kappa, \theta_2 = \alpha/\kappa, \theta_3 = \sigma)$, which we used to build `mu1` and `mu2`. We simulate from the CIR with $(\alpha = 0.2, \kappa = 0.4, \sigma = 0.15)$ and plot the density of α conditionally on $\kappa = 0.4$ and $\sigma = 0.15$.

```
> # ex3.10.R (cont)
> set.seed(123)
> X <- sde.sim(X0=1, model="CIR", theta=c(.2, .4, .15),delta=1e-3,N=100)
>
> xx <- seq(0,4,length=100)
> a <- system.time(sapply(xx, function(x) HPloglik(X, c(.4,x/.4,.15),
+              mu1, F, S))-> px1)
> b <- system.time(sapply(xx, function(x) HPloglik2(X, c(.4,x/.4,.15),
+            mu2, F, S))-> px2)
>
> # should be zero
> sum(abs(px1-px2))
[1] 0
>
> b/a
    user   system  elapsed
31.23019 34.33333 31.29146
```

The next listing constructs the true and the approximated likelihoods and calculates maximum likelihood estimators for α.

```
> # ex3.10.R (cont)
> # true CIR log-likelihood
> CIR.lik <- function(alpha,kappa,sigma) {
+   n <- length(X)
+   dt <- deltat(X)
+   sum(dcCIR(x=X[2:n], Dt=dt, x0=X[1:(n-1)], theta=c(alpha,kappa,sigma),
+     log=TRUE))
+ }
>
> CIR.negloglik <- function(ALPHA=1, KAPPA=1, SIGMA=1)
+    -CIR.lik(ALPHA,KAPPA,SIGMA)
>
> HP.negloglik <- function(THETA1=1, THETA2=1, THETA3=1)
+    -HPloglik(X,c(THETA1,THETA2,THETA3),mu1,F,S)
>
> mle(HP.negloglik,lower=c(0.01,0.01,0.01),
+     fixed=list(THETA1=.4,THETA3=.15), method="L-BFGS-B") -> fit.approx
> THETA <- coef(fit.approx)
> THETA
```

```
   THETA1     THETA2     THETA3
0.4000000 0.2147702 0.1500000
> # alpha = theta1*theta2
> cat(sprintf("alpha=%f\n",THETA[1]*THETA[2]))
alpha=0.085908
>
>
> mle(CIR.negloglik,lower=c(0.01,0.01,0.01),
+      fixed=list(KAPPA=.4,SIGMA=.15), method="L-BFGS-B") -> fit.CIR
> PAR <- coef(fit.CIR)
> PAR
    ALPHA      KAPPA      SIGMA
0.0859112 0.4000000 0.1500000
```

The maximum likelihood estimators on the true and the approximated likelihoods are quite close, even though the estimates are far from the true value. Figure 3.6, obtained with the R code that follows, shows the true and approximated log-likelihoods and their point of maximum.

```
> xx <- seq(0.0858,0.086,length=100)
> sapply(xx, function(x) HPloglik(X, c(.4,x/.4,.15), mu1, F, S))-> px1
> sapply(xx, function(x) CIR.lik(x,.4,.15))-> px3
> par(mar=c(5,5,1,1))
> plot(xx,px1,type="l",xlab=expression(alpha),ylab="log-likelihood")
> lines(xx,px3,lty=2)
> abline(v=PAR[1],lty=3)
> abline(v=prod(THETA[1:2]),lty=3)
```

To conclude this section, we recall that the code based on `HPloglik2` is quite generic and requires only the specification of μ_Y as an expression in the object `m0`. To understand why evaluation of symbolic derivatives generated by R is so slow, one should look inside the generated objects. For example, $\mu^{(3)}$, which corresponds to `m3`, contains

```
> m3
(4 * theta1 * theta2 - theta3^2) * (2 * theta3^2) * (2 * (2 *
    theta3^2 * (2 * theta3^2)))/((2 * x * theta3^2)^2)^2 - (4 *
    theta1 * theta2 - theta3^2) * (2 * theta3^2) * (2 * (2 *
    theta3^2 * (2 * x * theta3^2))) * (2 * (2 * (2 * theta3^2 *
    (2 * x * theta3^2)) * ((2 * x * theta3^2)^2)))/(((2 * x *
    theta3^2)^2)^2)^2
```

which evaluates exactly to

$$\mu^{(3)}(y,\theta) = -\frac{3(4\theta_1\theta_2 - \theta_3^2)}{y^4\theta_3^2}\,.$$

Indeed,

```
> # ex3.10.R (cont)
> theta <- c(.2, .4, .15)
> sapply(1:3, function(x) assign(params[x], theta[x], .GlobalEnv))
[1] 0.20 0.40 0.15
> mu1[[4]]
function(t, x, theta)
 -3*(4*theta[1]*theta[2]-theta[3]^2)/(x^4*theta[3]^2)
> mu2[[4]]
function(t,x,theta) eval(m3)
> mu1[[4]](0,1,theta)
[1] -39.66667
> mu2[[4]](0,1,theta)
[1] -39.66667
```

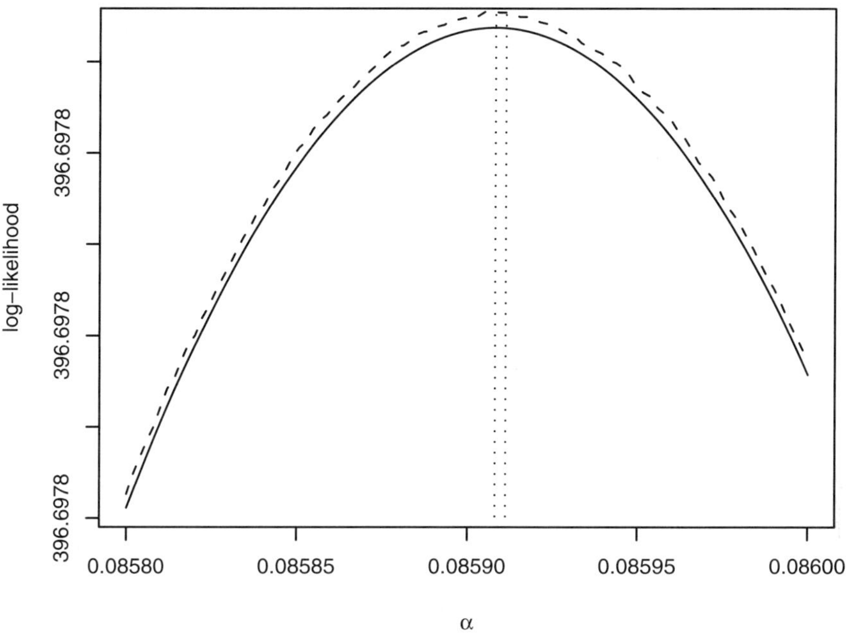

Fig. 3.6. True likelihood (solid line) against Hermite polynomial approximation (dashed line) for $\theta_2 \in (0, 4)$ and $(\theta_1, \theta_3) = (0.4, 0.15)$ on some simulated data for the CIR model with parameters $(\theta_1, \theta_2, \theta_3) = (0.4, 0.2, 0.15)$.

3.4 Bayesian estimation

In this section, we just mention the idea behind the Bayesian approach. In the general Bayesian paradigm, a Bayesian estimator of a parameter θ is obtained as the expected value of the posterior probability distribution of θ,

$$p(\theta|x^{obs}) = \frac{L_n(\theta)p(\theta)}{\int L_n(\theta)p(\theta)\mathrm{d}\theta},$$

where $p(\theta)$ is a prior distribution for θ and x^{obs} denotes the discrete-time observations from the diffusion process (see, e.g., [100]). When the likelihood $L_n(\theta)$ is known in explicit form, different results are available (see, e.g., [192]), but we already know that is quite a rare case. Recently a new stream of results based on Markov chain Monte Carlo (MCMC) algorithms have been proposed (see [80], [78], and [196]).

The MCMC technique may require an entire new book (see, e.g., [11]); hence we limit ourselves to mentioning the basic idea without touching on the details. This presentation follows [128]. The main idea behind MCMC algorithms is to sample θ from a convenient approximation of the posterior distribution $p(\theta|x^{obs})$. Let x_{i-1} and x_i be two consecutive observations and let $x_{i,k} = (t_{i-1} + k\delta_N)$ be the N unobservable values between x_{i-1} and x_i with $k = 1, \dots, N$, $\delta_N = \Delta/(N+1)$, and N a given integer. Set $x_{i,0} = x_{i-1}$ and $x_{i,N+1} = x_i$ and further introduce the set of unobserved data points $\tilde{x} = \{\tilde{x}_1, \dots, \tilde{x}_n\}$ with $\tilde{x}_i = (x_{i,1}, \dots, x_{i,N}\}$. For δ_N small enough, the transition density $p_\theta(\delta_N, x_{i,k}|x_{i-1,k})$ can be obtained by simulation as in Section 3.3.2. Note that this is another case where the exact algorithm in Section 2.12 or the method of simulation of diffusion bridges in section 2.13 is particularly handy. The intermediate points $\tilde{x}$ are regarded as missing values in the bayesian approach, the marginal distribution $p(\theta, \tilde{x}|x^{obs})$ is used to obtain $p(\theta|x^{obs})$, and the problem is reduced to sampling some θ's from $p(\theta, \tilde{x}|x^{obs})$ and averaging the values. The MCMC method is based on the construction of a Markov chain $(\tilde{x}^j, \theta^j)$ for $j = 1, 2, \dots$, whose limiting distribution is $p(\theta, \tilde{x}|x^{obs})$. This is also called the Gibbs sampler and proceeds as follows. The jth iteration is conducted in two steps:

(S1) sample $\tilde{x}^j$ from $p(\tilde{x}|x^{obs}, \theta^{j-1})$.
(S2) sample θ^j from $p(\theta|x^{obs}, \tilde{x}^j)$.

The Markov chain is started by sampling θ^0 from the prior distribution $p(\theta)$, and $\tilde{x}^0$ can be taken as the linear interpolation of the observed points x_i's. The final Bayes estimator is computed as

$$\hat{\theta} = \frac{1}{J - J_0} \sum_{j=J_0}^{J} \theta^j,$$

where J_0 is some burn-in time of the Markov chain and J is the total number of replications. But steps (S1) and (S2) are based on two unknown distributions, and hence some different route is taken to circumvent the issue. This is the Metropolis-Hastings algorithm combined with a Gaussian approximation. By the Markov property of the diffusion process, it follows that

$$p(\tilde{x}|x^{obs}, \theta) = \prod_{i=1}^{n} p(\tilde{x}_i|x_{i-1}, x_i, \theta)$$

and the approximation

$$p(\tilde{x}_i|x_{i-1}, x_i, \theta) \approx p^N(\tilde{x}_i|x_{i-1}, x_i, \theta)$$

is reasonable, with

$$p^N(\tilde{x}_i|x_{i-1}, x_i, \theta) \propto p^N(\tilde{x}_i, x_{i-1}, x_i, \theta) = \prod_{k=0}^{N} p^N(\delta_N, x_{i,k}, x_{i,k+1}|\theta),$$

where $\propto$ means proportionality. In general, the normalizing constant cannot be found. Hence the Metropolis-Hastings algorithm is used to sample from

$p^N(\tilde{x}_i|x_{i-1}, x_i, \theta)$ on the basis of $p^N(\tilde{x}_i, x_{i-1}, x_i|\theta)$ given some proposal distribution $q_1(\cdot|\tilde{x}^{j-1})$ (usually a Gaussian). This replaces step (S1). Similarly, taking some proposal distribution $q_2(\cdot|\theta^{j-1})$ for θ (for example, the prior $p(\theta)$), step (S2) is replaced by the Metropolis-Hastings step. This method is obviously computationally demanding, and convergence of the Markov chain is to be checked empirically case by case. Many variants of this method have been proposed so far along with others such as the particle filters approach (or sequential Monte Carlo sampling). The reader may want to investigate further.

3.5 Estimating functions

Estimating functions (see [106] for a complete reference on the subject) are functions, say F_n, that take as arguments both the parameters $\theta \in \Theta$ and the observed data X^{obs} with values in $\mathbb{R}$. An estimator is obtained as the solution in θ of the equation $F_n(X^{obs}, \theta) = 0$. What drives the estimating function is of course the score function, leading to the maximum likelihood estimator. We work in a large sample setup and hence need to ask a few properties of these functions: they should be asymptotically unbiased and capable of discriminating between different values of θ (*identifiability condition*); i.e., $\mathbb{E}F_n(X^{obs}, \theta) = 0$ if and only if $\theta = \theta_0$ (see also [215] for results on asymptotic theory). Estimating functions should be developed ad hoc for each observational model, and there might exist different kinds of them. We review some relevant cases.

3.5.1 Simple estimating functions

The class of *simple estimating functions* (see, e.g., [138]) is made of estimating functions based on the marginal or joint distribution of the process in some ways. These estimating functions are mainly of two forms,

$$F_n(\theta) = \sum_{i=1}^{n} f(X_i, \theta) \qquad \text{(type I)} \tag{3.35}$$

or

$$F_n(\theta) = \sum_{i=1}^{n} f(X_{i-1}, X_i, \theta) \qquad \text{(type II).} \tag{3.36}$$

The function $f(x, \theta)$ or $f(y, x, \theta)$ can take different aspects, such as $f = \partial_\theta \log \pi_\theta$ (i.e., the score function based on the invariant measure), $f_j(x, \theta) = x^j - \mathbb{E}_\theta X_0^j$, $j = 1, \ldots, p$ (i.e., based on the moment generating function), or any other suitable function on $X_{i\Delta}$. For simple estimating functions,[6] the

[6] The denominations "type I" and "type II" are only used in this book to make the treatment as easy as possible. There is currently no standard way of discriminating the two kinds of estimating functions in the literature.

identifiability conditions are

$$\mathbb{E}_{\theta_0} f(X_0, \theta) = 0 \quad \text{if and only if} \quad \theta = \theta_0 \qquad \text{(type I)}$$

and

$$\mathbb{E}_{\theta_0} f(X_0, X_\Delta, \theta) = 0 \quad \text{if and only if} \quad \theta = \theta_0 \qquad \text{(type II)}.$$

Estimating functions of type II contain the relevant case of the score function. Indeed, for discretely observed diffusion processes, the score function is

$$S_n(\theta) = \partial_\theta \log L_n(\theta) = \sum_{i=1}^{n} \partial_\theta \log p_\theta(\Delta, X_{i-1}|X_i).$$

For this model, $S_n(\theta)$ depends on the data through the terms X_{i-1} and X_i, and hence

$$f(y, x, \theta) = \partial_\theta \log p_\theta(\Delta, y|x).$$

When the transition density is known, the solution of $F_n(\theta) = 0$ is the maximum likelihood estimator. As we will see below, least squares estimators lead to estimating functions of type II. Among estimating functions of type I, the class

$$F_n(\theta) = \sum_{i=1}^{n} \left\{ b(X_i, \theta) h'(X_i) + \frac{1}{2} \sigma^2(X_i, \theta) h''(X_i) \right\} \tag{3.37}$$

occupies a central role, where $h(\cdot)$ is a twice continuously differentiable function. This estimating function is in fact built on the infinitesimal generator of the diffusion; i.e., $f(x, \theta) = \mathcal{L}_\theta h(x)$, $f : \Theta \times \mathbb{R} \to \mathbb{R}^p$, where $\mathcal{L}_\theta$ is as in (3.3). It was first introduced in [104] in the construction of the generalized method of moments. For a wide class of processes, the analytic solution of $F_n(\theta) = 0$ leads to estimators in explicit form, which makes this approach computationally faster then solving $F_n(\theta) = 0$ numerically , as we will see in several examples below. Under quite general conditions (see Theorem 3.1 in [138]), the resulting estimators are also consistent and asymptotically Gaussian. We will describe a very general implementation of estimating functions such as (3.37) in Section 3.5.2.

Simple estimating functions for the Ornstein-Uhlenbeck process

Consider the Ornstein-Uhlenbeck process solution to

$$\mathrm{d}X_t = -\theta X_t \mathrm{d}t + \mathrm{d}W_t$$

starting at x_0 at time $t_0 = 0$. We know that the transition density $p_\theta(\Delta, y|x)$ is Gaussian with parameters

$$\mathbb{E}_\theta(X_i|X_{i-1} = x) = x e^{-\theta\Delta}, \qquad \mathrm{Var}_\theta(X_i|X_{i-1}) = \frac{1 - e^{-2\theta\Delta}}{2\theta}.$$

Least squares estimator

As in any Gaussian case, the estimator of θ can be found by minimizing the quadratic distance

$$K_n(\theta) = \sum_{i=1}^{n} \left(X_i - e^{-\theta\Delta} X_{i-1}\right)^2$$

either via least squares or minimum contrast estimation. The simple estimating function corresponding to this example, derived from $\frac{\partial}{\partial\theta} K_n(\theta)$, is

$$F_n(\theta) = \sum_{i=1}^{n} X_{i-1}\left(X_i - e^{-\theta\Delta} X_{i-1}\right), \tag{3.38}$$

and the estimator $\hat{\theta}$ is obtained as the value of θ that solves $F_n(\theta) = 0$. In this case, $f(y, x, \theta) = x(y - e^{-\theta\Delta}x)$ and the estimating function is of type II as in (3.36). Easy calculations show that the resulting estimator has the explicit form

$$\hat{\theta}_n = -\frac{1}{\Delta} \log \left(\frac{\sum\limits_{i=1}^{n} X_{i-1} X_i}{\sum\limits_{i=1}^{n} X_{i-1}^2} \right), \tag{3.39}$$

and this estimator exists only if $\sum_{i=1}^{n} X_{i-1} X_i > 0$. It can be shown that the estimating function $F_n(\theta)$ above is also a martingale estimating function (see Section 3.5.4) and the estimator is consistent.

Kessler's estimator

If $\theta > 0$, the process is ergodic and it is also possible to show asymptotic normality. If we choose $h(x) = x^2$ in (3.37), we obtain Kessler's estimator (see [138]) from

$$F_n(\theta) = \sum_{i=1}^{n} \left(2\theta X_{i-1}^2 - 1\right).$$

Kessler's estimator is optimal in the sense that it has the lowest asymptotic variance v_0^*,

$$v_0^* = 2\theta_0^2 \frac{1 + e^{2\theta_0\Delta}}{1 - e^{2\theta_0\Delta}},$$

in the class of all estimators derived from estimating functions of the form (3.35). The estimator has the simple explicit form

$$\tilde{\theta} = \frac{n}{2\sum\limits_{i=1}^{n} X_{i-1}^2}. \tag{3.40}$$

Maximum likelihood estimator

For this model there is also available the maximum likelihood estimator $\bar{\theta}$, which is the solution in θ of the equation

$$\frac{1}{2\,(-1+e^{2\,\Delta\,\theta})^2\,\theta}\Bigg\{1+2\,x^2\,\theta+2\,\Delta\,\theta+e^{4\,\Delta\,\theta}\,\left(1-2\,y^2\,\theta\right) \\ -4\,e^{3\,\Delta\,\theta}\,x\,y\,\theta\,\left(-1+\Delta\,\theta\right)-4\,e^{\Delta\,\theta}\,x\,y\,\theta\,\left(1+\Delta\,\theta\right) \\ +2\,e^{2\,\Delta\,\theta}\Bigg(-1-\Delta\,\theta+x^2\,\theta\,\left(-1+2\,\Delta\,\theta\right) \\ +y^2\,\theta\,\left(1+2\,\Delta\,\theta\right)\Bigg)\Bigg\}=0, \tag{3.41}$$

which is the score function. We will now implement the three estimators for this model to show their empirical behavior. We will use the function `sde.sim` to simulate the trajectory of the process and test the three estimators over simulated paths. The experiment is as follows[7]: we simulate 1000 paths of lengths $n = 200$, $n = 500$, and $n = 1000$ from the Ornstein-Uhlenbeck process with parameter $\theta_0 = 1$. We will use Δ steps of amplitude 0.4, 1.0, and 5.0. For each trajectory, we calculate the three estimators `K.est`, `LS.est`, and `MLE.est`. Average values of the estimators and their standard deviations are reported in Table 3.2. One note for the maximum likelihood estimator is that instead of using the `uniroot` function to solve (3.41), our `MLE.est` uses `optim` to minimize the negative log-likelihood. For numerical efficiency, we optimize $-\sum_i \log p_\theta(\Delta, x_{i+1}|x_i)$ instead of $-\log(\prod_i p_\theta(\Delta, x_{i+1}|x_i))$. We will discuss in detail the use of R command `optim` in Section 3.5.2.

```
# ex3.11.R
K.est <- function(x) {
  n.obs <- length(x)
  n.obs/(2*(sum(x^2)))
}

LS.est <- function(x) {
  n <- length(x) -1
  k.sum <- sum(x[1:n]*x[2:(n+1)])
  dt <- deltat(x)
  ifelse(k.sum>0, -log(k.sum/sum(x[1:n]^2))/dt, NA)
}

MLE.est <- function(y, lower=0, upper=Inf){
 n <- length(y) - 1
 Dt <- deltat(y)
 Y <- y[2:(n+1)]
 g <- function(theta){
  ss <- sqrt((1-exp(-2*Dt*theta))/(2*theta))
  X <- y[1:n]*exp(-theta*Dt)
  lik <- dnorm(Y,mean=X,sd=ss)
  -sum(log(lik))
 }
 tmp <- try(optim(runif(1), g, method="L-BFGS-B", lower=lower,
                     upper=upper)$par)
```

[7] We try to replicate the experiment in [138]; see Table 1 there.

```
  if(class(tmp)=="try-error") tmp <- NA
  tmp
}
```

The reader should note that in `MLE.est` we have limited the search for the minimum of the negative log-likelihood in the interval $[0, +\infty)$ because we know that θ_0 is in that interval. In practical applications (i.e., with real data), one should first plot the likelihood function of the observed data to decide where to search for the extremal points or use some other adaptive approach.

As for Kessler's estimator, instead of using only the observations from X_0 to X_{n-1} (see (3.40)), the routine `K.est` uses all the observations up to X_n. In this way, no information is lost in finite (yet big) samples as in our experiment. The function `LS.est` returns `NA` if $\sum_{i=1}^n X_{i-1}X_i \leq 0$.

Table 3.2. Empirical comparison of the simple estimating function estimator $\hat\theta$ in (3.39), Kessler's estimator $\tilde\theta$ in (3.40), and the MLE estimator for the Ornstein-Uhlenbeck process (see the text). Average values over $K = 1000$ replications and standard deviation for different values of Δ and different number of observations n are given. The true $\theta_0 = 1$, and v_0^* is the best asymptotic variance for estimators in the class of simple estimating functions of the type (3.35). For the $\hat\theta$ estimator and $n = 1000$, valid cases (i.e., $\sum_{i=1}^n X_{i-1}X_i > 0$) are 543, 553, and 578 for $\Delta = 0.4$, 1.0, and 5.0, respectively.

Δ	n	$\sqrt{v_0^*/n}$	$\hat\theta$	$\tilde\theta$	$\bar\theta$
0.4	200		1.03	1.03	1.04
		0.16	(0.20)	(0.17)	(0.16)
	500		1.01	1.01	1.02
		0.10	(0.19)	(0.12)	(0.11)
	1000		0.66	1.01	1.01
		0.07	(0.23)	(0.10)	(0.10)
1.0	200		1.01	1.01	1.01
		0.11	(0.13)	(0.10)	(0.10)
	500		1.00	1.01	1.01
		0.07	(0.12)	(0.07)	(0.07)
	1000		0.74	1.01	1.01
		0.05	(0.24)	(0.06)	(0.06)
5.0	200		1.00	1.01	1.01
		0.10	(0.09)	(0.07)	(0.07)
	500		1.00	1.01	1.01
		0.06	(0.08)	(0.05)	(0.05)
	1000		0.79	1.00	1.00
		0.04	(0.21)	(0.04)	(0.04)

The next code calculates the results reported in Table 3.2. We run `K=1000` replications of the Ornstein-Uhlenbeck process for $\delta = 0.1 < \Delta$ and resample it for $\Delta = 0.4$, $\Delta = 1.0$, and $\Delta = 0.5$. This allows for finer simulations.[8] In each replication, we generate $5.0/0.1 \cdot 1000 = 50000$ observations in order to have a trajectory long enough to extract $n = 200$, 500, and 1000 observations for every value of Δ. We collect the estimates in three matrices, `kessler`, `mle`, and `simple` with nine columns each. Columns 1 to 3 are for estimates in the case of $\Delta = 0.4$ and $n = 200$, 500, and 1000, columns 4 to 6 for $\Delta = 1.0$ and different values of n, and the remaining columns for $\Delta = 5.0$. One thing worth noting in the following R code is the way we subsample the trajectories: the object `X` contains the whole trajectory generated for $\delta = 0.1$. The `Delta` varies in the R loop in the set `c(0.4,1,5)`. To pick up the times corresponding to the desired Δ step, we use the `window` function.[9] For example, `window(X,deltat=0.4)` will subsample the time series X for the specified $\Delta = 0.4$.

```
# ex3.11.R (cont)
theta0 <- 1
d <- expression( -1*x )
s <-  expression( 1 )
K <- 1000 # 1000 MC replications

set.seed(123)
kessler <- matrix(NA,K,9)
mle <- matrix(NA,K,9)
simple <- matrix(NA,K,9)

x0 <- rnorm(K,sd=sqrt(1/(2*theta0)))
sde.sim(X0=x0, drift=d, sigma=s, N=50000, delta=0.1, M=K)->X

for(k in 1:K){
 cat(".")
 m <- 0
 for(Delta in c(0.4,1,5)){
  m <- m+1
  j <- 0
  for(n in c(200,500,1000)){
   j <- j+1
   X.win <- window(X[,k], start=0, end=n*Delta, deltat=Delta)
   kessler[k,m+3*(j-1)] <- K.est(X.win)
   simple[k,m+3*(j-1)] <- LS.est(X.win)
   mle[k,m+3*(j-1)] <- MLE.est(X.win)
  }
 }
 cat(sprintf(" %3.3d / %3.3d completed\n",k,K))
}
```

The next code is used to obtain average values of the estimators, their standard deviations, the theoretical values of v_0^*, and the number of valid cases for the `LS.est` estimator.

```
# ex3.11.R (cont)
S1 <- apply(simple,2,function(x) mean(x,na.rm=T))
K1 <- apply(kessler,2,function(x) mean(x,na.rm=T))
M1 <- apply(mle,2,function(x) mean(x,na.rm=T))
A <- cbind(S1,K1,M1)
matrix(as.numeric(sprintf("%3.2f",A)),9,3)
```

[8] It is always desirable to simulate trajectories with a δ smaller than the Δ needed to test the estimators.

[9] Please note that up to version 2.1.1 of R, a bug prevented the use of `window` with noninteger values of `deltat`.

```
S2 <- apply(simple ,2,function(x) sd(x,na.rm=T))
K2 <- apply(kessler ,2,function(x) sd(x,na.rm=T))
M2 <- apply(mle ,2,function(x) sd(x,na.rm=T))
B <- cbind(S2,K2,M2)
matrix(as.numeric(sprintf("%3.2f",B)),9,3)

Delta <- c(0.4,1,5)
v0 <- 2*theta0^2 * (1+exp(-2*theta0*Delta))/(1-
  exp(-2*theta0*Delta))

sprintf("%3.2f",sqrt(v0[1]/c(200,500,1000)))
sprintf("%3.2f",sqrt(v0[2]/c(200,500,1000)))
sprintf("%3.2f",sqrt(v0[3]/c(200,500,1000)))

# valid cases for the LS estimator
apply(simple, 2, function(x) length(which(!is.na(x))))
```

Simple estimating function for the Cox-Ingersoll-Ross process

We introduced this process in Section 2.3.2 solution of the stochastic differential equation

$$\mathrm{d}X_t = (\alpha + \theta X_t)\mathrm{d}t + \sigma\sqrt{X_t}\mathrm{d}W_t\,,$$

$\alpha > 0$, $\theta < 0$ and $\sigma > 0$. Assume σ is known, and choose $h(x) = (x, x^2)^t$ as in [138] in the estimating function (3.37). Then, solving for $F_n(\theta) = 0$ yields the two explicit estimators for α and θ

$$\begin{aligned}
\tilde{\alpha}_n &= \frac{\left(\sum\limits_{i=1}^n X_{i-1}\right)^2}{2\left(n\sum\limits_{i=1}^n X_{i-1}^2 - \left(\sum\limits_{i=1}^n X_{i-1}\right)^2\right)}\,,\\
\tilde{\theta}_n &= \frac{-n\sum\limits_{i=1}^n X_{i-1}}{2\left(n\sum\limits_{i=1}^n X_{i-1}^2 - \left(\sum\limits_{i=1}^n X_{i-1}\right)^2\right)}\,,
\end{aligned} \tag{3.42}$$

which can be easily implemented as follows.

```
> # ex3.11.R (cont)
> # CIR-Model
> # theta = -1
> # alpha = 10
> # sigma = 1
> # x0 = 10
> set.seed(123);
> d <- expression(10 - x)
> s <- expression(sqrt(x))
> x0 <- 10
> sde.sim(X0=x0,drift=d, sigma=s,N=1000,delta=0.1) -> X
>
> # estimator for alpha
> (sum(X)^2)/(2*(length(X)*sum(X^2)-sum(X)^2))
[1] 11.52346
```

```
>
> # estimator for theta
> (-length(X)*sum(X))/(2*(length(X)*sum(X^2)-sum(X)^2))
[1] -1.136781
```

Linear case

In the linear case, as for Ornstein-Uhlenbeck or CIR processes, some general results are available. In particular, if the parameter to be estimated is only in the drift and $b(\theta,x) = \theta \cdot b(x)$ and $\sigma(\theta,x) = \sigma(x)$, for any[10] function h in the estimating function (3.37), the solution of $F_n(\theta) = 0$ leads to the following explicit form of the estimator

$$\tilde{\theta} = -\frac{1}{2}\frac{\sum_{i=1}^{n} \sigma^2(X_{i-1})h''(X_{i-1})}{\sum_{i=1}^{n} b(X_{i-1})h'(X_{i-1})}.$$

3.5.2 Algorithm 1 for simple estimating functions

We now describe the algorithm that can be used to estimate parameters using the generic estimating functions of type I (3.35) and type II (3.36). The algorithm is quite flexible in that the user can choose any way to describe the parametric model; i.e., use any name for the parameters and the algorithm will figure out the dimension on the parametric space. The user is asked to specify as input to the command `simple.ef` the observed data `X` and the function f as parameter `f`, which has to be an R list of expressions defined in terms of variables x and y and of course the parameters. If both x and y appear in the definition of f, then f is interpreted as a function of both; i.e., the algorithm solves the problem $F_n = 0$ for (3.36). If only x appears in the definition, then the solution is searched for an estimating function of type (3.35). Variable x cannot be omitted, and variables must be specified strictly with names '`x`' and '`y`' .

To solve $F_n(\theta) = 0$, we use the box-constrained optimization facility `optim` with method '`L-BFGS-B`'. This is a box-constrained minimization algorithm. Thus, to find the zeros of $F_n(\theta)$, we minimize $|F_n(\theta)|^2$ and we square $|F_n(\theta)|$ to penalize the extremes more. If the identifiability condition holds, this leads to the true zero of $F_n(\theta)$. Most of the following code is dedicated to extracting the parameter names from variables x and y into two vectors `f.pars` (the parameters) and `f.vars` (the two variables x and y). This code is also optimized for speed in the sense that all the internal functions are written to accept and return vector objects, avoiding R loops. At first, the code extracts parameter names from the coefficients using the `all.vars` command and checks if at

[10] Actually, $h(x) = x$ cannot be used because hypothesis A5 in [138] is not satisfied.

least variable 'x' has been specified; if it is missing, the command returns an error.

Then function `f` is transformed from an R expression to an R function and this function now depends on both x and y and all the parameters. This generalization makes the code quite general in accepting parameters. All evaluation is done in a new environment `e1`, which lives only inside command `simple.ef` to avoid any interference with the current R workspace. The function $F_n = (F_n^1, F_n^2, \ldots, F_n^p)$ is coded as `Fn` and has the same dimension p as the parameter space and function f. The optimizer calls function $G_n(\theta)$, which is in fact $G_n(\theta) = \sum_{i=1}^p |F_n^i(\theta)|^2$.

```
simple.ef <- function(X, f, guess, lower, upper){

 if(!is.ts(X))
  stop("Please provide a `ts' object")

 n.f <- length(f)
 f.vars <- NULL
 for(i in 1:n.f)
  f.vars <- c(f.vars,all.vars(f[[i]]))
 f.vars <- unique(f.vars)
 n.vars <- length(f.vars)
 match("x", f.vars) -> idx.x
 match("y", f.vars) -> idx.y
 has.x <- !is.na(idx.x)
 has.y <- !is.na(idx.y)
 if(!has.x){
  if(!has.y)
   stop("Variables `x' and `y' both missing")
  else
   stop("Variable `x' missing")
 }
 idx <- c(idx.x, idx.y)
 idx <- as.integer(na.omit(idx))
 f.pars <- f.vars[(1:n.vars)[-idx]]
 n.pars <- length(f.pars)
 f.vars <- f.vars[idx]

 if(!is.list(f) | (n.f!=n.pars) |
    !all(unlist(lapply(f,mode))=="expression"))
  stop("`f' must be a list of expressions of length equal to the \
        dimension of the parameter space")

 new.env() -> e1

 f1 <- vector(n.f, mode="list")
 for(i in 1:n.f){
  f1[[i]] <- function(j) {
   val <- eval(f[[j]],e1)
   if(length(val) != lx)
    val <- rep(val, lx)
   val
  }
 }

 X.data <- NA
 Y.data <- NA
 n.obs <- length(X)
 if(has.y)
  Y.data <- X[2:n.obs]
 if(has.x){
  if(has.y)
   X.data <- X[1:(n.obs-1)]
```

```
  else
   X.data <- X[1:n.obs]
 }
 assign("x",X.data, e1)
 assign("y",Y.data, e1)
 lx <- length(X.data)

 Fn <- function(theta){
  for(i in 1:n.pars)
   assign(f.pars[i], theta[i], e1)
  val <- numeric(n.f)
  for(i in 1:n.f){
   ff1 <-  f1[[i]](i)
   val[i] <- sum(ff1)
  }
  return(val)
 }

 Gn <- function(theta)  sum(abs(Fn(theta))^2)


 if(missing(guess))
  start <- runif(n.pars)
 else
  start <- guess

 if(missing(lower))
  lower <- rep(-Inf, n.pars)

 if(missing(upper))
  upper <- rep(Inf, n.pars)

 st <- start
 names(st) <- f.pars
 cat("\nInitial values for the optimization algorithm ")
 if(missing(guess))
  cat("(random)\n")
 else
  cat("\n")
 print(st)
 cat("\nOptimization constraints\n")
 ct <- as.matrix(cbind(lower, upper))
 rownames(ct) <- f.pars
 colnames(ct) <- c("lower", "upper")
 print(ct)

 cat("\nRunning optimizer...\n")
 mn <- optim(start, Gn, method="L-BFGS-B", lower=lower, upper=upper)
 names(mn$par) <- f.pars

 estimate <- mn$par
 fn <-  Fn(mn$par)

 return( list(estimate=estimate, Fn=fn) )
}
```

Listing 3.8. Simple estimating function, Algorithm 1.

The use of this function is quite easy, as the following code shows.

```
> # ex3.12.R
> set.seed(123);
> d <- expression(-1 * x)
> s <- expression(1)
> x0 <- rnorm(1,sd=sqrt(1/2))
> sde.sim(X0=x0,drift=d, sigma=s,N=2500,delta=0.1) -> X
```

```
>
> # Kessler's estimator revisited
> f <- list(expression(2*theta*x^2-1))
> simple.ef(X, f, lower=0, upper=Inf)

Running optimizer...
$estimate
   theta
1.122539

> K.est(X)
[1] 1.122539
>
> # Least Squares estimator revisited
> f <- list(expression(x*(y-x*exp(-0.1*theta))))
> simple.ef(X, f, lower=0, upper=Inf)

Running optimizer...
$estimate
   theta
1.100543

> LS.est(X)
[1] 1.100543
>
> # MLE estimator revisited, f is based on
> # the score function of the process
> f <-list(expression((1 + 2 * (x^2) * theta + 2 * 0.1 * theta +
+     exp(4 * 0.1 * theta) * (1 - 2 * (y^2) * theta)
+     - 4 * exp(3 * 0.1 * theta) * x * y * theta * (-1 + 0.1 * theta)
+     - 4 * exp(theta * 0.1) * x * y * theta * (1 + theta * 0.1)
+     + 2 * exp(2 * 0.1 * theta) * (-1 - 0.1 * theta + (x^2) * theta *
+     (-1 + 2 * 0.1 * theta) + (y^2) * theta * (1 + 2 *
+     0.1 * theta)))/((2 * (-1 + exp(2 * 0.1 * theta))^2) * theta)))
>
> simple.ef(X, f, lower=0, upper=Inf)

Running optimizer...
$estimate
   theta
1.119500

> MLE.est(X)
[1] 1.119500
```

3.5.3 Algorithm 2 for simple estimating functions

This second algorithm solves the problem for estimating functions of type I in the particular case of equation (3.37). The user is allowed to specify the full model in terms of drift and diffusion parameters and is asked to specify the function $h(\cdot)$. It eventually calculates first and second derivatives of the function $h(\cdot)$ in equation (3.37). It is also possible to specify lower and upper bounds of the parameter space. The only requirement is that the user specify the space variable x as 'x' in the coefficients. As before, the function $h(\cdot)$ is expected to be a list of R expressions, one for each parameter to be estimated, and we use a constrained optimization approach as before. Most of the following code is dedicated to extracting the parameter names from the coefficients of the stochastic differential equation and to preparing the functions (drift, diffusion, $h(\cdot)$ and its derivatives) to be passed to the optimizer. Variable 'x' will be interpreted as variable x in $b(x, \theta)$, $\sigma(x, \theta)$, and $h(x)$. If derivatives of function $h(\cdot)$ are missing, then symbolic differentiation is attempted as in the `sde.sim` function. Option `hessian=TRUE` in the function `deriv` is used to get second-order derivatives.

```
simple.ef2 <- function(X, drift, sigma, h, h.x, h.xx,
                       guess, lower, upper){

 if(!is.ts(X))
  stop("Please provide a `ts' object")

 if(!is.expression(drift) | !is.expression(sigma))
  stop("Coefficients `drift' and `sigma' must be expressions")

 d.vars <- all.vars(drift)
 s.vars <- all.vars(sigma)
 match("x", d.vars) -> d.has.x
 match("x", s.vars) -> s.has.x
 par.vars <- unique(c(d.vars, s.vars))
 n.vars <- length(par.vars)
 match("x", par.vars) -> idx
 if(is.na(idx))
  stop("One variable should be named `x'")
 par.vars <- par.vars[c(idx,(1:n.vars)[-idx])]
 n.h <- length(h)
 if(!is.list(h) | (n.h!=n.vars-1) |
   !all(unlist(lapply(h,mode))=="expression"))
  stop("`h' must be a list of expressions of length equal to the \
      dimension of the parameter space")

 new.env() -> e1

 self.hx <- FALSE
 if(missing(h.x)){
  cat("h.x not provided, attempting symbolic derivation.\n")
  h.x <- vector(n.h, mode="list")
  for(i in 1:n.h)
   h.x[[i]] <- deriv(h[[i]],"x")
  self.hx <- TRUE
 }


 self.hxx <- FALSE
 if(missing(h.xx)){
  cat("h.xx not provided, attempting symbolic derivation.\n")
  h.xx <- vector(n.h, mode="list")
  for(i in 1:n.h)
   h.xx[[i]] <- deriv(h[[i]],"x",hessian=TRUE)
  self.hxx <- TRUE
 }


 h1.x <- vector(n.h, mode="list")
 for(i in 1:n.h){
  h1.x[[i]] <- function(x,j) {
   lx <- length(x)
   assign("x",x, e1)
   val <- eval(h.x[[j]],e1)
   if(self.hx)
    val <- as.numeric(attr(val,"gradient"))
   if(length(val) != lx)
    val <- rep(val, lx)
   val
  }
 }
```

```
h1.xx <- vector(n.h, mode="list")
for(i in 1:n.h){
 h1.xx[[i]] <- function(x,j) {
  lx <- length(x)
  assign("x",x, e1)
  val <- eval(h.xx[[j]],e1)
  if(self.hxx)
   val <- as.numeric(attr(val,"hessian")[1,1,1])
  if(length(val) != lx)
   val <- rep(val, lx)
  val
 }
}

Dd <- deriv(d, par.vars)
Ds <- deriv(s, par.vars)

D1 <- function(x){
 assign("x",x, e1)
 val <- as.numeric(eval(Dd,e1))
 if(is.na(d.has.x))
  val <- rep(val, length(x))
 val
}

S1 <- function(x){
 assign("x",x, e1)
 val <- as.numeric(eval(Ds,e1))
 if(is.na(s.has.x))
  val <- rep(val, length(x))
 val
}

Fn <- function(theta){
 for(i in 2:n.vars)
  assign(par.vars[i], theta[i-1],e1)
 dd1 <-  D1(X)
 ss1 <- S1(X)
 val <- numeric(n.h)
 for(i in 1:n.h){
  hh1.x <-  h1.x[[i]](X,i)
  hh1.xx <- h1.xx[[i]](X,i)
  val[i] <- sum(dd1*hh1.x + 0.5*(ss1^2)*hh1.xx)
 }
 return(val)
}

Gn <- function(theta)  sum(abs(Fn(theta))^2)


if(missing(guess))
 start <- runif(n.vars-1)
else
 start <- guess

if(missing(lower))
 lower <- rep(-Inf, n.vars-1)

if(missing(upper))
 upper <- rep(Inf, n.vars-1)

st <- start
names(st) <- par.vars[-1]
cat("\nInitial values for the optimization algorithm ")
if(missing(guess))
 cat("(random)\n")
```

```
  else
   cat("\n")
  print(st)
  cat("\nOptimization constraints\n")
  ct <- as.matrix(cbind(lower, upper))
  rownames(ct) <- par.vars[-1]
  colnames(ct) <- c("lower", "upper")
  print(ct)

  cat("\nRunning optimizer...\n")
  mn <- optim(start, Gn, method="L-BFGS-B", lower=lower, upper=upper)
  names(mn$par) <- par.vars[-1]

  estimate <- mn$par
  fn <-  Fn(mn$par)

  return( list(estimate=estimate, Fn=fn) )
}
```

Listing 3.9. Simple estimating function, Algorithm 2.

We now show examples using this function and we suggest that the reader study the code a bit on its own.

Example 3.1 (Ornstein-Uhlenbeck). As before, we compare the performance of estimators for this process. Recall that Kessler's estimator `K.est` should coincide numerically with the estimator in (3.37) choosing $h(x) = x^2$. In the following, some output has been dropped.

```
> # ex3.12.R (cont)

> set.seed(123);
> d <- expression(-1 * x)
> s <- expression(1)
> x0 <- rnorm(1,sd=sqrt(1/2))
> sde.sim(X0=x0,drift=d, sigma=s,N=1500,delta=0.1) -> X
> d <- expression(-theta* x)
> s <- expression(1)
> h <- list(expression(x^2))
> simple.ef2(X, d, s, h, lower=0, upper=Inf)

Running optimizer...
$estimate
   theta
1.141632

> K.est(X)
[1] 1.141632
> MLE.est(X)
[1] 1.138554
> LS.est(X)
[1] 1.128546
```

As can be seen Kessler's explicit estimator and the one obtained via constrained optimization match. We have specified as input h as a `list` of expressions of length 1. We also specified the lower and upper bounds for the parameter θ, which we decided to call `theta`.

Example 3.2 (Cox-Ingersoll-Ross model: the linear case). For this bidimensional case, we have already mentioned and calculated the explicit estimators in (3.42). These estimators are the solution of $F_n = 0$ in (3.37) with

$h = (x, x^2)^t$. We will specify $h(\cdot)$ as a list of length 2 and we will also specify different bounds for the α and θ. Please note that in the next code we use generic names for the parameters.

```
> # ex3.12.R (cont)

> set.seed(123);
> d <- expression(10 - x)
> s <- expression(sqrt(x))
> x0 <- 10
> sde.sim(X0=x0,drift=d, sigma=s,N=1500,delta=0.1) -> X
> d <- expression(alpha +theta* x)
> s <- expression(sqrt(x))
> h <- list(expression(x),expression(x^2))
> simple.ef2(X, d, s, h, lower=c(0,-Inf), upper=c(Inf ,0))

Running optimizer...
$estimate
    alpha      theta
11.613569 -1.144988

> # explicit estimator for alpha
> (sum(X)^2)/(2*(length(X)*sum(X^2)-sum(X)^2))
[1] 11.61357

> # explicit estimator for theta
> (-length(X)*sum(X))/(2*(length(X)*sum(X^2)-sum(X)^2))
[1] -1.144988
```

Example 3.3 (Cox-Ingersoll-Ross model: the nonlinear case). An interesting estimation problem for a nonlinear model is derived from finance, and it is a generalization of the previous CIR model. The process is a solution of the stochastic differential equation

$$\mathrm{d}X_t = (\alpha + \theta X_t)\mathrm{d}t + \sigma X_t^\gamma \mathrm{d}W_t\,,$$

$\alpha > 0$, $\theta < 0$, $\sigma > 0$, and $\gamma \in [0, 1]$. Assume as in [138] $\alpha = 10$, $\theta = -1$, and $\sigma = 1$, and we want to estimate γ when the true parameter is $\gamma_0 = \frac{1}{2}$. We use $h(x) = x^2$.

```
> # ex3.12.R (cont)

> set.seed(123);
> d <- expression(10 - x)
> s <- expression(sqrt(x))
> x0 <- 10
> sde.sim(X0=x0,drift=d, sigma=s,N=1500,delta=0.1) -> X
> d <- expression(10- x)
> s <- expression(x^gamma)
> h <- list(expression(x^2))
> simple.ef2(X, d, s, h, lower=0, upper=1)

Running optimizer...
$estimate
    gamma
0.5315879
```

Limits of an optimizer

It is well-known that the optimizer in high dimensions can lead to local minima/maxima instead of a global one. Even solutions provided by robust algorithms, such as the one used in **simple.ef**, depend on the initial point used to start the search. The command **simple.ef** allows us to specify an initial point in Θ used to start the search via the parameter **guess**; otherwise it chooses at random. We show an example of such use in the case of estimation of the three-dimensional parameter (α, θ, γ) for the previous model.

```
> # ex3.13.R

> set.seed(123)
> d <- expression(10 - x)
> s <- expression(sqrt(x))
> x0 <- 10
> sde.sim(X0=x0,drift=d, sigma=s,N=1500,delta=0.1) -> X
> d <- expression(alpha + theta*x)
> s <- expression(x^gamma)
> h <- list(expression(x), expression(x^2), expression(x^2))
> simple.ef2(X, d, s, h, lower=c(0,-Inf,0), upper=c(Inf,0,1))

Running optimizer...
$estimate
     alpha      theta      gamma
 1.1449879 -0.1128850  0.0000000

> # user defined guess
> simple.ef2(X, d, s, h, lower=c(0,-Inf,0), upper=c(Inf,0,1),
+     guess=c(1,-.9,.7))

Running optimizer...
$estimate
     alpha      theta      gamma
 9.3275921 -0.9195144  0.4525340

> # another guess
> simple.ef2(X, d, s, h, lower=c(0,-Inf,0), upper=c(Inf,0,1),
+     guess=c(1,-1.2,.3))

Running optimizer...
$estimate
     alpha      theta      gamma
11.0891507 -1.0929864  0.4886947
```

To conclude, as has already been noted elsewhere (see, e.g., [231]), in high dimensions one should start by solving simpler problems in lower dimensions to estimate the initial guess of the parameters and then attack the global problem and, as usual in optimization, play around with different initial starting points by doing some sensitivity analysis.

3.5.4 Martingale estimating functions

We denote by $G_n(\theta)$ estimating functions that are also martingales with respect to the σ-field $\mathcal{F}_n$; i.e., $\mathbb{E}(G_n(\theta)|\mathcal{F}_{n-1}) = G_{n-1}(\theta)$. There exist several types of *martingale estimating functions*. There are good reasons to use this type of estimating function such as that the score function is a martingale and that a huge amount of machinery for martingales is already available. An estimating function is said to be *unbiased* if $\mathbb{E}(G(\theta)) = 0$. This property is also called Fisher consistency. The more general form of a martingale estimating function is

$$G_n(\theta) = \sum_{i=1}^{n} g(\Delta, X_{i-1}, X_i; \theta) \tag{3.43}$$

with

$$g(\Delta, x, y; \theta) = \sum_{j=1}^{N} a_j(\Delta, x; \theta) h_j(\Delta, x, y; \theta)\,, \tag{3.44}$$

where $h_j(\Delta, x, y; \theta)$, $j = 1, \ldots, N$, are given real-valued functions satisfying

$$\int_l^r h_j(\Delta, x, y; \theta) p_\theta(\Delta, y|x) \mathrm{d}y = 0$$

for all $\Delta > 0$, $x \in (l, r)$, and $\theta \in \Theta$. The functions $a_j(\cdot)$ are called *weights* (or *instruments* in econometrics) and can be chosen in an optimal way with functions $h_j(\cdot)$ in hand using the theory of optimal estimating functions that we are not going to treat (see [106], [121], [122]). Martingale estimating functions have been studied in [31], [32], [214], [140], [138], [33], and [120], [48] and [139] contain comparative simulation studies.

3.5.5 Polynomial martingale estimating functions

Another form of estimating function leading to explicit estimators are polynomial estimating functions. In particular, linear and quadratic estimating functions will be treated here. They rely on the knowledge of the first conditional moments instead of the whole transition density. When conditional moments are not known, they can be estimated using the Monte Carlo approach, and this task, though computationally intensive, is simpler than the estimation of the whole transition density. When conditional moments are known, the estimators are also consistent. In general, this type of estimating function provides estimators that are very robust to model misspecification, as shown for example in [31].

Linear estimating functions

A simple type of estimating function that can be used when the parameter θ is only in the drift coefficient is the linear estimating function obtained for $N = 1$ with

$$h_1(\Delta, x, y; \theta) = y - H_\theta(\Delta|x)\,,$$

where

$$H_\theta(\Delta|x) = \mathbb{E}_\theta(X_\Delta|X_0 = x) = \int_l^r y p_\theta(\Delta, y|x) \mathrm{d}y\,.$$

They were studied for diffusion models in [31] and derived as an approximation to the continuous-time likelihood function. In some models (e.g. the Ornstein-Uhlenbeck model), the conditional expectation $H(\Delta, x; \theta)$ is known. If it isn't, one needs to estimate it using a Monte Carlo method. One of the martingale estimating functions proposed in [31] has the particular form

$$G_n(\theta) = \sum_{i=1}^{n} \frac{\partial_\theta b(X_{i-1},\theta)}{\sigma^2(X_{i-1},\theta)} \left\{X_i - \mathbb{E}_\theta(X_i|X_{i-1})\right\}. \tag{3.45}$$

This estimating function is derived as an approximation to the *optimal estimating function*[11] and it also arises in the problem of approximating the continuous score function in [211]. These martingale linear estimating functions are particular cases of the simple estimating functions (3.36) of type II when Θ is one-dimensional, so the same R command `simple.ef` can be used. Also notice again that in the Ornstein-Uhlenbeck example, when the conditional moments are known in explicit form, the least squares estimator in (3.39) is indeed obtained via a martingale estimating function (3.38), which is just a particular case of (3.43).

Quadratic estimating functions

If the parameter θ belongs to the diffusion coefficient, then the linear estimating function might be too "simple" for the task. Thus a second term $h_2(\Delta, x, y; \theta)$ has to be used. This function takes the form

$$h_2(\Delta, x, y; \theta) = h_1(\Delta, x, y; \theta)^2 - \Phi_\theta(\Delta|x),$$

where

$$\Phi_\theta(\Delta|x) = \mathrm{Var}_\theta(X_\Delta|X_0 = x) = \int_l^r h_1(\Delta, x; \theta)^2 p_\theta(\Delta, y|x)\mathrm{d}y.$$

Example 3.4 (Mean-reverting diffusion). Consider the mean-reverting diffusion

$$\mathrm{d}X_t = -\beta(X_t - \alpha)\mathrm{d}t + \tau\sqrt{X_t}\mathrm{d}W_t, \qquad \beta, \tau > 0.$$

For this model

$$H_\theta(\Delta|x) = xe^{-\beta t} + \alpha\left(1 - e^{-\beta t}\right)$$

and

$$\Phi(\Delta|x, \alpha, \beta, \tau) = \frac{\tau^2}{\beta}\left(\left(\frac{1}{2}\alpha - x\right)e^{-2\beta} - (\alpha - x)e^{-\beta} + \frac{1}{2}\alpha\right).$$

Monte Carlo evaluation of conditional moments

When the conditional mean and variance are not known, they can be estimated as follows. Fix one value of θ and simulate numerically M independent trajectories $\{X^j_{\delta i}, i = 1, \ldots, K\}$, where $\delta = \Delta/K$ with K sufficiently large. Then reasonable estimators are

$$H_\theta(\Delta|x) = \frac{1}{M}\sum_{j=1}^{M} X^j_{K\delta}$$

[11] We don't treat this class of estimating functions here, but the reader can refer to [106].

and

$$\Phi_\theta(\Delta|x) = \frac{1}{M}\sum_{j=1}^{M}\left(X_{K\delta}^j\right)^2 - (H_\theta(\Delta|x))^2 .$$

Polynomial estimating functions

One can generalize quadratic estimating functions using the functions

$$h_j(\Delta, x, y; \theta) = \phi_j(y;\theta) - \Pi_\Delta^\theta(\phi_j(\theta))(x),$$

where the function $\phi_j(y) = y^j$ and Π_Δ^θ is the transition operator such that

$$\Pi_\Delta^\theta(u)(x) = \mathbb{E}_\theta(u(X)|X_0 = x) .$$

Polynomial estimating functions have been studied in [179] and [136].

```
linear.mart.ef <- function(X, drift, sigma, a1, a2, guess,
  lower, upper, c.mean, c.var){

 if(!is.ts(X))
  stop("Please provide a `ts' object")

 N <- 1
 if(missing(a1))
  stop("`a1' is missing")
 if(!missing(a2)){
  if(length(a1) != length(a2))
   stop("`a1' and `a2' not of the same dimension")
  N <- 2
 }

 if(!is.expression(a1))
  stop("`a1' must be a vector of expressions")

 if(N==2)
  if(!is.expression(a2))
  stop("`a2' must be a vector of expressions")

 if(!is.expression(drift) | !is.expression(sigma))
  stop("Coefficients `drift' and `sigma' must be expressions")

 if(!is.expression(c.mean))
  stop("Conditional mean must be an expression")

 if(!is.expression(c.var))
  stop("Conditional variance must be an expression")

 d.vars <- all.vars(drift)
 s.vars <- all.vars(sigma)
 match("x", d.vars) -> d.has.x
 match("x", s.vars) -> s.has.x
 par.vars <- unique(c(d.vars, s.vars))
 n.vars <- length(par.vars)
 match("x", par.vars) -> idx
 if(is.na(idx))
  stop("One variable should be named `x'")
 par.vars <- par.vars[c(idx,(1:n.vars)[-idx])]
 n.pars <- n.vars - 1

# We check the list of expressions needed for the weights a(x,theta)
# each a_i(x,theta) must be a vector of expressions of the same
```

```
# dimension of the parameter space
 a.vars <- all.vars(a1)
 if(N==2)
  a.vars <- c(a.vars, all.vars(a2))

 a.vars <- unique(a.vars)
 n.a.vars <- length(a.vars)
 match("x", a.vars) -> a.idx
 a.has.x <- !is.na(a.idx)
 a.idx <- as.integer(na.omit(a.idx))
 a.pars <- a.vars[(1:n.a.vars)[-a.idx]]
 n.a.pars <- length(a.pars)

 if(n.pars != length(a1))
  stop("weight functions `a' must have the same dimension of the\
  parameter space")

 new.env() -> e1
 # when missing c.mean and c.var need to be estimated
 # via MC methods. Easy.

 F.XT <- function(){
  val <- eval(c.mean,e1)
  if(length(val) != lx)
   val <- rep(val, lx)
  val
 }

 PHI.XT <- function(){
  val <- eval(c.var,e1)
  if(length(val) != lx)
   val <- rep(val, lx)
  val
 }

# vectorized version of a: A[i,j,] = a_{ij}()
  A <- function() {
   val <- array(0, c(N, n.pars, lx))
   for(k in 1:n.pars){
    val[1,k,] <- eval(a1[k],e1)
    if(N==2)
     val[2,k,] <- eval(a2[k],e1)
   }
   val
  }


 n.obs <- length(X)
 Y.data <- X[2:n.obs]
 X.data <- X[1:(n.obs-1)]


 assign("x",X.data, e1)
 assign("y",Y.data, e1)
 lx <- length(X.data)

 Fn <- function(theta){
  for(i in 2:n.vars){
   assign(par.vars[i], theta[i-1],e1)
   assign(par.vars[i], theta[i-1], .GlobalEnv)
  }

  aa <-  A() # this contains the weights

  H1 <- Y.data - F.XT()
  H2 <- H1^2 -  PHI.XT()
```

```
  val <- 0
  for(i in 1:n.pars)
   val <- val + sum(aa[1,i,]*H1)
  if(N==2)
   for(i in 1:n.pars)
    val <- val + sum(aa[2,i,]*H2)
  return(val)
 }

 Gn <- function(theta)  sum(abs(Fn(theta))^2)


 if(missing(guess))
  start <- runif(n.vars-1)
 else
  start <- guess

 if(missing(lower))
  lower <- rep(-Inf, n.vars-1)

 if(missing(upper))
  upper <- rep(Inf, n.vars-1)

 st <- start
 names(st) <- par.vars[-1]
 cat("\nInitial values for the optimization algorithm ")
 if(missing(guess))
  cat("(random)\n")
 else
  cat("\n")
 print(st)
 cat("\nOptimization constraints\n")
 ct <- as.matrix(cbind(lower, upper))
 rownames(ct) <- par.vars[-1]
 colnames(ct) <- c("lower", "upper")
 print(ct)

 cat("\nRunning optimizer...\n")
 mn <- optim(start, Gn, method="L-BFGS-B", lower=lower, upper=upper)
 names(mn$par) <- par.vars[-1]

 estimate <- mn$par
 fn <- Fn(mn$par)

 return( list(estimate=estimate, Fn=fn) )
}
```

Listing 3.10. Polynomial estimating function.

```
> # ex3.14.R

> set.seed(123)
> d <- expression(-1 * x)
> s <- expression(1)
> x0 <- rnorm(1,sd=sqrt(1/2))
> sde.sim(X0=x0,drift=d, sigma=s,N=1000,delta=0.1) -> X
> d <- expression(-theta * x)

> linear.mart.ef(X, d, s, a1=expression(-x), lower=0, upper=Inf,
+    c.mean=expression(x*exp(-theta*0.1)),
+    c.var=expression((1-exp(-2*theta*0.1))/(2*theta)))

Running optimizer...
$estimate
   theta
1.144589
```

```
> # the linear mart. e.f. coincides with the
> # least square estimator
> LS.est(X)
[1] 1.144589
```

3.5.6 Estimating functions based on eigenfunctions

Another way to build estimating functions for diffusions was introduced in [140] and is based on the infinitesimal generator of the diffusion $\mathcal{L}_\theta$ in (3.3). A twice-differentiable function $\phi(x;\theta)$ is called an *eigenfunction* of $\mathcal{L}_\theta$ if it satisfies

$$\mathcal{L}_\theta\phi(x;\theta) = -\lambda(\theta)\phi(x;\theta),$$

where $\lambda(\theta)$ is a nonnegative real number called the *eigenvalue*, corresponding to $\phi(x;\theta)$. Under weak regularity conditions (see [140]), it holds true that

$$\Pi_\Delta^\theta(\phi(\theta))(x) = \mathbb{E}_\theta(\phi(X_\Delta;\theta)|X_0 = x) = e^{-\lambda(\theta)\Delta}\phi(x;\theta).$$

Thus, a martingale estimating function of the form (3.43) can be defined using

$$h_j(\Delta, x, y;\theta) = \phi_j(y;\theta) - e^{-\lambda_j(\theta)\Delta}\phi_j(x;\theta),$$

where $\phi_1(\cdot;\theta),\ldots,\phi_N(\cdot;\theta)$ are the eigenfunctions for $\mathcal{L}_\theta$ with eigenvalues $\lambda_1(\theta),\ldots,\lambda_N(\theta)$. This method is quite general for one-dimensional diffusions but multidimensional processed eigenvalues are not necessarily real values.

Example 3.5 (Cox-Ingersoll-Ross). For the Cox-Ingersoll-Ross model, the eigenfunctions are the Laguerre polynomials. Hence the resulting martingale estimating function in this case is another polynomial estimating function.

Example 3.6 (Ornstein-Uhlenbeck on the finite interval). As in [140], consider the one-dimensional diffusion with drift $b(x,\theta) = -\theta\tan(x)$ and $\sigma(x,\theta) = 1$ defined on $(-\pi/2, +\pi/2)$. For $\theta \geq \frac{1}{2}$, the process is ergodic. The eigenfunctions in this case are $\phi_j(x;\theta) = C_j^\theta(\sin(x))$ and eigenvalues $j(\theta+j/2)$ for $j = 0, 1, \ldots$, where C_j^θ are the Gegenbauer polynomials of order j. Considering only the first eigenfunction, one obtains

$$G_n(\theta) = \sum_{i=1}^n \sin(X_{i-1})\left\{\sin(X_i) - e^{-(\theta+\frac{1}{2})}\sin(X_{i-1})\right\}, \tag{3.46}$$

which clearly has an explicit solution of $G_n(\theta) = 0$. This estimating function is also an approximation of the optimal estimating function when using $h(x, y, \theta) = \sin(y) - e^{-(\theta+0.5)}\sin(x)$.

Example 3.7 (The Pearson diffusions). For the Pearson diffusion solution to (1.59), when the process has all finite moments up to order N, the eigenfunctions are just the first N eigenpolynomials. The asymptotic properties and optimality for this class of processes have been studied in [89].

More examples with discussion can be found in Section 3 of [30], but it is clear that previous algorithms for martingale estimating functions can be applied in this framework as well.

3.5.7 Estimating functions based on transform functions

Recently, in [133] a new method of creating estimating functions was proposed that does not need the explicit knowledge of conditional moments or distributions. This method relies on the Wagner-Platen expansion (see [141]) of a suitable transformation U of the original process. The authors prove that for affine diffusion and power transform function $U(x) = x^\lambda$, explicit solutions of the estimating functions can be found. Even if the calculations are easy to deal with, the formulas might be too lengthy and do not fit the purposes of this text; hence we direct the reader to the cited reference. One last thing worth mentioning is that this approach works well in high dimensions and also for nonergodic and (time)-inhomogeneous diffusions.

Which one to choose?

As seen from the previous material, there is no general method that can be applied to any kind of model without ad hoc adaptation to the specific situation (parameters in the drift and/or the diffusion part, high dimension, nonergodic case, etc.), so we prefer to show at least a selection of approaches. We mention without going into detail that, for example, combinations of estimating functions when the dimension of the parameter space is high proved to be quite effective (see, e.g., [33], where the authors propose to use a simple estimating function of type I for the parameters in the drift and a martingale estimating function for the diffusion part). A good account on estimating functions and their statistical properties can also be found in [30].

3.6 Discretization of continuous-time estimators

We mentioned at the beginning of this chapter that for continuous-time observations standard likelihood estimation can be done using the Girsanov theorem directly. In very early works on inference for stochastic differential equations from discrete observations, one approach was to consider discretized versions of continuous-time estimators (see, e.g., [151], [142]). This approach usually involves the discretization of stochastic integrals. Recently, in [52] and [53], performance results of these estimates were considered in comparison with estimates obtained from exact and approximated likelihood methods. Although we do not suggest this approach as a generic approach to obtain good estimates, for some specific models for which the hypotheses of other approximate methods do not hold, these estimates are still of some interest.

Table 3.3. Integrals I_j in maximum likelihood estimators of k and θ (see the text).

Model	I_1	I_2	I_3	I_4	I_5
CIR	$\int_0^T \frac{\mathrm{d}X_s}{X_s}$	$\int_0^T \mathrm{d}X_s$	$\int_0^T \frac{\mathrm{d}s}{X_s}$	$\int_0^T X_s \mathrm{d}s$	$\int_0^T \mathrm{d}s$
GBM	$\int_0^T \frac{\mathrm{d}X_s}{X_s^2}$	$\int_0^T \frac{\mathrm{d}X_s}{X_s}$	$\int_0^T \frac{\mathrm{d}s}{X_s^2}$	$\int_0^T \mathrm{d}s$	$\int_0^T \frac{\mathrm{d}s}{X_s}$

We discuss here an application for a couple of models just to show some details of the implementations. Consider the model

$$\mathrm{d}X_t = k(\theta - X_t)\mathrm{d}t + \sigma X_t^\beta \mathrm{d}W_t \,.$$

Usually, σ is estimated using the quadratic variation estimator (3.17),

$$\hat{\sigma}^2 = \frac{1}{n\Delta}\sum_{i=1}^{n}(X_i - X_{i-1})^2,$$

or (see, e.g., [186], [206]) by

$$\hat{\sigma}^2 = \frac{1}{n\Delta}\sum_{i=1}^{n}\frac{(X_i - X_{i-1})^2}{X_{i-1}^{2\beta}},$$

and β is assumed to be specified by the researcher. We consider here $\beta = 1$, which gives geometric Brownian motion, and $\beta = \frac{1}{2}$, which gives the Cox-Ingersoll-Ross model. Direct maximization of the continuous likelihood (3.4) gives (see, e.g., [159] or [149]) the continuous-time estimators

$$\hat{k} = \frac{I_1 I_5 - I_2 I_3}{I_3 I_4 - I_5^2}, \qquad \theta = \frac{I_1 I_4 - I_2 I_5}{I_1 I_5 - I_2 I_3},$$

where I_j are (eventually stochastic) integrals. Table 3.3, taken from [53], gives the actual form of the I_j. Usually, the nonstochastic integrals are evaluated using one of the following methods:

$$\int_0^T f(X_s)\mathrm{d}s = \begin{cases} \Delta \sum_{i=i}^{n} f(X_{i-1}), & \text{Cauchy formula;} \\ \Delta \sum_{i=1}^{n} \frac{f(X_{i-1}) + f(X_i)}{2}, & \text{trapezoidal;} \\ \Delta \sum_{i=1}^{n-1} \frac{f(X_{i-1}) + 4f(X_i) + f(X_{i+1})}{3}, & \text{Simpson.} \end{cases}$$

In the first case, the approximation error is of order $O(\Delta)$, in the second case it is $O(\Delta^2)$, and in the latter, it is $O(\Delta^4)$. Usually, the trapezoidal rule is

Table 3.4. Discretized versions of the integrals I_j in Table 3.3.

Model	I_1	I_2
CIR	$\log \frac{X_n}{X_0} + \frac{\sigma^2}{2} I_3$	$X_n - X_0$
GBM	$\sigma^2 I_5 - \left(\frac{1}{X_n} - \frac{1}{X_0}\right)$	$\log \frac{X_n}{X_0} + \frac{\sigma^2}{2} n\Delta$

Model	I_3	I_4	I_5
CIR	$\Delta \sum_{i=1}^{n} \frac{X_{i-1}^{-1} + X_i^{-1}}{2}$	$\Delta \sum_{i=1}^{n} \frac{X_{i+1} + X_i}{2}$	$n\Delta$
GBM	$\Delta \sum_{i=1}^{n} \frac{X_{i-1}^{-2} + X_i^{-2}}{2}$	$n\Delta$	$\Delta \sum_{i=1}^{n} \frac{X_{i-1}^{-1} + X_i^{-1}}{2}$

considered a good compromise between numerical efficiency and ease of implementation. When concerned with stochastic integrals, first the Itô formula is used to eliminate the stochastic integral and then one of the approximation formulas above is applied to the resulting terms. Rarely, the following direct approximation is considered:

$$\int_0^T f(X_s) \mathrm{d}X_s = \sum_{i=1}^{n} f(X_{i-1})(X_i - X_{i-1}) .$$

For the processes considered in this section, the discretized integrals take the form in Table 3.4. We implement in the next code only the algorithm for the Cox-Ingersoll-Ross case and run a small simulation study to test the performance of the estimators for a given σ and when σ has to be estimated using the quadratic variation estimator. The next experiment consists of a 1000 Monte Carlo replications in which a Cox-Ingersoll-Ross model is simulated with the initial value sampled from the stationary law of the process (`rsCIR`) and then the integrals I_1 to I_5 are calculated appropriately following Table 3.4. As I_1 depends on σ, we compute two estimators for k and θ, one time using the value of $\sigma = 0.15$ (the true value) and one other time using the quadratic variation estimator. At the end, we calculate both the average and standard deviation of the estimates over the 1000 simulations. The time horizon T is set to 100, the time increment Δ is chosen as 0.01, and hence a total of $N = T/\Delta$ observations are available for the estimation.

```
> # ex3.15.R

> k <- NULL
> theta <- NULL
```

```
> k1 <- NULL
> theta1 <- NULL
> sigma.hat <- NULL
> sigma <- 0.15
> pars <- c(0.5, 0.2, sigma)
> n.sim <- 1000
> Delta <- 0.01
> set.seed(123)
> x0 <- rsCIR(n.sim, pars)
> T <- 100
>
> for(i in 1:n.sim){
+   X <- sde.sim(X0=x0[i], model="CIR", theta=pars, N=T/Delta, delta=Delta)
+
+   n <- length(X)
+
+ # CIR
+  I3 <- Delta * sum(1/X[1:(n-1)] + 1/X[2:n])/2
+  I1 <- log(X[n]/X[1]) + 0.5*sigma^2 * I3
+  I2 <- X[n] - X[1]
+  I4 <- Delta * sum(X[1:(n-1)] + X[2:n])/2
+  I5 <- n*Delta
+
+  k <- c(k, (I1*I5-I2*I3)/(I3*I4-I5^2))
+  theta <- c(theta, (I1*I4-I2*I5)/(I1*I5-I2*I3))
+  sigma.est <- sqrt(sum((X[2:n]-X[1:(n-1)])^2/X[1:(n-1)])/(n*Delta))
+  sigma.hat <- c(sigma.hat, sigma.est)
+
+  I1 <- log(X[n]/X[1]) + 0.5*sigma.est^2 * I3
+  k1 <- c(k1, (I1*I5-I2*I3)/(I3*I4-I5^2))
+  theta1 <- c(theta1, (I1*I4-I2*I5)/(I1*I5-I2*I3))
+
+ }
> cat(sprintf("kappa=%f, theta=%f, sigma=%f : kappa1=%f, theta1=%f\n",
+ mean(k*theta),  mean(k), mean(sigma.hat), mean(k1*theta1), mean(k1)))
kappa=0.524426, theta=0.211023, sigma=0.150207 : kappa1=0.524789,
theta1=0.211176

> cat(sprintf("SD: kappa=%f, theta=%f, sigma=%f : kappa1=%f, theta1=%f\n",
+  sd(k*theta), sd(k), sd(sigma.hat), sd(k1*theta1), sd(k1)))
SD: kappa=0.075822, theta=0.033268, sigma=0.001058 : kappa1=0.075948,
theta1=0.033317

> cat(sprintf("kappa=%f, theta=%f, sigma=%f\n", pars[1], pars[2], pars[3]))
kappa=0.500000, theta=0.200000, sigma=0.150000
```

From the above, it emerges that there is no substantial difference in terms of precision of the estimates if σ is estimated using the quadratic variation estimator. The reader might want to verify that these results are highly dependent on the discretization step. Indeed, more bias in the estimates of k and θ emerges as Δ increases. For extensive accounts of similar simulation studies, see [53] and [52].

3.7 Generalized method of moments

The generalized method of moments (GMM) dates back to 1982 (see [103]). It is a generalization of the method of moments that is based on the matching of theoretical moments and sample moments. Let X_i be the stationary sequence and θ the $m \times 1$ vector of parameters characterizing the stationary law of the

X_i's. Let $u_i = u(X_i;\theta)$ be an $r \times 1$-valued function, $r \geq m$, such that

$$\mathbb{E}\{u_i\} = \mu, \quad \forall\, i,$$

and

$$\mathrm{Cov}(u_i, u_{i+j}) = \mathbb{E}\{(u_i - \mu)(u_{i+j} - \mu)^T\} = S_j, \quad \forall\, i, j,$$

and define $S_{-j} = \mathrm{Cov}(u_i, u_{i-j})$. It is assumed that the following moment condition (or orthogonality condition) holds if θ_0 is the true value of the parameter θ:

$$\mathbb{E}\{u(X_i;\theta)\} = 0 \quad \text{only if} \quad \theta = \theta_0\,. \tag{3.47}$$

Usually, the function $u(\theta)$ is the difference between the exact kth moment and X_i^k for some powers k. Let

$$g_n(\theta) = \frac{1}{n}\sum_{i=1}^{n} u(X_i;\theta)$$

be the sample counterpart of condition (3.47). We expect that $\mathbb{E}\{g_n(\theta_0)\} = 0$. It is assumed further that the strong law of large numbers holds; i.e.,

$$g_n(\theta) \overset{a.s.}{\to} \mathbb{E}\{u(X_i;\theta)\}$$

uniformly in θ. The main point is that moments and, in general, the structure of the function $u(\cdot)$ should be known for each model[12] (see also [92] and [90]). If the dimension m of θ and the number of conditions r defined by u match, there is one single solution to the problem, but in general there might be more moment conditions than parameters to estimate ($r \geq m$), i.e., the model is *overidentified.* The optimization problem is transformed as

$$\hat{\theta} = \arg\min_{\theta} Q(\theta) = \arg\min_{\theta} g_n(\theta)^T W g_n(\theta),$$

where W is a given positive definite matrix of weights. A particular choice of W is S^{-1}, where

$$S = \mathbb{E}\{uu^T\}$$

is the long-run covariance matrix. This choice of W guarantees the smallest asymptotic covariance matrix of the GMM estimator. The matrix S is estimated from the data as follows. Let

$$\hat{S}_j = \frac{1}{n}\sum_{i=j+1}^{n} u_i u_{i-j}^T, \quad j = 0, 1, \ldots, \ell,$$

where ℓ is the maximum lag chosen a priori.[13] Then, the estimate is

---

[12] Unless some nonparametric approach such as SNP by Gallant and Tauchen [91] is used to calculate the moments numerically in place of exact formulas. See also `http://www.econ.duke.edu/~get/snp.html`.

[13] This is not actually a trivial choice (see [166]), but we do not discuss it here.

$$\hat{S} = \hat{S}_0 + \sum_{j=1}^{\ell} w_j(\hat{S}_j + \hat{S}_j^T). \tag{3.48}$$

The set of weights w_j are such that $\hat{S}$ is a positive definite matrix. If the w_j's are all equal to one, all lags are considered to contribute equally. One usual choice is the *Bartlet weights*, defined as

$$w_j = 1 - \frac{j}{\ell + 1}.$$

Several authors (see, e.g., [166]) have shown that the choice of the weights is not really an issue, so we do not present other alternatives here. The interested reader may find more details on the estimation of S in Section 3.5.3 of [101].

3.7.1 The GMM algorithm

Hansen [103] proposed a two-step procedure.

Step 1: Set $W = I$ with I the identity matrix, and solve the nonlinear least squares problem

$$\hat{\theta}^{(1)} = \arg\min_{\theta} g_n(\theta) g_n(\theta)^T.$$

Step 2: Compute

$$\hat{u}_i = u\left(x_i; \hat{\theta}^{(1)}\right),$$

estimate S_j as

$$\hat{S}_j = \frac{1}{n} \sum_{i=j+1}^{n} \hat{u}_i \hat{u}_{i-j}^T, \quad j = 0, 1, \ldots, \ell,$$

and then calculate $\hat{S}$ as in (3.48). Now set $W = \hat{S}^{-1}$ and estimate a new value for θ as follows:

$$\hat{\theta}^{(2)} = \arg\min_{\theta} g_n(\theta)^T W g_n(\theta).$$

Step 2 can be iterated until, at the $k+1$st step, one obtains $\hat{\theta}^{(k+1)} \simeq \hat{\theta}^{(k)}$. The GMM estimators are asymptotically normal with asymptotic variance $V/n = (DS^{-1}D)^{-1}/n$, where D is the gradient of g_n evaluated at θ_0. The asymptotic variance can of course be estimated by the sample counterpart of V and D calculated at $\hat{\theta}$. Under additional regularity conditions, the estimator is also consistent (see [103]).

Further, it is also possible to construct a χ^2 test to verify if the model is overidentified. Indeed, under the null hypothesis $H_0 : \mathbb{E}\{u_i(\theta)\} = 0$, the following statistic J is such that

$$J = g_n(\hat{\theta}) W g_n(\hat{\theta}) \sim \chi^2_{r-m} .$$

There exists a nice link between likelihood inference, estimating functions, and the generalized method of moments (see, e.g., [48]). By choosing u as the score function,

$$u(X_i; \theta) = \frac{\partial}{\partial \theta} \log p_\theta(t, X_i; X_{i-1}) ,$$

the GMM estimator is just the maximum likelihood estimator, and the limiting variance V takes the usual form.

For diffusion processes observed at discrete times, the moment conditions have to be replaced by *conditional* moment conditions,

$$\mathbb{E}\{u_i(\theta_0)|\mathcal{F}_i\} = 0 ,$$

provided that they exist in explicit form or otherwise are obtained with the simulated moment method as proposed in [51].

3.7.2 GMM, stochastic differential equations, and Euler method

One application that increased the popularity of GMM in inference for stochastic differential equations was the paper [49], where the famous CKLS model for the interest rate was proposed for the first time

$$\mathrm{d}X_t = (\alpha + \beta X_t)\mathrm{d}t + \sigma X_t^\gamma \mathrm{d}W_t . \tag{3.49}$$

At least in the case $\gamma = \frac{1}{2}$ (i.e., the Cox-Ingersoll-Ross model), the first two conditional moments exist in explicit form (see formulas (1.49) and (1.50)). The usual approach in econometrics (see [43], [41], [40]) is to discretize the stochastic differential equation above in a way that is easy to treat but not the exact (or the best) discretization of equation (3.49),

$$X_{i+1} - X_i = (\alpha + \beta X_i)\Delta + \epsilon_{i+1} . \tag{3.50}$$

It is then assumed that, for the discretized model, the following are true:

$$\mathbb{E}\{\epsilon_{i+1}|\mathcal{F}_i\} = 0, \quad \mathbb{E}\{\epsilon_{i+1}^2|\mathcal{F}_i\} = \Delta\sigma^2 X_i^{2\gamma}, \quad \mathbb{E}\{\epsilon_i\epsilon_j|\mathcal{F}_i\} = 0 .$$

Hence $\mathbb{E}(X_{i+1}|\mathcal{F}_i) = X_i + (\alpha + \beta X_i)\Delta$ and $X_{i+1} - \mathbb{E}(X_{i+1}|\mathcal{F}_i) = \epsilon_{i+1}$. The function $u(\cdot)$ for the moment condition is consequently constructed as

$$u_i(\theta) = \begin{bmatrix} X_{i+1} - \mathbb{E}(X_{i+1}|\mathcal{F}_i) \\ X_i(X_{i+1} - \mathbb{E}(X_{i+1}|\mathcal{F}_i)) \\ \mathrm{Var}(X_{i+1}|\mathcal{F}_i) - (X_{i+1} - \mathbb{E}(X_{i+1}|\mathcal{F}_i))^2 \\ X_i\{\mathrm{Var}(X_{i+1}|\mathcal{F}_i) - (X_{i+1} - \mathbb{E}(X_{i+1}|\mathcal{F}_i))^2\} \end{bmatrix} ,$$

and the function $g_n(\theta)$ is then

$$g_n(\theta) = \frac{1}{n}\sum_{i=1}^{n} u_i(\theta)\,.$$

Unfortunately, conditional moments of the continuous-time model (3.49) and conditional moments of the discretization (3.50) do not coincide. Hence, in this case, using conditional moments may lead to inconsistent estimates. In particular (see, e.g., [48], [5]), let us denote by α_ρ, β_ρ, σ_ρ, and γ_ρ the true parameters of (3.50) and by α, β, σ, and γ the true parameters of (3.49). Easy calculations show that if we check the moment condition (3.47) for the first two elements in u (i.e., $\mathbb{E}(\epsilon_{i+1}) = 0$ and $\mathbb{E}(X_i\epsilon_{i+1}) = 0$), we obtain

$$\alpha_\rho = \frac{\alpha}{\beta\Delta}\left(e^{\beta\Delta} - 1\right) \quad \text{and} \quad \beta_\rho = \frac{e^{\beta\Delta} - 1}{\Delta}.$$

Therefore, the GMM estimator is a consistent estimator of α_ρ and β_ρ and not α and β. Similar results hold for σ_ρ and γ_ρ. We have already noticed similar facts in Section 3.2.1, and of course α_ρ and β_ρ are nearly equal to their continuous-time counterparts α and β as $\Delta \to 0$. Many other solutions have been proposed so far such as moment approximation of specific model discretizations (as in [48]), the simulated method of moments [51], the efficient method of moments (see [93]), and indirect inference [99], but we do not discuss these methods here. Conversely, we present an example of implementation in which we compare exact maximum likelihood and GMM estimators. We make use of the `gmm` function in the package `sde`. The interface of `gmm` is similar to that of `simple.ef`, with the only difference that this function needs the $u(\cdot)$ function specified as `u <- function(x,y,theta)`, where `theta` is a vector and the dimension of `theta` is specified via the parameter `dim` or obtained by the function from the length of the `guess` vector of initial values. An application of `gmm` is the easiest way to show how it functions. In order to obtain reasonable working GMM estimates, we need to have stationary observations from the CIR process and small Δ. Hence we simulate a path of the CIR process of length 500000 with $\Delta = 0.001$ with stationary initial value `x0`. Then, we retain only the second half of the trajectory and subsample it further with $\Delta = 0.01$. In the end, we are left with 5000 observations.

```
> # ex3.16.R
> alpha <- 0.5
> beta <- 0.2
> sigma <- sqrt(0.05)
> true <- c(alpha, beta, sigma)
> names(true) <- c("alpha", "beta", "sigma")
>
> set.seed(123)
> x0 <- rsCIR(1,theta=true)
> sde.sim(X0=x0,model="CIR",theta=true,N=500000,delta=0.001) -> X
> X <- window(X, deltat=0.1)
> DELTA = deltat(X)
> n <- length(X)
> X <- window(X, start=n*DELTA*0.5)
> plot(X)
```

For the Cox-Ingersoll-Ross model, we know that

$$\mathbb{E}\{X_{i+1}|X_i = y\} = \frac{\theta_1}{\theta_2} + \left(y - \frac{\theta_1}{\theta_2}\right) e^{-\theta_2 \Delta},$$

$$\text{Var}\{X_{i+1}|X_i = y\} = y\frac{\theta_3^2(e^{-\theta_2\Delta} - e^{-2\theta_2\Delta})}{\theta_2} + \frac{\theta_1\theta_3^2(1 - e^{-2\theta_2\Delta})}{2\theta_2^2},$$

where $(\alpha, \beta, \sigma) = (\theta_1, \theta_2, \theta_3)$. Then, we can easily construct the function $u_i(\theta)$ = `u(x,y,theta)` as the code below shows.

```
> # ex3.16.R (cont)
> u <- function(x, y, theta, DELTA){
+   c.mean <- theta[1]/theta[2] + (y-theta[1]/theta[2])*exp(-theta[2]*DELTA)
+   c.var <- ((y*theta[3]^2 *
+     (exp(-theta[2]*DELTA)-exp(-2*theta[2]*DELTA))/theta[2] +
+       +theta[1]*theta[3]^2*(1-exp(-2*theta[2]*DELTA))/(2*theta[2]^2)))
+   cbind(x-c.mean,y*(x-c.mean), c.var-(x-c.mean)^2, y*(c.var-(x-c.mean)^2))
+ }
```

Further, we estimate the parameters using the true maximum likelihood approach, and we pass these estimates to the **gmm** function.

```
> # ex3.16.R (cont)
> CIR.lik <- function(theta1,theta2,theta3) {
+   n <- length(X)
+   dt <- deltat(X)
+   -sum(dcCIR(x=X[2:n], Dt=dt, x0=X[1:(n-1)],
+   theta=c(theta1,theta2,theta3),  log=TRUE))
+ }
>
> fit <- mle(CIR.lik, start=list(theta1=.1,  theta2=.1,theta3=.3),
+      method="L-BFGS-B",lower=c(0.001,0.001,0.001), upper=c(1,1,1))
> # maximum likelihood estimates
> coef(fit)
   theta1    theta2    theta3
0.5232152 0.2096333 0.2273695
>
> gmm(X,u, guess=as.numeric(coef(fit)), lower=c(0,0,0), upper=c(1,1,1))

Dimension of parameter space set to 3
Initial values for the optimization algorithm
[1] 0.5232152 0.2096333 0.2273695

Optimization constraints
       lower upper
theta1     0     1
theta2     0     1
theta3     0     1

Running optimizer...

First stage estimates:
   theta1    theta2    theta3
0.5232972 0.2094014 0.1631663

Starting second stage...
         theta1          theta2          theta3            Q1  |theta1-theta2|
   0.5299241256    0.2127399561    0.1623548320  0.0005173358     0.0107770245
$par
   theta1    theta2    theta3
0.5299241 0.2127400 0.1623548

$val
```

```
[1] 0.0005173358

$hessian
           theta1    theta2    theta3
theta1   7.002527 -14.16295 -22.52342
theta2 -14.162946  33.48923  63.42024
theta3 -22.523415  63.42024 493.56295

> true
    alpha      beta     sigma
0.5000000 0.2000000 0.2236068
```

The initial guess value was good enough, but the reader may want to try other starting points in the parameter space or change the tolerances `tol1` and/or `tol2` to see how unstable the optimization step might be. As usual, reducing the number of parameters to be estimated increases the probability that the optimizer will find a good solution. In the code below, we only impose the first two moment conditions, which are enough to obtain estimates of θ_1 and θ_2.

```
> # ex3.16.R (cont)
> u2 <- function(x, y, theta, DELTA){
+   c.mean <- theta[1]/theta[2] + (y-theta[1]/theta[2])*exp(-theta[2]*DELTA)
+    cbind(x-c.mean,y*(x-c.mean))
+ }
>
> set.seed(123)
> gmm(X, u2, dim=2, lower=c(0,0), upper=c(1,1))

Initial values for the optimization algorithm (random)
[1] 0.2875775 0.7883051

Optimization constraints
       lower upper
theta1     0     1
theta2     0     1

Running optimizer...

First stage estimates:
   theta1    theta2
0.5178472 0.2073139

Starting second stage...
          theta1            theta2                Q1   |theta1-theta2|
    5.076414e-01      2.034049e-01      2.567041e-13      1.411470e-02
$par
   theta1    theta2
0.5076414 0.2034049

$val
[1] 2.567041e-13

$hessian
          theta1    theta2
theta1   5.85667 -11.54539
theta2 -11.54539  26.07322

> true
    alpha      beta     sigma
0.5000000 0.2000000 0.2236068
```

Listing 3.11 contains the code of the gmm function used above.

```
gmm <- function(X, u, dim, guess, lower, upper,maxiter=30,
            tol1=1e-3,tol2=1e-3){
  if(!is.ts(X))
    stop("Please provide a `ts' object")
  DELTA = deltat(X)
  n <- length(X)

  if(!missing(guess)){
   if(length(guess)>0){
    dim <- length(guess)
    cat(sprintf("\nDimension of parameter space set to %d", dim))
   }
  }

  if(missing(dim))
   stop("Please specify dimension of parameter space")

  H <-function(theta)
   apply(u(X[2:n], X[1:(n-1)], theta, DELTA), 2, mean)

  Q <-function(theta) sum(H(theta)^2)

  S <- function(j, theta)
   ( (t(u(X[(j+2):n],X[(j+1):(n-1)], theta, DELTA)) %*%
      u(X[2:(n-j)],X[1:(n-j-1)], theta, DELTA))/n )

  ell <- n-2
  w <- 1-(1:ell)/(ell+1) # Bartlet weights

  cat("\nInitial values for the optimization algorithm ")
  if(missing(guess)){
    guess <- runif(dim)
    cat("(random)\n")
  } else {
   cat("\n")
  }
  print(guess)

  if(missing(lower))
    lower <- rep(-Inf, dim)
  if(missing(upper))
    upper <- rep(Inf, dim)

  cat("\nOptimization constraints\n")
  ct <- as.matrix(cbind(lower, upper))
    rownames(ct) <- paste("theta",1:dim,sep="")
    colnames(ct) <- c("lower", "upper")
    print(ct)
    cat("\nRunning optimizer...\n")


  theta1 <- optim(guess,Q,method="L-BFGS-B",
                  upper=upper, lower=lower)$par
  for(i in 1:dim){
   names(theta1)[i] <- sprintf("theta%d",i)
  }
  cat("\nFirst stage estimates:\n")
  print(theta1)

  cat("\nStarting second stage...\n")
  goOn <- TRUE
  iter <- 0
  while(goOn){
   iter <- iter + 1
   S.hat <- S(0, theta1)
   for(i in 1:ell)
```

```
    S.hat = S.hat + w[i]*(S(i,theta1)+t(S(i,theta1)))
   W <- solve(S.hat)
   Q1 <-function(theta) H(theta) %*% W %*% H(theta)
   fit <- optim(theta1,Q1,method="L-BFGS-B", upper=upper,
                 lower=lower, hessian=TRUE)
   theta2 <- fit$par
   val <- fit$value
   hes <- fit$hessian
   out <- c(theta2, val, sum(abs(theta1-theta2)))
   names(out) <- c(names(theta1),"Q1","|theta1-theta2|")
   print(out)
   names(theta2) <- names(theta1)
   if(sum(abs(theta1-theta2))<tol1 || val<tol2 || iter>maxiter)
    goOn <- FALSE
   theta1 <- theta2
 }

  list(par=theta1, val=val, hessian=hes)
}
```

Listing 3.11. Generalized method of moments.

3.8 What about multidimensional diffusion processes?

As mentioned in the early pages of this text, we do not deal with the multidimensional case. In general, methods based on simulation work equally well for the multidimensional case. These include the simulated likelihood method and the MCMC approach. Estimating functions and the generalized method of moments also have variants for the multidimensional case. Finally, Hermite polynomial expansion is available for multidimensional diffusion processes. For all the above mentioned methods, the references cited in this book also contain the multidimensional result. The reader is invited to read them as a starting point.

4

Miscellaneous Topics

We now review some inference problems that do not directly relate to the parametric estimation methods presented in the previous chapter even though part of the results already presented will be useful here. We will describe the problem of model identification via Akaike's information criterion, the problem of nonparametric estimation for diffusion processes, and a change-point problem for the volatility of a diffusion process.

4.1 Model identification via Akaike's information criterion

Consider a diffusion process solution to the stochastic differential equation

$$\mathrm{d}X_t = b(X_t, \alpha)\mathrm{d}t + \sigma(X_t, \beta)\mathrm{d}W_t \tag{4.1}$$

with some initial condition $X_0 = x_0$, where the parameter $\theta = (\alpha, \beta)$ is such that $\theta \in \Theta_\alpha \times \Theta_\beta = \Theta$, $\Theta_\alpha \subset \mathbb{R}^p$, $\Theta_\beta \subset \mathbb{R}^q$, and Θ convex. As usual, $b(\cdot, \cdot)$ and $\sigma(\cdot, \cdot)$ are two known (up to α and β) regular functions such that a solution of (4.1) exists. The process X_t is also assumed to be ergodic for every θ with invariant distribution $\pi_\theta(\cdot)$. Observations are assumed to be equally spaced and such that the discretization step Δ_n shrinks as the number of observations increases: $\Delta_n \to 0$, $n\Delta_n = T \to \infty$ under the rapidly increasing design; i.e., $n\Delta_n^2 \to 0$ as $n \to \infty$. The aim is to try to identify the underlying continuous model on the basis of discrete observations using an information criterion that is a function of the dimension of the parameter space [223]. The Akaike information criterion (AIC) [9], [10] dates back to 1973 and is constructed in such a way that it searches the best model embedded in a wider class of models. It is a likelihood-based method that, roughly speaking, is defined as minus twice the log-likelihood plus twice the dimension of the parameter space. So it is based on the idea that an overspecified model (high dimension of parameter space = too many parameters in the stochastic differential equation) is less

S.M. Iacus, *Simulation and Inference for Stochastic Differential Equations*,
doi: 10.1007/978-0-387-75839-8_4,

valuable than a correctly specified one. Given a class of competing models, the best model is the one that minimizes the AIC criterion. The main assumption is that the true model is currently included among the competing ones; otherwise there is a misspecification problem (see, e.g., [220], [144], [145] for other generalized information criteria). Let $\ell_n(\theta)$ be the log-likelihood of the process. Then the AIC statistic is defined as

$$\mathrm{AIC} = -2\ell_n\left(\hat{\theta}_n^{(ML)}\right) + 2\dim(\Theta),$$

where $\hat{\theta}_n^{(ML)}$ is the true maximum likelihood estimator. Since, as we have seen, there are only a few models for which the explicit expression of $\ell_n(\theta)$ is known, in most of the cases one of the approximated likelihood methods presented in Chapter 3 is needed. The solution proposed in [223] is to consider the approximated log-likelihood function due to Dacunha-Castelle and Florens-Zmirou [62],

$$u_n(\theta) = \sum_{k=1}^{n} u(\Delta_n, X_{i-1}, X_i, \theta)\,, \tag{4.2}$$

where

$$u(t,x,y,\theta) = -\frac{1}{2}\log(2\pi t) - \log\sigma(y,\beta) - \frac{S^2(x,y,\beta)}{2t} + H(x,y,\theta) + t\tilde{g}(x,y,\theta)\,,$$

with

$$\begin{aligned} S(x,y,\beta) &= \int_x^y \frac{\mathrm{d}u}{\sigma(u,\beta)}\,, \\ H(x,y,\theta) &= \int_x^y \frac{B(u,\theta)}{\sigma(u,\beta)}\mathrm{d}u\,, \\ \tilde{g}(x,y,\theta) &= -\frac{1}{2}\left\{C(x,\theta) + C(y,\theta) + \frac{1}{3}B(x,\theta)B(y,\theta)\right\}\,, \\ C(x,\theta) &= \frac{1}{2}B^2(x,\theta) + \frac{1}{2}B_x(x,\theta)\sigma(x,\beta)\,, \\ B(x,\theta) &= \frac{b(x,\alpha)}{\sigma(x,\beta)} - \frac{1}{2}\sigma_x(x,\beta)\,. \end{aligned}$$

Moreover, the following contrast function is defined in order to obtain an asymptotically efficient estimator to plug into the AIC statistic:

$$g_n(\theta) = \sum_{k=1}^{n} g(\Delta_n, X_{i-1}, X_i, \theta)\,,$$

where

$$g(t,x,y,\theta) = -\frac{1}{2}\log(2\pi t) - \log\sigma(x,\beta) - \frac{(y-x-tb(x,\alpha))^2}{2t\sigma^2(x,\beta)}\,.$$

The minimum contrast estimator is then defined as

$$\hat{\theta}_n^{(C)} = \arg\sup_{\theta} g_n(\theta).$$

Further, define the functions

$$\begin{aligned}
s(x, \beta) &= \int_0^x \frac{\mathrm{d}u}{\sigma(u, \beta)}, \\
\tilde{B}(x, \theta) &= B(s^{-1}(x, \beta), \theta), \\
\tilde{h}(x, \theta) &= \tilde{B}^2(x, \theta) + \tilde{B}_x(x, \theta),
\end{aligned}$$

and denote by $\theta_0 = (\alpha_0, \beta_0)$ the true value of the parameter θ. We now introduce the set of assumptions that should be verified by the model in order to obtain the good properties for the AIC statistic.

Assumption 4.1 *The coefficients are such that*

(i) equation (4.1) has a unique strong solution on $[0, T]$*;*
(ii) $\inf_{x,\beta} \sigma^2(x, \beta) > 0$*;*
(iii) X *is ergodic for every* θ *with invariant law* μ_θ *and all moments of* μ_θ *are finite;*
(iv) for all $m \geq 0$ *and for all* θ, $\sup_t \mathbb{E}_\theta |X_t|^m < \infty$*; and*
(v) for every θ, $b(x, \alpha)$ *and* $\sigma(x, \beta)$ *are twice continuously differentiable with respect to* x *and the derivatives are of polynomial growth in* x *uniformly in* θ*;*
(vi) $b(x, \alpha)$ *and* $\sigma(x, \beta)$ *and all their partial derivatives with respect to* x *up to order 2 are three times differentiable with respect to* θ *for all* x *and are of polynomial growth in* x*, uniformly in* θ*.*

Assumption 4.2 *The function* $\tilde{h}(\cdot)$ *is such that*

(i) $\tilde{h}(x, \theta) = O(|x|^2)$ *as* $x \to \infty$*;*
(ii) $\sup_\theta \sup_x |\tilde{h}^3(x, \theta)| \leq M < \infty$*;*
(iii) there exists $\gamma > 0$ *such that for every* θ *and* $j = 1, 2$, $|\tilde{B}^j(x, \theta)| = O(|\tilde{B}(x, \theta)|^\gamma)$ *as* $|x| \to \infty$.

Assumption 4.3 *Almost surely with respect to* $\pi_\theta(\cdot)$ *and for all* x, $b(x, \alpha) = b(x, \alpha_0)$ *implies* $\alpha = \alpha_0$ *and* $\sigma(x, \beta) = \sigma(x, \beta_0)$ *implies* $\beta = \beta_0$.

These assumptions imply the existence of a good estimator and the validity of the approximation of the log-likelihood function (see [137] and [62]), but they also imply that the estimator $\hat{\theta}_n^{(C)}$ is asymptotically efficient [223] and that the following version of the AIC, which will be used in practice, converges to the true AIC statistic (based on the true likelihood and calculated at the true maximum likelihood estimator):

$$\mathrm{AIC} = -2u_n\left(\hat{\theta}_n^{(C)}\right) + 2\dim(\Theta). \tag{4.3}$$

The same result holds true if $\hat{\theta}_n^{(C)}$ is replaced by the approximated maximum likelihood estimator, say $\hat{\theta}_n^{(AML)}$, obtained by direct maximization of (4.2),

$$\text{AIC} = -2u_n\left(\hat{\theta}_n^{(AML)}\right) + 2\dim(\Theta)\,.$$

In most of the cases, though, the estimator $\hat{\theta}_n^{(C)}$ is easier to obtain numerically than $\hat{\theta}_n^{(AML)}$ because g_n is simpler than u_n. Conversely, it is not a good idea to use g_n instead of u_n to build the AIC statistic because the simple Gaussian contrast is in general too rough an approximation of the conditional density as we discussed in Chapter 3. Numerical evidence about the discrepancy of g_n and u_n from the true likelihood was shown in [223]. In principle, any other good approximation of the likelihood can be used in (4.3) and, as suggested by the authors of this method, a good candidate is the Aït-Sahalia method based on Hermite polynomial expansion discussed in Section 3.3.3.

The package `sde` contains the `sdeAIC` function which evaluates the AIC statistics but also estimates the underlying model using the function $g_n(\cdot)$. The code of the `sdeAIC` function is given in Listing 4.1. Notice that with `optim` we minimize $-g_n(\cdot)$ instead of maximizing $g_n(\cdot)$. The function needs a specification of the model and the data `X`. The model is specified giving the functions $b(\cdot,\cdot)$ and $\sigma(\cdot,\cdot)$ as functions of `theta` and `x`, where `theta` is a vector of the parameters. The minimum setup to use this function is the specification of $b(\cdot,\cdot)$, $\sigma(\cdot,\cdot)$, and their derivatives or, as an alternative to explicit derivatives, the functions `B` and `B.x` (respectively $B(\cdot,\cdot)$ and $B_x(\cdot,\cdot)$). The functions `H` and `S`, if missing, are evaluated numerically. This is the easiest way to specify all the ingredients to evaluate the AIC statistics according to previous theory. The following example clarifies the use of the `sdeAIC` function. The following code calculates AIC after estimation of parameters of the process

$$\mathrm{d}X_t = -(X_t - 10)\mathrm{d}t + 2\sqrt{X_t}\mathrm{d}W_t, \quad X_0 = 10\,.$$

Please note that the functions passed to `sdeAIC` should be able to accept vectors and return vectors to speed up the execution of the code. The reader should check the definition of `b.x`.

```
> # ex4.01.R
> set.seed(123)
>
> dri <- expression(-(x-10))
> dif <- expression(2*sqrt(x))
> sde.sim(X0=10,drift=dri, sigma=dif,N=1000,delta=0.1) -> X
>
> b <- function(x,theta) -theta[1]*(x-theta[2])
> b.x <- function(x,theta)  -theta[1]+0*x
>
> s <- function(x,theta) theta[3]*sqrt(x)
> s.x <- function(x,theta) theta[3]/(2*sqrt(x))
> s.xx <- function(x,theta) -theta[3]/(4*x^1.5)
>
> # we let sdeAIC calculate the estimates and the AIC statistics
> sdeAIC(X, NULL, b, s, b.x, s.x, s.xx, guess=c(1,1,1),
```

```
+              lower=rep(1e-3,3), method="L-BFGS-B")
estimating the model...
[1] 1.123802 9.171913 0.847953
AIC value:
[1] 3364.784
```

We now run a Monte Carlo experiment that is borrowed from [223]. We simulate 15 trajectories from the same process as before,

$$dX_t = -(X_t - 10)dt + 2\sqrt{X_t}dW_t\,,$$

with initial value $X_0 = 10$. We use $n = 1000$ and $\Delta = 0.1$. We test the performance of the AIC statistics for the three competing models

$$\begin{aligned} dX_t &= -\alpha_1(X_t - \alpha_2)dt + \beta\sqrt{X_t}dW_t && \text{(true model)},\\ dX_t &= -\alpha_1(X_t - \alpha_2)dt + \sqrt{\beta_1 + \beta_2 X_t}dW_t && \text{(competing model 1)},\\ dX_t &= -\alpha_1(X_t - \alpha_2)dt + (\beta_1 + \beta_2 X_t)^{\beta_3}dW_t && \text{(competing model 2)}, \end{aligned}$$

Then, at each replication, we let the `sdeAIC` function estimate the parameters of each model and calculate the corresponding AIC statistics. After the replications, we count how many times the true model has been selected using AIC. Some of the output has been skipped.

```
> # ex4.02.R
> set.seed(123)
> n.sim <- 15
> aic <-matrix(, n.sim, 3)
> for(i in 1:n.sim){
+ dri <- expression(-(x-10))
+ dif <- expression(2*sqrt(x))
+ sde.sim(X0=10,drift=dri, sigma=dif,N=1000,delta=0.1) -> X
+
+ b <- function(x,theta) -theta[1]*(x-theta[2])
+ b.x <- function(x,theta)  -theta[1]+0*x
+
+ s <- function(x,theta) theta[3]*sqrt(x)
+ s.x <- function(x,theta) theta[3]/(2*sqrt(x))
+ s.xx <- function(x,theta) -theta[3]/(4*x^1.5)
+ aic[i,1] <- sdeAIC(X, NULL, b, s, b.x, s.x, s.xx, guess=c(1,1,1),
+                            lower=rep(1e-3,3), method="L-BFGS-B")
+
+ s <- function(x,theta) sqrt(theta[3]*+theta[4]*x)
+ s.x <- function(x,theta) theta[4]/(2*sqrt(theta[3]+theta[4]*x))
+ s.xx <- function(x,theta) -theta[4]^2/(4*(theta[3]+theta[4]*x)^1.5)
+ aic[i,2] <- sdeAIC(X, NULL, b, s, b.x, s.x, s.xx, guess=c(1,1,1,1),
+                         lower=rep(1e-3,4), method="L-BFGS-B")
+
+ s <- function(x,theta) (theta[3]+theta[4]*x)^theta[5]
+ s.x <- function(x,theta)
+       theta[4]*theta[5]*(theta[3]+theta[4]*x)^(-1+theta[5])
+ s.xx <- function(x,theta)
+    (theta[4]^2*theta[5]*(theta[5]-1)*(theta[3]+theta[4]*x)^(-2+theta[5]))
+ aic[i,3] <- sdeAIC(X, NULL, b, s, b.x, s.x, s.xx, guess=c(1,1,1,1,1),
+                            lower=rep(1e-3,5), method="L-BFGS-B")
+ }
```

Now the matrix `aic` contains three columns (one for each model) and 15 rows (one for each replication). Each element of the matrix `aic` is one of the AIC statistics.

```
> print(aic)
          [,1]      [,2]      [,3]
 [1,]  3364.784  3365.884  3449.916
 [2,]  7539.537  7540.566  7299.706
 [3,]  2706.390  2708.315  2743.830
 [4,]  1650.316  1652.973  1677.536
 [5,]  1858.206  1861.913  1859.676
 [6,]  3160.675  3161.724  3184.985
 [7,]  3807.816  3809.665  3886.972
 [8,] 15163.537 15164.077 14947.754
 [9,]  3024.119  3028.236  3076.922
[10,] 23573.548 23582.335 27507.025
[11,] 32712.191 32723.685 36261.744
[12,]  1953.586  1957.112  1912.955
[13,]  1432.877  1436.096  1451.208
[14,] 12802.626 12802.338 12457.432
[15,]  6113.929  6114.784  6243.610
> table(apply(aic,1,function(x) which(x==min(x))))

 1  3
11  4
```

For the last `table` command, it emerges that the first model has been selected 11 times over 15. The original simulation study was made on a reasonable number of replications and for different values of Δ_n. Here we just wanted to show that the AIC criterion is reasonably good even if, trajectory to trajectory, it may select the wrong model.

```
sdeAIC <- function(X, theta, b, s, b.x, s.x, s.xx,
            B, B.x, H, S, guess, ...){
        n <- length(X)
        DELTA <- deltat(X)

        if(missing(theta) || is.null(theta)){
         if(missing(guess))
          stop("cannot estimate the model. \
             Specify initial guess values for theta")
         g <- function(theta,X,drift,sigma){
            sum(log(sigma(X[-n],theta) +
              (diff(X)-DELTA*drift(X[-n],theta))^2/
              (2*DELTA*sigma(X[-n],theta)^2)))
          }
         cat("estimating the model...\n")
         est <- optim(guess,g,drift=b,sigma=s,X=X,...)
         theta <- est$par
         print(theta)
        }

        if(missing(B))
          B <- function(x,theta) b(x,theta)/s(x,theta) - 0.5*s.x(x,theta)
         if(missing(B.x) &&
           !( missing(b.x) || missing(s.x) || missing(s.xx))){
          B.x <- function(x,theta) (b.x(x,theta)/s(x,theta)
             - b(x,theta)*s.x(x,theta)/(s(x,theta)^2)
              -0.5*s.xx(x,theta))
        } else {
          stop("error")
        }

        C1 <- function(x) (B(x,theta)^2)/3 + 0.5*B.x(x,theta)*s(x,theta)
        g <- function(x,y) -0.5*(C1(x)+C1(y)+B(x,theta)*B(y,theta)/3)
        h <- function(x) B(x,theta)/s(x,theta)
        if(missing(H))
         H <- function(x,y) integrate(h, x, y)$value
        s1 <- function(x) 1/s(x,theta)
```

```
    if(missing(S))
     S <- function(x,y) integrate(s1, x, y)$value

    u <- function(x,y,theta) (-0.5*log(2*pi*DELTA) - log(s(y,theta))
        - S(x,y)^2/(2*DELTA)  + H(x,y) + DELTA*g(x,y))
    cat("AIC value:\n")
    -2*sum(u(X[1:(n-1)],X[2:n],theta)) + 2*length(theta)
}
```

Listing 4.1. Akaike's information criterion for diffusion processes.

4.2 Nonparametric estimation

When there is no specific reason to specify a parametric form for either the diffusion or the drift coefficient or both, nonparametric methods help in the identification of the diffusion model. In this section, we review, without going too much into details, some nonparametric techniques that are easy to use inside R. The main inference problems are related to invariant density function estimation and/or drift and diffusion coefficients. Nice reviews on the subject can be found in [218], [81], and [128]. Justification of the nonparametric approach to inference for diffusion processes can also be found in [4], which contains an extensive analysis of how poorly standard parametric models for the interest rate fit actual historical data. In finance, misspecification of the underlying interest rate model can lead to serious pricing and hedging errors[1] (see, e.g., [46]). Let us consider the ergodic diffusion process X solution to

$$\mathrm{d}X_t = b(X_t)\mathrm{d}t + \sigma(X_t)\mathrm{d}W_t,$$

where $b(\cdot)$ and $\sigma(\cdot)$ satisfy the usual assumptions of regularity and Assumption 1.5 holds true. Our primary interest is now the invariant density $\pi(x)$. As in the i.i.d. case, a simple kernel type estimator can be used. Let K be a nonnegative function such that

$$\int K(u)\mathrm{d}u = 1$$

and K is bounded and twice continuously differentiable on $\mathbb{R}$. K and its derivatives are supposed to be in $L^2(\mathbb{R})$. Such a function K is called a *kernel* of order $r > 1$ if there exists an integer r such that

$$\int_{-\infty}^{+\infty} x^i K(x)\mathrm{d}x = 0, \quad i = 1, \dots, r-1,$$

and

$$\int_{-\infty}^{+\infty} x^r K(x)\mathrm{d}x \neq 0, \quad \int_{-\infty}^{+\infty} |x|^r |K(x)|\mathrm{d}x < \infty\,.$$

We assume K to be of order 2 and we further define

[1] In this section, we do not focus on generic functionals of the diffusion process.

$$K_h(u) = \frac{1}{h} K\left(\frac{u}{h}\right)$$

and notice that

$$\lim_{h\to\infty} K_h(u) = \delta(u)\,,$$

where δ is the Dirac delta.

4.2.1 Stationary density estimation

The estimator

$$\hat{\pi}_n(x) = \frac{1}{nh_n}\sum_{i=1}^{n} K\left(\frac{x-X_i}{h_n}\right) = \frac{1}{n}\sum_{i=1}^{n} K_{h_n}(x-X_i) \tag{4.4}$$

is the kernel estimator of $\pi(x)$ and will be used in this framework. Usually the Gaussian kernel

$$K(u) = \frac{1}{\sqrt{2\pi}}\exp\left(-\frac{1}{2}u^2\right)$$

is used, but any other reasonable kernel can be considered without loss in the optimality results (see, e.g., [79] and [210]). The more critical choice is known to be the bandwidth h_n. The *bandwidth* h_n is a shrinking sequence with n; i.e., $h_n \to 0$ as $n \to \infty$. For general m-dimensional densities, the bandwidth is usually chosen according to Scott's rule [203], which assumes h_n to be proportional to $d\cdot n^{-\frac{1}{m+4}}$, where d is the standard deviation of the time series (in our case $m = 1$). This seems to be the standard choice in econometrics according to [218]. In [3], the choice of the bandwidth is $h_n = c_n n^{-1/4.5}$, where c_n is c times the standard deviation of the data divided by $\log(n)$ and c is chosen to minimize the mean integrated square error (MISE) of the estimator of the density. Other approaches to bandwidth selection are available, such as the cross-validation strategy [199]. An extensive guide on the bandwidth selection problem can be found in [105]. Under the assumption $\lim_{n\to\infty} nh^{4.5} = 0$ and mild regularity conditions (see [3] and [192]), the stationary density estimator $\hat{\pi}_n$ behaves as in the i.i.d. setting. In particular, we have

$$\sqrt{nh_n}\,(\hat{\pi}_n(x) - \pi(x)) \xrightarrow{d} \mathcal{N}(0, V_x),$$

where

$$V_x = \pi(x)\int_{-\infty}^{+\infty} K^2(u)\mathrm{d}u.$$

Moreover, for $x_1 \neq x_2$, the estimators $\hat{\pi}_n(x_1)$ and $\hat{\pi}_n(x_2)$ are asymptotically independent. This approach does not require $\Delta_n \to 0$. The next listing simulates a Cox-Ingersoll-Ross model,

$$\mathrm{d}X_t = (\theta_1 - \theta_2 X_t)\mathrm{d}t + \theta_3\sqrt{X_t}\mathrm{d}W_t\,,$$

with $\theta = (6, 2, 1)$, and estimates the corresponding stationary density of the CIR model, which can be obtained using the `dsCIR` function of the `sde` package. R implements many different kernel functions K with several automatic bandwidth selection criteria via the `density` function. For didactic purposes, we show how to construct a kernel function in a vectorized form and then we plot the true stationary density against the estimated one.

```
> # ex4.03.R
> set.seed(123)
> theta <- c(6,2,1)
> X <- sde.sim(X0=rsCIR(1, theta),model="CIR",theta=theta,N=1000,delta=1)
> f <-function(x) dsCIR(x, theta)
```

We now choose the bandwidth according to Scott's rule.

```
> h <- length(X)^(-1/5)*sd(X)
> K <-function(x) exp(-0.5*x^2)/sqrt(2*pi)
> p <-function(x) sapply(x, function(x) mean(K((x-X)/h)))/h
```

Then we plot the true density, our version of the kernel estimator and R's implementation of kernel estimators. The last two commands return the same result.

```
> curve(f,0,8)
> curve(p,0,8, col="red", add=TRUE,lty=2)
> lines(density(X,bw=h),col="green",lty=3)
```

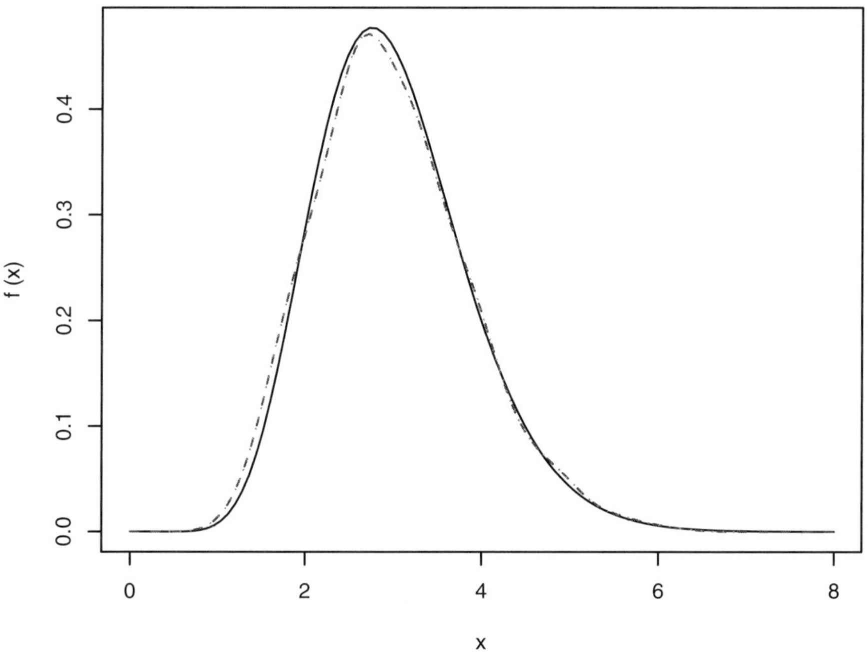

Fig. 4.1. Implementation of the kernel estimator for the stationary distribution of the CIR model.

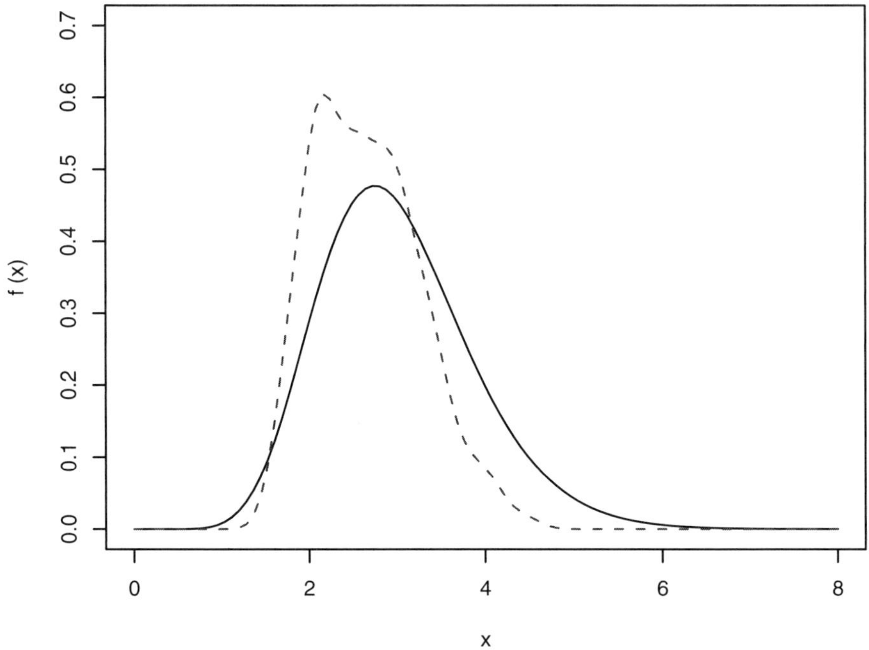

Fig. 4.2. Effect of the bandwidth h_n and the mesh Δ_n on the kernel estimator.

Figure 4.1 shows the good performance of the simple kernel estimator, for $\pi(x)$. Even if in principle the approach does not require $\Delta_n = 1$, empirically it can be seen that the quality of the estimates is less accurate for values of $\Delta_n < 1$. The next example shows such an effect.

```
> # ex4.03.R (cont)
> set.seed(123)
> X <- sde.sim(X0=rsCIR(1,theta),model="CIR",theta=theta,N=1000,delta=0.01)
> h <- length(X)^(-1/5)*sd(X)
> curve(f,0,8,ylim=c(0,0.7))
> curve(p,0,8, col="red", add=TRUE,lty=2)
```

This is not really an issue if the observations are not collected in a very high frequency scheme or if the sample size is relatively high, but still it is always good to play with the bandwidth a bit to figure out what the underlying unknown stationary density looks like. In many cases, nonparametric density (or regression) estimation is a preliminary (explorative data) analysis that helps in finding a possible parametric model for the data when no prior information on the model that generates the data is not available. Just to show that the previous effect shown by Figure 4.2 is not due to the simulation part of the code, the user can run the following example for the same Δ_n with an increased number of observations; e.g., $n = 15000$. The corresponding plot of the estimated density will look like Figure 4.1.

```
> # ex4.03.R (cont)
```

```
> set.seed(123)
> X <- sde.sim(X0 = rsCIR(1,theta), model = "CIR", theta = theta,
+              N = 15000, delta = 0.01)
> h <- length(X)^(-1/5)*sd(X)
> curve(f,0,8,ylim=c(0,0.7))
> curve(p,0,8, col="red", add=TRUE,lty=2)
```

4.2.2 Local-time and stationary density estimators

A relationship between h_n and Δ_n was established in [18]. The result is given in terms of *local-time* estimation also for nonstationary processes. In this case, the local-time estimator generalizes the stationary density estimator. The result in [18] is given for general càdlàg processes[2] X_t. Here, we consider only diffusion processes without jumps, and hence the local time is defined as

$$L_X(T,x) = \lim_{\epsilon\to 0}\frac{1}{\epsilon}\int_0^T \mathbf{1}_{[x,x+\epsilon)}(X_s)\mathrm{d}{<}X,X{>}_s, \tag{4.5}$$

where ${<}X,X{>}_s$ is the quadratic variation process and x is the state space of the process. The local time is intuitively the amount of time a process sojourns in a neighborhood of x between time 0 and T. We know from Chapter 3 that, for continuous diffusion processes, $\mathrm{d}{<}X,X{>}_t = \sigma^2(X_t)\mathrm{d}t$. From this it follows that the local time can be transformed into the so-called *chronological* local time (see, e.g., [39], [185]), defined as

$$\bar{L}_X(T,x) = \frac{1}{\sigma^2(x)}L_X(T,x)\,. \tag{4.6}$$

The difference between L_X and $\bar{L}_X$ is that the local time in (4.5) is the amount of time expressed in time units of the quadratic variation process, while the chronological local time in (4.6) is expressed in terms of real time units, which is the time we deal with in estimation. The relationship in (4.6) is interesting because it is related to the occupation measure of the process X,

$$\eta_A^T = \int_0^T \mathbf{1}_A(X_s)\mathrm{d}s = \int_A \bar{L}_X(T,x)\mathrm{d}x\,.$$

Therefore, $\bar{L}_X(T,x)$ is a version of the Radon-Nikodým derivative of this measure. Moreover, from the preceding formula, we have that

$${<}X,X{>}_t = \int_{-\infty}^{+\infty} L_X(t,x)\mathrm{d}x\,,$$

which closes the circle.

---

[2] A càdlàg process is a process such that its trajectories are continuous to the right with left limit.

Fact 4.1 ([87] and [18]) *If $h_n \to 0$ and $n \to \infty$ with fixed $T = \bar{T}$ in such a way that*

$$\frac{1}{h_n}\sqrt{\Delta_n \log \frac{1}{\Delta_n}} = o(1),$$

then

$$\hat{\hat{L}}_X(T,x) = \frac{\Delta_n}{h_n}\sum_{i=1}^n K\left(\frac{X_i - x}{h_n}\right) \stackrel{a.s.}{\to} \bar{L}_X(T,x)\,.$$

The result above shows the limiting quantity is a random object, which is not what happens in standard kernel density estimation, and this is not surprising. The relation with kernel density estimation is given in the next result.

Fact 4.2 ([18]) *If $h_n \to 0$, $T = n\Delta_n \to \infty$ as $n \to \infty$ such that*

$$\frac{T}{h_n}\sqrt{\Delta_n \log \frac{1}{\Delta_n}} = o(1)\,.$$

Then

$$\frac{\hat{\hat{L}}_X(T,x)}{T} = \frac{\hat{\hat{L}}_X(T,x)}{n\Delta_n} = \frac{1}{nh_n}\sum_{i=1}^n K\left(\frac{X_i - x}{h_n}\right) = \hat{\pi}_n(x) \stackrel{a.s.}{\to} \pi(x)\,.$$

Notice that $\hat{\hat{L}}_X(T,x)/T$ is also an estimator of the expected local time.

4.2.3 Estimation of diffusion and drift coefficients

We saw in Section 1.13 that the drift and diffusion coefficients are related to the stationary density π via the forward and backward Kolmogorov equations. In particular, from the Kolmogorov forward equation (1.26), by subsequent integrations, it is possible to derive the following relationship: [19]

$$\sigma^2(x) = \frac{2}{\pi(x)}\int_0^x b(u)\pi(u)\mathrm{d}u\,.$$

Given as the estimator the kernel estimator (4.4) of π, if $b(\cdot)$ is known or has a parametric form for which consistent estimators for the parameters exist, it is possible to use the estimator

$$\hat{\sigma}_n^2(x) = \frac{2}{\hat{\pi}_n(x)}\int_0^x b(u)\hat{\pi}_n(u)\mathrm{d}u\,,$$

where $b(x)$ can eventually be replaced by $b(x;\hat{\theta}_n)$ if it has the parametric form $b = b(x;\theta)$, where $\hat{\theta}_n$ is a $\sqrt{n}$-consistent estimator of θ.

Assumption 4.4

$$\lim_{x\to 0}\sigma(x)\pi(x)=0 \qquad \textit{or} \qquad \lim_{x\to\infty}\sigma(x)\pi(x)=0$$

and

$$\lim_{x\to 0}\left|\frac{\sigma(x)}{2b(x)-\sigma(x)\sigma_x(x)}\right|<\infty \qquad \textit{or} \qquad \lim_{x\to\infty}\left|\frac{\sigma(x)}{2b(x)-\sigma(x)\sigma_x(x)}\right|<\infty\,.$$

The conditions in Assumption 4.4 imply geometric ergodicity [104], which in turn implies the following mixing condition on the observed data (see, e.g., [191] and [3]).

Assumption 4.5 *The observed data X_i, $i=1,\dots,n$, is a strictly stationary β-mixing sequence satisfying $k^\delta\beta_k\to 0$ and $k\to\infty$ for some $\delta>1$.*

Assumption 4.6 *As $n\to\infty$ and $h_n\to 0$, we have*

$$\sqrt{nh_n^{2r+1}}\to 0 \quad \textit{and} \quad nh_n\to\infty \quad \textit{and} \quad nh_n^3\to\infty\,,$$

where r is the order of the kernel K.

Fact 4.3 ([3]) *Suppose Assumptions 4.5 and 4.6 hold true and $\sigma^2(x)>0$. Assume that the drift $b(x)$ is known (or $b(x;\theta)$ unknown up to a finite-dimensional parameter θ) and $\sigma(x)$ is differentiable with continuous derivatives on $(0,\infty)$ of order greater than or equal to 2. Then*

$$\sqrt{nh_n}\left(\hat\sigma_n^2(x)-\sigma^2(x)\right)\xrightarrow{d}\mathcal{N}(0,V_x)\,,$$

where

$$V_x=\frac{\sigma^4(x)}{\pi(x)}\int_{-\infty}^{+\infty}K^2(u)\mathrm{d}u\,.$$

Moreover, a consistent estimator of V_x is given by

$$\hat V_x=\frac{\left(\hat\sigma_n^2(x)\right)^2}{\hat\pi_n(x)}\int_{-\infty}^{+\infty}K^2(u)\mathrm{d}u,$$

and for any two values $x\neq y$ the estimates $\hat\sigma_n^2(x)$ and $\hat\sigma_n^2(y)$ are independent.

The result above is interesting in real-life applications only when the drift $b(x)$ has at least a parametric form $b(x)=b(x;\theta)$, but it is hardly reasonable to assume that $\sigma(x)$ is unknown and $b(x)$ is completely known. On the contrary, one can imagine as in [3] or [218] that the process is, for example, mean reverting and hence assume a model such as

$$\mathrm{d}X_t=\theta_1(\theta_2-X_t)\mathrm{d}t+\sigma(X_t)\mathrm{d}W_t$$

with $b(x;\theta)=\theta_1(\theta_2-x)$. Therefore, it is possible to consistently estimate $\theta=(\theta_1,\theta_2)$, use the kernel estimator $\hat\pi_n(x)$, and finally estimate the diffusion coefficient with $\hat{\sigma}_n^2(x)$. But in all other cases in which such simple forms for the drift are not available, it is possible to estimate the diffusion coefficient directly without information on the drift b or by nonparametric estimation of the drift. This approach requires a high-frequency asymptotic. We start with the Florens-Zmirou [87] estimator

$$\hat{\sigma}_n^2(x) = \frac{\sum_{i=0}^{n-1} K\left(\frac{x-X_i}{h_n}\right)(X_{i+1}-X_i)^2}{\Delta_n \sum_{i=0}^{n-1} K\left(\frac{x-X_i}{h_n}\right)} . \tag{4.7}$$

In the original paper, K is the uniform kernel. Then Jiang and Knight [127] extended the Florens-Zmirou results to the Gaussian kernel, and finally Bandi and Phillips [18] completed the theory by extending the results to general stationary processes and general kernels. Many other works have considered the problem of diffusion coefficient estimation by different techniques, including wavelet methods, nearest neighbor, simulated annealing, adaptive estimation, etc. We cannot discuss all these approaches, but the reader might want to consider at least the following references as a starting point: [44], [94], [95],[96], [45], [108], [123]. The Florens-Zmirou estimator (4.7), and its variations, require $\Delta_n \to 0$ at some proper rate in order to have consistency and asymptotic normality. In the same way, a nonparametric drift estimator can be obtained as follows:

$$\hat{b}_n(x) = \frac{\sum_{i=0}^{n-1} K\left(\frac{x-X_i}{h_n}\right)(X_{i+1}-X_i)}{\Delta_n \sum_{i=0}^{n-1} K\left(\frac{x-X_i}{h_n}\right)} . \tag{4.8}$$

In Stanton's approach [218], the two estimators are nothing but Nadaraya-Watson kernel regression estimators of the following conditional expectations

$$b(x) = \lim_{t\to 0} \frac{1}{t}\mathbb{E}\{X_t - x | X_0 = x\},$$
$$\sigma^2(x) = \lim_{t\to 0} \frac{1}{t}\mathbb{E}\{(X_t - x)^2 | X_0 = x\}.$$

In this approach, $b(x)$ and $\sigma^2(x)$ are seen as *instantaneous* conditional means and variances of the process when $X_0 = x$. The two quantities can be rewritten, for fixed Δ_n, as

$$\begin{aligned} b(x) &= \frac{1}{\Delta_n}\mathbb{E}\{X_{i+1} - X_i | X_i = x\} + \frac{o(\Delta_n)}{\Delta_n}, \\ \sigma^2(x) &= \frac{1}{\Delta_n}\mathbb{E}\{(X_{i+1} - X_i)^2 | X_i = x\} + \frac{o(\Delta_n)}{\Delta_n}. \end{aligned} \tag{4.9}$$

From these expressions and given the estimator $\hat{\pi}_n$, the two estimators (4.7) and (4.8) are readily obtained. The discretizations above are obtained using the Itô-Taylor expansion (3.19),

$$\mathbb{E}\{\phi(X_{i+1})|\mathcal{F}_i\} = \phi(X_i) + \Delta_n \mathcal{L}\phi(X_i) + \frac{1}{2}\Delta_n^2 \mathcal{L}^2\phi(X_i) + \cdots ,$$

where $\mathcal{L}$ is the infinitesimal generator of the diffusion and $\phi(x)$ an appropriate function (see, e.g., [107]). Contrary to what we saw in Chapter 3, in this case the unknowns are the coefficients $b(\cdot)$ and $\sigma(\cdot)$, and the function $\phi(\cdot)$ is chosen in order to obtain estimators of them. The Itô-Taylor expansion can be rewritten as

$$\mathcal{L}\phi(X_i) = \frac{1}{\Delta_n}\mathbb{E}\{\phi(X_{i+1}) - \phi(X_i)|\mathcal{F}_i\} - \frac{1}{2}\Delta_n \mathcal{L}^2\phi(X_i) - \frac{1}{6}\Delta_n^2 \mathcal{L}^3\phi(X_i) + \cdots .$$

At first order, we have

$$\mathcal{L}\phi(X_i) = \frac{1}{\Delta_n}\mathbb{E}\{\phi(X_{i+1}) - \phi(X_i)|\mathcal{F}_i\} + O(\Delta_n) .$$

From the definition of $\mathcal{L}\phi(x) = \phi_x(x)b(x) + \frac{1}{2}\phi_{xx}(x)\sigma^2(x)$, we have that, choosing

$$\phi_1(x) = x, \qquad \phi_2(x) = (x - X_i)^2,$$

we obtain

$$\mathcal{L}\phi_1(X_i)b(X_i) , \qquad \mathcal{L}\phi_2(X_i) = \sigma^2(X_i),$$

from which (4.9) follows. But the Itô-Taylor expansion approach allows for further refinements in the approximation. If we consider a time step of $2\Delta_n$ and perform the expansion up to order 2, after simple calculations we obtain

$$\mathcal{L}\phi(X_i) = \frac{1}{2\Delta_n}\Big(4\mathbb{E}\{\phi(X_{i+1})-\phi(X_i)|\mathcal{F}_i\}-\mathbb{E}\{\phi(X_{i+2})-\phi(X_i)|\mathcal{F}_i\}\Big)+O(\Delta_n^2) ,$$

from which we derive as second-order approximations for $b(\cdot)$ and $\sigma^2(\cdot)$

$$\begin{aligned} b(x) &= \frac{1}{2\Delta_n}\Big(4\mathbb{E}\{X_{i+1} - X_i|X_i = x\} - \mathbb{E}\{X_{i+2} - X_i|X_i = x\}\Big) + O(\Delta_n^2) , \\ \sigma^2(x) &= \frac{1}{2\Delta_n}\Big(4\mathbb{E}\{(X_{i+1} - X_i)^2|X_i = x\} - \mathbb{E}\{(X_{i+2} - X_i)^2|X_i = x\}\Big) \\ &\quad + O(\Delta_n^2), \end{aligned}$$

and similarly, using third-order approximations, we derive

$$\begin{aligned} b(x) =& \frac{1}{6\Delta_n}\Big(18\mathbb{E}\{X_{i+1} - X_i|X_i = x\} - 9\mathbb{E}\{X_{i+2} - X_i|X_i = x\} \\ &+ 2\mathbb{E}\{X_{i+3} - X_i|X_i = x\}\Big) + O(\Delta_n^3) , \\ \sigma^2(x) =& \frac{1}{6\Delta_n}\Big(18\mathbb{E}\{(X_{i+1} - X_i)^2|X_i = x\} - 9\mathbb{E}\{(X_{i+2} - X_i)^2|X_i = x\} \\ &+ 2\mathbb{E}\{(X_{i+3} - X_i)^2|X_i = x\}\Big) + O(\Delta_n^3) \end{aligned}$$

and kernel estimators for $\mathbb{E}\{(X_{i+j} - X_i)^k | X_i = x\}$ can be used. More frequently, $\mathbb{E}\{(X_{i+j} - X_i)^2 | X_i = x\}$ are replaced by the conditional variance of X_{i+j} without changing the order of the approximation. Therefore, we have

$$
\begin{aligned}
\sigma^2(x) &= \frac{1}{\Delta_n} \mathrm{Var}\{X_{i+1}|X_i = x\} + O(\Delta_n), \\
\sigma^2(x) &= \frac{1}{2\Delta_n} \Big(4\mathrm{Var}\{X_{i+1}|X_i = x\} - \mathrm{Var}\{X_{i+2}|X_i = x\} \Big) + O(\Delta_n^2), \\
\sigma^2(x) &= \frac{1}{6\Delta_n} \Big(18\mathrm{Var}\{X_{i+1}|X_i = x\} - 9\mathrm{Var}\{X_{i+2}|X_i = x\} \\
&\qquad + 2\mathrm{Var}\{X_{i+3}|X_i = x\} \Big) + O(\Delta_n^3).
\end{aligned}
$$

The code in Listing 4.2 implements the nonparametric estimation of drift and diffusion coefficients using the estimators (4.8) and (4.7), respectively, and also an optimized version of the stationary density estimator (4.4). It makes use of the functions `ksdrift`, `ksdiff`, and `ksdens` in package `sde`. These functions are self-explanatory and are coded to be compatible with the standard functions `density` and `ksmooth` in the base R package. In particular, `ksdens` returns an object of class `density`, and the two other functions return a list of `x` and `y` coordinates that can be printed on the graphic device using the functions `plot`, `lines`, or `points` as for the output of `density` and `ksmooth`. The bandwidth is calculated using Scott's rule.

```
> # ex4.04.R
> set.seed(123)
> theta <- c(6,2,1)
> X <- sde.sim(X0 = rsCIR(1, theta), model="CIR", theta = theta,
+              N = 1000, delta = 0.1)
>
> f <-function(x) dsCIR(x, theta)
> b <- function(x) theta[1]-theta[2]*x
> sigma <- function(x) theta[3]*sqrt(x)
>
> minX <- min(X)
> maxX <- max(X)
>
> par(mfrow=c(2,1))
> curve(b,minX,maxX,main="drift coefficient")
> lines(ksdrift(X),lty=3,col="red",lwd=2)
>
> curve(sigma,minX, maxX,main="diffusion coefficient")
> lines(ksdiff(X),lty=3,col="red",lwd=2)
```

Figure 4.3 shows the performance of the nonparametric estimators of the drift and diffusion coefficients for the Cox-Ingersoll-Ross model simulated in the previous listing.

```
ksdrift <- function(x,bw,n=512){
 len <- length(x)
 xval <- seq(min(x), max(x), length=n)
 if(missing(bw))
  bw <- len^(-1/5)*sd(x)
  y <- sapply(xval, function(xval) {
  tmp <- dnorm(xval, x[1:(len-1)], bw)
```

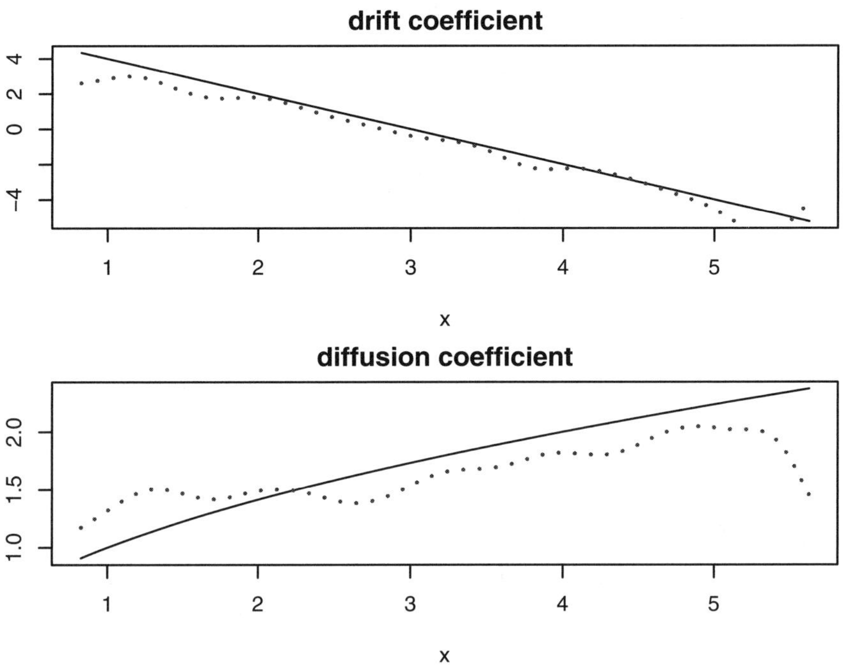

Fig. 4.3. Nonparametric estimators of the drift and diffusion coefficients for the CIR process.

```
  sum(tmp * diff(x)) / (deltat(x) * sum(tmp))})
  invisible(list(x=xval, y=y))
}

ksdiff <- function(x,bw,n=512){
 len <- length(x)
 xval <- seq(min(x), max(x), length=n)
 if(missing(bw))
  bw <- len^(-1/5)*sd(x)
  y <- sapply(xval, function(xval) {
  tmp <- dnorm(xval, x[1:(len-1)], bw)
  sum(tmp * as.numeric(diff(x))^2) / (deltat(x) * sum(tmp))})
  invisible(list(x=xval, y=sqrt(y)))
}

ksdens <- function(x,bw,n=512){
   len <- length(x)
   if(missing(bw))
    bw <- len^(-1/5)*sd(x)
   invisible(density(x,bw=bw,n=n))
}
```

Listing 4.2. Nonparametric estimator for the drift and diffusion coefficients and the stationary density of a diffusion.

4.3 Change-point estimation

Change-point estimation consists in the identification of the instant in which a change occurs in the parameter of some model. There are several approaches to the solution of this problem, and here we consider a least squares solution (see, e.g., [15], [119], [50]), but other approaches, such as maximum likelihood change-point estimation, are also possible (see, e.g., [61], [16]). We assume we have a diffusion process[3] solution to

$$\mathrm{d}X_t = b(X_t)\mathrm{d}t + \theta\sigma(X_t)\mathrm{d}W_t\,, \tag{4.10}$$

where $b(\cdot)$ and $\sigma(\cdot)$ are known functions and $\theta \in \Theta \subset \mathbb{R}$ is the parameter of interest. As in [64], given discrete observations from (4.10) on $[0, T = n\Delta_n]$, we want to identify retrospectively if and when a change in value of the parameter θ occurred and estimate consistently the parameter before and after the change point. The asymptotics is $\Delta_n \to 0$ as $n \to \infty$ and $n\Delta_n = T$ fixed.[4] For simplicity, we assume that the change occurs at instant k_0, which is one of the integers in $1, \ldots, n$. This is a problem of volatility change-point estimation that frequently occurs in finance applications. We assume that $\theta = \theta_1$ before the time change and $\theta = \theta_2$ after the time change with $\theta_1 < \theta_2$ (but this does not matter in the final results).

In order to obtain a simple least squares estimator, we use Euler approximation. So, from now on, we assume all the hypotheses necessary to have the Euler approximation in place. Namely, we can write the Euler scheme as

$$X_{i+1} = X_i + b(X_i)\Delta_n + \theta\sigma(X_i)(W_{i+1} - W_i)$$

and introduce the standardized residuals

$$Z_i = \frac{(X_{i+1} - X_i) - b(X_i)\Delta_n}{\sqrt{\Delta_n}\sigma(X_i)} = \theta\frac{(W_{i+1} - W_i)}{\sqrt{\Delta_n}}\,.$$

The Z_i's are i.i.d. Gaussian random variables. The change-point estimator is obtained as

$$\hat{k}_0 = \arg\min_k \left(\min_{\theta_1,\theta_2} \left\{ \sum_{i=1}^{k} (Z_i^2 - \theta_1^2)^2 + \sum_{i=k+1}^{n} (Z_i^2 - \theta_2^2)^2 \right\} \right) \tag{4.11}$$

with , $k = 2, \ldots, n-1$. We denote by $[x]$ the integer part of the real x, and sometimes we write $k_0 = [n\tau_0]$ and $k = [n\tau]$, $\tau, \tau_0 \in (0,1)$ to indicate the change point in the continuous timescale. Define the partial sums

---

[3] For continuous-time observations this problem was studied in [152]. A bayesian approach for discrete-time observations can be found in [158].

[4] For ergodic diffusion processes and $n\Delta_n = T \to \infty$, under additional mild regularity conditions, the results mentioned here are still valid.

$$S_n = \sum_{i=1}^{n} Z_i^2, \quad S_k = \sum_{i=1}^{k} Z_i^2, \quad S_{n-k} = \sum_{i=k+1}^{n} Z_i^2,$$

and denote by $\bar{\theta}_1^2$ and $\bar{\theta}_2^2$ the *initial* least squares estimators of θ_1^2 and θ_2^2 for any given value of k in (4.11),

$$\bar{\theta}_1^2 = \frac{S_k}{k} = \frac{1}{k}\sum_{i=1}^{k} Z_i^2$$

and

$$\bar{\theta}_2^2 = \frac{S_{n-k}}{n-k} = \frac{1}{n-k}\sum_{i=k+1}^{n} Z_i^2.$$

These estimators will be refined once a consistent estimator of k_0 is obtained. Denote by U_k^2 the quantity

$$U_k^2 = \sum_{i=1}^{k}(Z_i^2 - \bar{\theta}_1^2)^2 + \sum_{i=k+1}^{n}(Z_i^2 - \bar{\theta}_2^2)^2.$$

Then, $\hat{k}_0$ is defined as

$$\hat{k}_0 = \arg\min_k U_k^2 .$$

To study the asymptotic properties of U_k^2, it is better to rewrite it as

$$U_k^2 = \sum_{i=1}^{n}(Z_i^2 - \bar{Z}_n)^2 - nV_k^2 ,$$

where

$$\bar{Z}_n = \frac{1}{n}\sum_{i=1}^{n} Z_i^2$$

and

$$V_k = \left(\frac{k(n-k)}{n^2}\right)^{\frac{1}{2}} (\bar{\theta}_2^2 - \bar{\theta}_1^2) = \frac{S_n D_k}{\sqrt{k(n-k)}}$$

with

$$D_k = \frac{k}{n} - \frac{S_k}{S_n} .$$

This representation of U_k^2 is obtained by lengthy but straightforward algebra, and it is rather useful because minimization of U_k^2 is equivalent to the maximization of V_k and hence of D_k. So it is easier to consider the following estimator of k_0

$$\hat{k}_0 = \arg\max_k |D_k| = \arg\max_k (k(n-k))^{\frac{1}{2}} |V_k| . \tag{4.12}$$

As a side remark, it can be noted that, for fixed k (and under suitable hypotheses), D_k is also an approximate likelihood ratio statistic for testing the null hypothesis of no change in volatility (see, e.g., [119]). Once $\hat{k}_0$ has been obtained, the following estimators of the parameters θ_1 and θ_2 can be used:

$$\hat{\theta}_1^2 = \frac{S_{\hat{k}_0}}{\hat{k}_0}, \tag{4.13}$$

$$\hat{\theta}_2^2 = \frac{S_{n-\hat{k}_0}}{n-\hat{k}_0}. \tag{4.14}$$

Next results provide consistency of $\hat{k}_0$, $\hat{\theta}_1^2$, and $\hat{\theta}_2^2$ as well as their asymptotic distributions.

Fact 4.4 ([64]) *Under H_0: $\theta_1 = \theta_2 = 1$, we have that*

$$\sqrt{\frac{n}{2}}|D_k| \xrightarrow{d} |W^0(\tau)|, \tag{4.15}$$

where $\{W^0(\tau), 0 \le \tau \le 1\}$ is a Brownian bridge.

The asymptotic result above is useful to test if a change point doesn't exist. In particular, it is possible to obtain the asymptotic critical values for the distribution of the statistic by means of the same arguments used in [61], but we do not go into these details here.

Fact 4.5 ([64]) *The estimator $\hat{\tau}_0 = \frac{\hat{k}_0}{n}$ satisfies*

$$|\hat{\tau}_0 - \tau_0| = n^{-1/2}(\theta_2^2 - \theta_1^2)^{-1} O_p(\sqrt{\log n})\,. \tag{4.16}$$

Moreover, for any $\beta \in (0, 1/2)$,

$$n^{\beta}(\hat{\tau}_0 - \tau_0) \xrightarrow{p} 0\,.$$

Finally,

$$\hat{\tau}_0 - \tau_0 = O_p\left(\frac{1}{n(\theta_2^2 - \theta_1^2)^2}\right). \tag{4.17}$$

It is also interesting to know the asymptotic distribution of $\hat{\tau}_0$ for small discrepancies between θ_1 and θ_2. The case $\vartheta_n = \theta_2^2 - \theta_1^2$ equal to a constant is less interesting because when ϑ_n is large the estimate of k_0 is quite precise.

Assumption 4.7 *$\vartheta_n \to 0$ in such a way that $\frac{\sqrt{n}\vartheta_n}{\sqrt{\log n}} \to \infty$.*

Assumption 4.7 and Fact 4.5 imply the consistency of $\hat{\tau}_0$.

Fact 4.6 ([64]) *Under Assumption 4.7, for $\Delta_n \to 0$ as $n \to \infty$, we have that*

$$\frac{n\vartheta_n^2(\hat{\tau}_0 - \tau_0)}{2(\tilde{\theta}_n^2)^2} \xrightarrow{d} \arg\max_v \left\{\mathcal{W}(v) - \frac{|v|}{2}\right\}, \tag{4.18}$$

where $\mathcal{W}(u)$ is a two-sided Brownian motion,

$$\mathcal{W}(u) = \begin{cases} W_1(-u), & u < 0 \\ W_2(u), & u \geq 0 \end{cases}, \tag{4.19}$$

with W_1 and W_2 two independent Brownian motions and $\tilde{\theta}_n^2$ a consistent estimator for θ_1^2 or θ_2^2.

Finally we have the asymptotic distributions for the estimators $\hat{\theta}_1^2, \hat{\theta}_2^2$, defined in (4.13) and (4.14). We denote by θ_0 the common limiting value of θ_1 and θ_2.

Fact 4.7 ([64]) *Under Assumption 4.7, we have that*

$$\sqrt{n}\begin{pmatrix} \hat{\theta}_1^2 - \theta_1^2 \\ \hat{\theta}_2^2 - \theta_2^2 \end{pmatrix} \stackrel{d}{\to} N(0, \Sigma), \tag{4.20}$$

where

$$\Sigma = \begin{pmatrix} 2\tau_0^{-1}\theta_0^4 & 0 \\ 0 & 2(1-\tau_0)^{-1}\theta_0^4 \end{pmatrix}. \tag{4.21}$$

We show an example of implementation. We assume we have the Cox-Ingersoll-Ross model

$$\mathrm{d}X_t = (6 - 2X_t)\mathrm{d}t + \theta\sqrt{X_t}\mathrm{d}W_t,$$

where $\theta = \theta_1 = 1$ for $t < \tau_0 = 0.6$ and $\theta = \theta_2$ for $t \geq \tau_0$ and $t \in (0,1)$. We test the estimator outside the asymptotics for small n and fixed T. We simulate two paths `X1` and `X2` of the CIR process, one in $[0, \tau_0]$ and the other in $[\tau_0, 1]$, under the condition $X2(\tau_0) = X1(\tau_0)$, and we collate the two into one single trajectory. We resample the trajectory to $\Delta_n = 0.01$, which will give $n = 100$ observations.

```
> # ex4.05.R
> tau0 <- 0.6
> k0 <- ceiling(1000*tau0)
> set.seed(123)
> X1 <- sde.sim(X0=1, N=2*k0, t0=0, T=tau0, model="CIR", theta=c(6,2,1))
> X2 <- sde.sim(X0=X1[2*k0+1], N=2*(1000-k0), t0=tau0,
+               T=1, model="CIR", theta=c(6,2,3))
>
> Y <- ts(c(X1,X2[-1]), start=0, deltat=deltat(X1))
> X <- window(Y,deltat=0.01)
> DELTA <- deltat(X)
> n <- length(X)
```

Now we construct the residuals Z_i and the statistic D in order to identify k_0.

```
> # ex4.05.R (cont)
> mu <- function(x) 6-2*x
> sigma <- function(x) sqrt(x)
> Z <- (diff(X) - mu(X[1:(n-1)])*DELTA)/(sqrt(DELTA)*sigma(X[1:(n-1)]))
>
> tau <- seq(0,1, length=length(Z))
```

```
> k <- ceiling(n*tau)
>
> Sn <- cumsum(Z^2)
> S <- sum(Z^2)
> D <- abs((2:n)/n - Sn/S)
>
> k0 <- which(D==max(D))
> tau[k0]
[1] 0.5959596
> sqrt(Sn[k0]/k0)
[1] 0.8559207
> sqrt((S-Sn[k0])/(n-k0))
[1] 3.038949
```

The estimated value $\hat{\tau} = 0.59$ is likely close to the real value $\tau_0 = 0.6$, and the two values of the volatility are estimated as $\hat{\theta}_1 = 0.85$ and $\hat{\theta}_2 = 3.04$. The next code plots the graph of the trajectory and the statistic $|D_k|$. The result is given in Figure 4.4.

```
> # ex4.05.R (cont)
> par(mar=c(3,3,1,1))
> par(mfrow=c(2,1))
> plot(X)
> abline(v=tau0,col="red",lty=3)
> plot(tau,D,type="l")
> abline(v=tau[k0],col="blue")
> abline(v=tau0,col="red",lty=3)
```

4.3.1 Estimation of the change point with unknown drift

When both $b(x)$ and $\sigma^2(x)$ are unknown, it is necessary to assume that at least $\sigma(x)$ is constant, and hence we consider the stochastic differential equation

$$\mathrm{d}X_t = b(X_t)\mathrm{d}t + \theta\mathrm{d}W_t\,. \tag{4.22}$$

Then $b(x)$ can be estimated nonparametrically with $\hat{b}_n(x)$ from (4.8), and the residuals Z_i are estimated as

$$\hat{Z}_i = \frac{X_{i+1} - X_i}{\sqrt{\Delta_n}} - \hat{b}_n(X_i)\sqrt{\Delta_n}\,.$$

In this case, we can use the following contrast to identify the change point:

$$\tilde{k}_0 = \arg\min_k \left\{ \sum_{i=1}^{k} \left(\hat{Z}_i^2 - \frac{\hat{S}_k}{k} \right)^2 + \sum_{i=k+1}^{n} \left(\hat{Z}_i^2 - \frac{\hat{S}_{n-k}}{n-k} \right)^2 \right\}, \tag{4.23}$$

where

$$\hat{S}_k = \sum_{i=1}^{k} \hat{Z}_i^2 \quad \text{and} \quad \hat{S}_{n-k} = \sum_{i=k+1}^{n} \hat{Z}_i^2\,.$$

We obtain the new statistic

$$\hat{V}_k = \left(\frac{k(n-k)}{n^2} \right)^{\frac{1}{2}} \left(\frac{\hat{S}_{n-k}}{n-k} - \frac{\hat{S}_k}{k} \right) = \frac{\hat{S}_n \hat{D}_k}{\sqrt{k(n-k)}}\,,$$

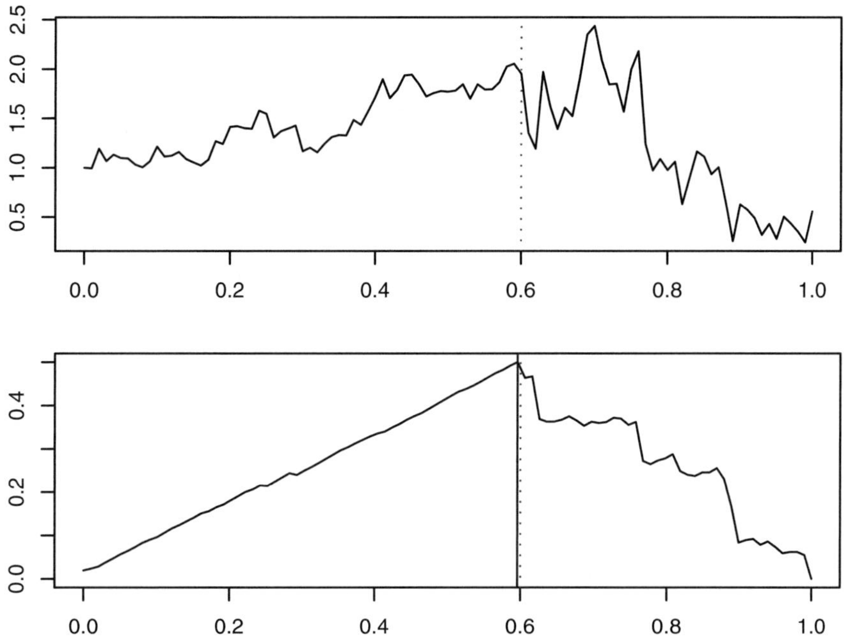

Fig. 4.4. Top: the simulated CIR process changes volatility at time $\tau_0 = 0.6$. Bottom: the shape of statistic $|D_k|$.

where

$$\hat{D}_k = \frac{k}{n} - \frac{\hat{S}_k}{\hat{S}_n}$$

and the change point is identified as the solution to

$$\hat{k}_0 = \arg\max_k |\hat{D}_k| \,.$$

Consistency and distributional results mentioned in the previous section hold (see [64]).

```
cpoint <- function(x, mu, sigma){
 DELTA <- deltat(x)
 n <- length(x)
 Z <- NULL
 if(!missing(mu) && !missing(sigma)){
  Z <- (diff(x) - mu(x[1:(n-1)])*DELTA)/(sqrt(DELTA)*sigma(x[1:(n-1)]))
 } else {
        bw <- n^(-1/5) * sd(x)
    y <- sapply(x[1:(n-1)], function(xval) {
        tmp <- dnorm(xval, x[1:(n - 1)], bw)
        sum(tmp * diff(x))/(DELTA * sum(tmp))
    })
   Z <- diff(x)/sqrt(DELTA) - y*sqrt(DELTA)
 }
```

```
  Sn <- cumsum(Z^2)
  S <- sum(Z^2)
  D <- abs((2:n)/n - Sn/S)
  k0 <- which(D==max(D))
  return(list(k0=k0, tau0=time(x)[k0],
      theta1=sqrt(Sn[k0]/k0), theta2=sqrt((S-Sn[k0])/(n-k0))))
}
```

Listing 4.3. Change-point estimator of the volatility of a diffusion process.

The package `sde` contains the function `cpoint`, whose code is shown in Listing 4.3. It is very simple to use it requires the data and the two drift and diffusion coefficients. If the drift coefficient is missing, it is estimated nonparametrically. An example of the function's application follows.

```
> # ex4.06.R
> tau0 <- 0.6
> k0 <- ceiling(1000*tau0)
> set.seed(123)
> X1 <- sde.sim(X0=1, N=2*k0, t0=0, T=tau0, model="CIR", theta=c(6,2,1))
> X2 <- sde.sim(X0=X1[2*k0+1], N=2*(1000-k0), t0=tau0,
+               T=1, model="CIR", theta=c(6,2,3))
>
> Y <- ts(c(X1,X2[-1]), start=0, deltat=deltat(X1))
> X <- window(Y,deltat=0.01)
> DELTA <- deltat(X)
> n <- length(X)
>
> mu <- function(x) 6-2*x
> sigma <- function(x) sqrt(x)
>
> cpoint(X,mu,sigma)
$k0
[1] 60

$tau0
[1] 0.59

$theta1
[1] 0.8559207

$theta2
[1] 3.038949

>
> # nonparametric estimation of the drift
> cpoint(X)
$k0
[1] 60

$tau0
[1] 0.59

$theta1
[1] 1.011887

$theta2
[1] 2.930061
```

4.3.2 A famous example

We analyze the DWJ dataset, which contains the weekly closings of the Dow-Jones industrial average in the period July 1971–August 1974. These data were proposed by Hsu [113, 114, 228] and used by many other authors to test change-point estimators. There are 162 data, and the main evidence found by several authors is that a change in the variance occurred around $k = 88$, which corresponds to the third week of March 1973. Instead of working on the values, we transform the data into returns as usual, $S(t_i) = (X(t_i) - X(t_{i-1}))/X(t_{i-1})$, $i = 1, \ldots, n$, with X the series of Dow-Jones closings and S the returns. We assume the drift coefficient to be unknown.

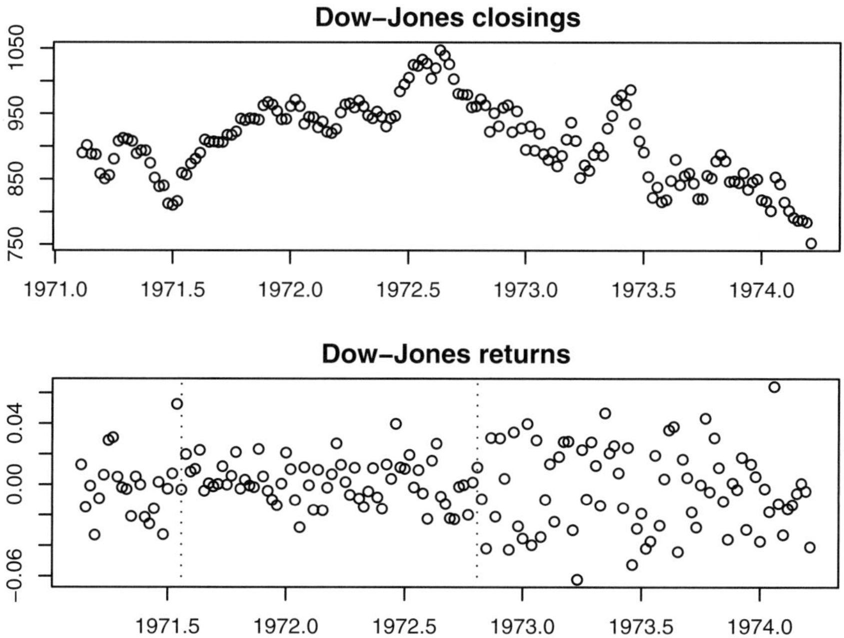

Fig. 4.5. Change-point analysis of the Dow-Jones weekly closings.

```
> # ex4.07.R
> data(DWJ)
> ret <- diff(DWJ)/DWJ[-length(DWJ)]
> par(mfrow=c(2,1))
> par(mar=c(3,3,2,1))
> plot(DWJ,main="Dow-Jones closings",ylab="",type="p")
> plot(ret,main="Dow-Jones returns",ylab="",type="p")
> cp <- cpoint(ret)
> cp
$k0
[1] 88
```

```
$tau0
[1] 1972.808

$theta1
[1] 0.1097246

$theta2
[1] 0.2022780
> abline(v=cp$tau0,lty=3)
```

Looking at Figure 4.5, it emerges that another change point may be present. So we reanalyze the first part of the series to spot the second change point.

```
> # ex4.05.R (cont)
> cp <- cpoint(window(ret,end=cp$tau0))
> cp
$k0
[1] 23

$tau0
[1] 1971.558

$theta1
[1] 0.1372465

$theta2
[1] 0.0944348
> abline(v=cp$tau0,lty=3)
```

Both change points correspond to important shocks in the U.S. market: the first one seems to be related to the announcement of the broken gold/dollar link, and the second one is in relation to the Watergate scandal.

Appendix A: A Brief Excursus into R

This appendix is not intended to be a guide to the R language because it focuses only on special aspects of the language that are used throughout the book. The reader is invited to read the brief R manual called "An Introduction to R" that comes with every installed version of R and use as reference both [63] for a primer on the R language and [226] for a more advanced guide on R programming. The reader might find it useful to read these notes when he or she encounters problems with the R code presented in the main body of the book.

A.1 Typing into the R console

R is mainly an interactive language with a simple command-line interface. This means that the user needs to type most of the commands with limited or no support from the graphical user interface (GUI). All the commands are given as inputs to R after the prompt > and are analyzed by the R parser after the user presses the "return"/"enter" key (or a newline character is encountered in the case of a script file).

```
> cat("Hello World!")
Hello World!
```

R inputs can be multiline; hence, if the R parser thinks that the user did not complete some command (because of unbalanced parentheses or quotation marks), on the next line a + symbol will appear instead of a prompt.

```
> cat("Hello World!"
+
```

This can be quite frustrating for novice users, so it is better to know how to exit from this impasse. Depending on the implementation of R usually pressing CTRL+C or ESC on the keyboard helps. Otherwise, for GUI versions of R, pushing the "stop" button of the R console will exit the parser. Of course, another solution is to complete the command.

```
> cat("Hello World!"
+ )
Hello World!
```

Everything in R is an object, and objects live in a workspace. The main workspace can be saved and loaded in R with the `save.image` and `load` commands, respectively. When exiting from the R console (e.g., with `q()`), the user is asked whether to save the workspace. If the user responds yes", the workspace is saved in the current directory as a hidden (on some operating systems) file named `.RData` and reloaded the next time R is started. Workspace or R objects can be saved and loaded with `load` and `save` commands, and almost all R GUIs have functionalities for these. The reader is invited to use the `help` command to retrieve information on how to use each R command. R documentation can be accessed in essentially two ways:

```
> help(load)
```

or

```
> ?load
```

For some special operators, the user should specify the argument like this: `?"for"`, `?"+"`, etc. There is more information on the help page for `help` (i.e., `help()`). If the help page for a topic contains examples of uses, the user can run them using `example(topic)`; for example, `help(plot)` and `example(plot)` run the examples for the `plot` function.

Usually R graphics are displayed on a device that corresponds to a window for a GUI version of R (for example, under MS-Windows, X11, or Mac OS X). Otherwise a Postscript file `Rplots.ps` is generated in the current working directory. Sometimes, in interactive uses of R, it is useful to set

```
> par(ask=TRUE)
```

so that R will ask for a confirmation before drawing a new plot. The call `par(ask=FALSE)` restores the original behavior.

A.2 Assignments

Many functions in R operate *vector wisely* on vectors of different nature. Nevertheless, as will be discussed later, the reader should not think of R vectors as usual linear algebra vectors. Indeed, R vectors are just indexed sequences of objects of the same type; for example, numbers, labels, functions, etc. The following command creates a scalar in R.

```
> x <- 4
```

`<-` is an operator that literally, behaves as “assign the right-hand side to the left-hand side,” and the following is equivalent.[1]

```
> 4 -> x
```

[1] The user is discouraged from using “=” for assignments.

Now `x` is a new object in the workspace, and it has been created as a vector of length 1 containing the real[2] number 4. To see what is inside an R object, usually typing its name in the console will help; otherwise `print(object)` might be another option. In our case,

```
> x
[1] 4
```

and the first `[1]` is indicating that the first element of `x` is 4. The following command creates a more interesting vector `y` containing the numbers 2, 7, 4, and 1 concatenated in a single object using the function `c()`.

```
> y <- c(2,7,4,1)
> y
[1] 2 4 7 1
```

The command `ls()` shows the current content of the workspace.

```
> ls()
[1] "x" "y"
```

We mentioned that R vectors are not equivalent to the vectors of linear algebra. Consider the example

```
> x*y
[1] 8 28 16 4
```

which seems a natural answer since x is scalar. But consider the following

```
> y*y
[1] 4 49 16 1
```

This is the term-by-term product of the elements in y (i.e., $y_i \cdot y_i$). If y was a vector of linear algebra, then one should expect two possible cases only, $y \cdot y^T$ or $y^T \cdot y$, where "T" is the transposition operator and we still did not mention whether y is a column or a row vector. To clarify what is R behavior, one should consider the following equivalent command.

```
> y^2
[1] 4 49 16  1
> log(y)
[1] 0.6931472 1.9459101 1.3862944 0.0000000
```

From this example, it is clear that R is applying the function $f(z) = z^2$ (or $f(z) = \log(z)$) to each element of the vector `y`. One way to think of this is in terms of statistical models and corresponding data analysis. If one has to model a response variable Y in terms of some covariates $X_1, \ldots, X_k$ with a model function f, say $Y \sim f(X_1, \ldots, X_k)$, and corresponding data come as n observations $(Y_i, X_{i1}, \ldots, X_{ik})$, $i = 1, \ldots, n$, one wants to apply the model f to each observation i and fit it accordingly.

[2] In R, there is no distinction between single- and double-precision real numbers as in other languages. All objects of class `real` are stored and treated as double-precision real numbers internally.

A.3 R vectors and linear algebra

To use linear algebra on vectors and matrices, one should use the linear algebra operator %*% as follows

```
> t(y) %*% y
     [,1]
[1,]   70
```

and

```
> z <- y %*% t(y)
> z
     [,1] [,2] [,3] [,4]
[1,]    4   14    8    2
[2,]   14   49   28    7
[3,]    8   28   16    4
[4,]    2    7    4    1
```

Conventionally, R vectors are considered as column vectors, and implicitly the product

```
> y %*% y
```

is treated as

```
> t(y) %*% y
```

Of course, `t()` is the transposition operator in R. Unless defined as a matrix with dimension $1 \times k$ of $k \times 1$, R vectors are always printed horizontally. For example,

```
> -1:30
 [1] -1  0  1  2  3  4  5  6  7  8  9 10
[13] 11 12 13 14 15 16 17 18 19 20 21 22
[25] 23 24 25 26 27 28 29 30
```

creates a vector using the sequence of numbers from -1 to 30 with step 1. Matrices can be created using the `matrix` command as follows.

```
> a <- matrix(1:30, 5,6)
> a
     [,1] [,2] [,3] [,4] [,5] [,6]
[1,]    1    6   11   16   21   26
[2,]    2    7   12   17   22   27
[3,]    3    8   13   18   23   28
[4,]    4    9   14   19   24   29
[5,]    5   10   15   20   25   30
```

In the code above, a matrix of five rows and six columns is created and filled with the numbers from 1 to 30. The matrix is filled columnwise using the elements of the vector 1:30. If the input vector is not long enough, its elements are recycled as in the following example.[3]

```
> b <- matrix(LETTERS, 5, 6)
> b
     [,1] [,2] [,3] [,4] [,5] [,6]
[1,] "A"  "F"  "K"  "P"  "U"  "Z"
[2,] "B"  "G"  "L"  "Q"  "V"  "A"
[3,] "C"  "H"  "M"  "R"  "W"  "B"
[4,] "D"  "I"  "N"  "S"  "X"  "C"
[5,] "E"  "J"  "O"  "T"  "Y"  "D"
```

[3] A warning message has been suppressed.

LETTERS is a predefined vector of the 26 capital letters of the English alphabet. The reader may want to explore different uses of the command matrix, looking at the corresponding help page. Two things are worth mentioning at this point. The first is that the argument "recycling" is a standard feature of many R functions, and the reader is invited to keep this in mind. The second is the way R represents matrices which helps in understanding how flexible R is in subsetting objects. We give some examples of subsetting of matrices without comments.

```
> a[1,] # extract 1st column
[1]  1  6 11 16 21 26
> a[1:2,3] # extracts a submatrix
[1] 11 12
> a[1:2,3:4]
     [,1] [,2]
[1,]   11   16
[2,]   12   17
> a[,4] # extract the 4th column
[1] 16 17 18 19 20
> a[c(1,3),] # a submatrix of noncontiguous elements
     [,1] [,2] [,3] [,4] [,5] [,6]
[1,]    1    6   11   16   21   26
[2,]    3    8   13   18   23   28
```

In the code above, we used the symbol #, which indicates a comment in the spirit of what REM does for BASIC, c for FORTRAN, etc. All characters following the # symbol are ignored by the R parser.

A.4 Subsetting

As already mentioned, subsetting is an important feature of the language and widely used in many R programs, including the one contained in this book. Subsetting is strictly related to the ability to create indexes programmatically. We have seen one way of creating sequences like the following code.

```
> x <- 3:8
> x
[1] -3 -2 -1  0  1  2  3  4  5  6  7  8
```

The more general way of doing this is the command seq (sequence). Consider the following

```
> seq(-3,6,2)
[1] -3 -1 1  3  5
> seq(-3,-1,.33)
[1] -3.00 -2.67 -2.34 -2.01 -1.68 -1.35 -1.02
```

In this code, we have created two sequences using different steps (2 and 0.33, respectively), but one should specify the total length of a sequence as follows.

```
> seq(-3, 1, length=10)
 [1] -3.0000000 -2.5555556 -2.1111111 -1.6666667 -1.2222222
 [6] -0.7777778 -0.3333333  0.1111111  0.5555556  1.0000000
```

Although this way of creating sequences is useful, for subsetting we need something different. The main command in this direction is `which`, which works on logical expressions. For example, let

```
> x <- -3:8
> x
[1] -3 -2 -1  0  1  2  3  4  5  6  7  8
```

If we want to know which elements of `x` are strictly less than 2, we will use

```
> which(x<2)
[1] 1 2 3 4 5
```

or, for the elements greater than or equal to -1 or strictly greater than 5, we will write

```
> which((x >= -1) & (x < 5))
[1] 3 4 5 6 7 8
```

Here "`&`" is the logical operator *and*. The *or* logical operator is "`|`". For example,

```
> which((x < -2) | (x > 1)) -> z
> z
[1]  1  6  7  8  9 10 11 12
```

returns the indexes of the elements of `x` such that `x` is strictly less than 2 or greater than 1. But what we really want is to extract such elements from `x`; i.e., we want to do subsetting.

```
> x[z]
[1] -3  2  3  4  5  6  7  8
```

The command `which` just returns the index of the logical elements equal to `TRUE` in a vector. R logical vectors can contain the symbols `TRUE` and `FALSE` (which are not equivalent to the numbers 1 and 0 as in other programming languages). For example,

```
> x < 3.2
 [1]  TRUE  TRUE  TRUE  TRUE  TRUE  TRUE  TRUE FALSE FALSE
[10] FALSE FALSE FALSE
```

is the vector of "`TRUE/FALSE`" above, which is the result of applying the function `is < 3.2` element by element to vector `x`. One such logical vector `which` returns the following.

```
> which(x < 3.2)
[1] 1 2 3 4 5 6 7
```

A.5 Different types of objects

In the R language it is also possible to initialize object without specifying the initial value.

```
> matrix(,2,3)
     [,1] [,2] [,3]
[1,]   NA   NA   NA
[2,]   NA   NA   NA
```

The matrix above is filled with missing values and represented by the symbol `NA`, literally `Not Available`. Many R functions are able to treat missing values automatically, but at this stage we only mention that a few special symbols in the R language are sealed in the sense that they cannot be redefined by the user. Indeed, one can decide to redefine all other R objects by overriding their definitions. The only limitations are naming conventions of the language itself. Object names are usually a sequence of letters and numbers but may also include special characters such as the dot "." or the underscore "_". Nevertheless, object names cannot start with a digit. This means that `x2`, `.x2`, or `x.2` are admissible names but `2x` is not. Objects whose name starts with a dot are usually masked by a simple use of the `ls()` command. The few names of objects that are sealed are the following.

```
FALSE TRUE Inf NA NaN NULL
break else for function if in next repeat while
```

Of course, it is common sense not to redefine functions such as `c()`, `ls()`, etc. All objects belong to some classes, such as the following basic ones:

- `character` : alphanumeric string of characters;
- `numeric` : real numbers, internally stored as double-precision floating-point numbers;
- `integer` : integer numbers, eventually with sign;
- `logical` : objects that take only values `TRUE` or `FALSE`;
- `complex` : complex numbers with real and imaginary parts;
- `function` : functions of named or variable number of arguments;
- `expression` : true mathematical expressions to be evaluated.

Objects of these classes can be organized differently in vectors, matrices, and arrays or lists and environments. Vectors, matrices, arrays and objects of class `list` are also indexable but not objects of class `environment`. Still, it is possible to subset the environment using the key-value convention, retrieving objects in an environment by name and obtaining the corresponding value. While vectors, matrices, and arrays contain elements of the same class, lists and environments are containers of different objects. For example, the R workspace is itself an environment called the "global environment."

```
> .GlobalEnv
<environment: R_GlobalEnv>
```

Vectors can be created in essentially two ways: by "initialization" or "filling" (or assignment). We have already used both approaches, so the following two lines of R code are given without comment, being self-explanatory.

```
> x <- 1:3
> x
[1] 1 2 3
> x <- numeric(3)
> x
[1] 0 0 0
> x <- vector(3, mode="numeric")
> x
[1] 0 0 0
```

The option `mode` can be any of the classes above and eventually many others, but at least one mode has to be specified: try `vector(3)`. Objects of class `list`, along with `environment`, are heavily used in the R language because they can contain exhaustive descriptions of, for example, models, data, etc. The following code creates a few objects of different types and collects them into a `list`.

```
> myCmp <- complex(real=1:10,imaginary=-1:9)
> myCmp # recycling is taking place
 [1]  1-1i  2+0i  3+1i  4+2i  5+3i  6+4i  7+5i  8+6i
 [9]  9+7i 10+8i  1+9i
> myStr <- c("up", "down")
> myLog <- c(TRUE, TRUE, FALSE, FALSE, FALSE)
> myMat <- matrix(1, 4, 2)
> myExpr <- c(expression(sin(x)), expression(cos(2*x)))
```

There are two ways of doing this:

```
> myList1 <- list(CPLX = myCmp, LAB = myStr, BOOL = myLog,
+                 MAT = myMat, EXPR = myExpr)
> myList2 <- list(myCmp, myStr, myLog,  myMat,  myExpr)
```

`myList1` and `myList2` are almost the same object, but the difference is that elements of `myList1` can be accessed by names (`CPLX`, `LAB`, etc.) but elements of `myList2` only by index.

```
> myList1
$CPLX
 [1]  1-1i  2+0i  3+1i  4+2i  5+3i  6+4i  7+5i  8+6i
 [9]  9+7i 10+8i  1+9i

$LAB
[1] "up"   "down"

$BOOL
[1]  TRUE  TRUE FALSE FALSE FALSE

$MAT
     [,1] [,2]
[1,]    1    1
[2,]    1    1
[3,]    1    1
[4,]    1    1

$EXPR
expression(sin(x), cos(2 * x))
```

but

```
> myList2
[[1]]
 [1]  1-1i  2+0i  3+1i  4+2i  5+3i  6+4i  7+5i  8+6i
 [9]  9+7i 10+8i  1+9i
```

```
[[2]]
[1] "up"   "down"

[[3]]
[1]  TRUE  TRUE FALSE FALSE FALSE

[[4]]
     [,1] [,2]
[1,]    1    1
[2,]    1    1
[3,]    1    1
[4,]    1    1

[[5]]
expression(sin(x), cos(2 * x))
```

So, for myList1, to access the second element of the container, we can use both

```
> myList1$LAB
[1] "up"   "down"
> myList1["LAB"]   # key-value access to the element of a list
$LAB
[1] "up"   "down"
> myList1[2]
$LAB
[1] "up"   "down"
```

but for myList2 only myList2[2]. The operator $ has a special meaning in R, and we will use it frequently to access elements of lists or environments. We should also mention that if one wants direct access to the elements of, say, the LAB element of myList1, one should use the subsetting operator [[. We give an example of this without comment.

```
> myList1$LAB[1]
[1] "up"
> myList1[[2]][1]
[1] "up"
> myList1[["LAB"]] # Note:  [[ is different from [  below
[1] "up"   "down"
> myList1["LAB"]
$LAB
[1] "up"   "down"
```

One useful command to inspect the nature of the object is str, "structure"; for example,

```
> str(myList1)
List of 5
 $ CPLX: cplx [1:11] 1-1i 2+0i 3+1i ...
 $ LAB : chr [1:2] "up" "down"
 $ BOOL: logi [1:5]  TRUE  TRUE FALSE FALSE FALSE
 $ MAT : num [1:4, 1:2] 1 1 1 1 1 1 1 1
 $ EXPR:  expression(sin(x), cos(2 * x))
```

We will discuss the peculiarity of environment objects after discussing a bit of objects of class function.

A.6 Expressions and functions

Mathematical expressions in R are objects of class expression. R expressions are intended to be evaluated and can be manipulated symbolically until the evaluation is needed. R has a limited set of primitive mathematical functions along with a table of derivatives for them. They can also be plotted using a LaTeX style visualization on graphs. R expressions must be constructed as such. For example, `cos(x)` is not interpreted as an expression by R but `expression(cos(x))` is

```
> ex <- expression(cos(x))
> str(ex)
  expression(cos(x))
```

and can be evaluated using the `eval` function

```
> eval(ex)
Error in eval(expr, envir, enclos) : object "x" not found
> x <- 2
> eval(ex)
[1] -0.4161468
```

From the example above, it is clear that in order to evaluate an expression everything needed for the evaluation should be available. Expressions can be differentiated, but the result is not immediately clear for the average R user.

```
> dx <- deriv(ex,"x")
> dx
expression({
    .value <- cos(x)
    .grad <- array(0, c(length(.value), 1), list(NULL, c("x")))
    .grad[, "x"] <- -sin(x)
    attr(.value, "gradient") <- .grad
    .value
})
> eval(dx)
[1] -0.4161468
attr(,"gradient")
              x
[1,] -0.9092974
```

We discuss manipulation and differentiation of R expressions in the body of the book, particularly in Chapter 2. R commands are objects of class `function`, and functions can have a fixed or variable number of arguments.[4] Writing a function in R is as easy as the following.

```
> f <- function(x=1,y){
+    if(y != 0)
+     x/y
+    else
+     stop("are you joking?")
+ }
```

`f` is constructed using the special R command `function` as a function of two arguments `x` and `y`. The first argument has a default value `x` set equal to `1`. The body of the function is enclosed within "{" and "}". All arguments must be named, and the order of the arguments matters only in calls in which the argument names are not used.

```
> f(1,2)
[1] 0.5
> f(y=2,x=1)
```

[4] We do not treat this second case.

```
[1] 0.5
> f(y=2)
[1] 0.5
> f(y=0)
Error in f(y = 0) : are you joking?
> f(0)
Error in f(0) : argument "y"  is not specified and has no default value
```

R expressions can be transformed into R functions as follows.

```
> ez <- expression(cos(z))
> fz <- function(z) eval(ez)
> fz(1)
[1] 0.5403023
```

Notice that, in the case above, the variable `z` is taken from the arguments passed to the function and not from the global environment.

```
> eval(ez)
Error in eval(expr, envir, enclos) : object "z" not found
> z <- 2
> eval(ez)
[1] -0.4161468
> fz(1)
[1] 0.5403023
> fz(2)
[1] -0.4161468
```

A.7 Loops and vectorization

Even if we cannot go into the details of the *object-oriented* nature of the R language, the reader should know some basic principles on how the R language works. In the function `f` above, a call to `f(x,y)` implies that copies of the objects `x` and `y` are passed to the function instead of just their pointers (i.e., memory addresses). This means that function calls usually generate memory copies that in iterative procedures might lead to inefficiency, in particular in cases where the objects are large in size. One should think that operators such as `+` are indeed functions: `a+b` should be interpreted as `+(a,b)`, where `+(,)` is just an R function. This is why, in the R language, loops are "slow" compared with some other widely used languages, and the best programming practice is to use vectorization instead.

A `for` loop in R is a construct defined as

`for(`**index** `in` **some set**`)`{ **execute these commands** }

The following `for` loop, which executes the task of summing the first 10000 integer numbers

```
> s <- 0
> for(i in 1:10000){
+   s <- s + i
+ }
> s
[1] 50005000
```

is not as efficient as

```
> sum(1:10000)
[1] 50005000
```

The apparently innocuous line `s <- s + i` takes a copy of `s` and `i`, passes them to the function `+`, gets the result, and copies it back onto `s`. If one thinks that this is iterated 10000 times, it is clear why such a `for` loop is not efficient. The line `sum(1:10000)` is instead a call to some quite efficient internal, primitive R function. Many R functions are vectorized; furthermore, not many algorithms require a true iterative scheme and can instead be vectorized. We discuss this in detail in Chapter 1.

Vectorization of functions is made easy by a series of `*apply` functions. For example, the `sapply` function applies some R function to each element of a vector and returns a vector of the same length as the original.

```
> x <- 1:3
> sapply(x, log) # this is equivalent to log(x)
[1] 0.0000000 0.6931472 1.0986123
```

Other examples are the functions `apply` and `lapply`. The first one operates along a specified dimension of an array, and the second one iterates the element of the list.

```
> a <- matrix(rep(1:5,6),5,6)
> a
     [,1] [,2] [,3] [,4] [,5] [,6]
[1,]    1    1    1    1    1    1
[2,]    2    2    2    2    2    2
[3,]    3    3    3    3    3    3
[4,]    4    4    4    4    4    4
[5,]    5    5    5    5    5    5
> apply(a, 1, sum) # applies `sum' iterating on the rows of a
[1]  6 12 18 24 30
> apply(a, 2, sum) # applies `sum' iterating on the columns of a
[1] 15 15 15 15 15 15
```

We further discuss vectorization for the functions introduced in the book when appropriate.

A.8 Environments

The only exceptions in the base R language of objects that are not copied in function calls are objects of class `environment` for which only the address is passed. An environment is created using the command `new.env()`. Once created, the user can store objects in it. The following listing creates an environment `e1` and stores objects in it. To show that the R workspace is just a particular environment, the command `ls()` is used on `e1`.

```
> rm(list=ls()) # removes everything from the workspace
> ls()
character(0)
> e1 <- new.env() # creates a new environment
> e1$a <- 10 # an object a is created in e1 and initialized with 10
> e1$a
[1] 10
> ls() # the R workspace contains only e1. Where is a?
```

```
[1] "e1"
> ls(e1) # here it is!
[1] "a"
> assign("b", 3, env=e1) # equivalent to e1$b <- 3
> ls()
[1] "e1"
> ls(e1)
[1] "a" "b"
```

Environments are useful in different scopes. In particular, we have seen that evaluation of R expressions needs to find objects in environments (either the global or the function environment in the examples above). It is possible to specify directly in `eval` the environment in which to work or to assign a predetermined environment to functions. With this approach, one is always sure about what is really evaluated and that objects created in some environment do not interfere with (override) objects in other environments. The following listing is an example of such an implementation.

```
> a+b
Error: object "a" not found
> eval(a+b)
Error in eval(a + b) : object "a" not found
> eval(expression(a+b), envir=e1)
[1] 13
> e1
<environment: 0x1c8df40> # this number may be different for the reader
> str(e1)
length 2 <environment>
```

The following is an example of a function that operates in a separate environment.

```
> rm(list=ls())
> e2 <- new.env()
> e2
<environment: 0x1feecec> # this might be different for the reader
> e2$a <- 1
> new.f <- function(u) u+a
> new.f
function(u) u+a
> environment(new.f)
<environment: R_GlobalEnv>
> new.f(5)
Error in new.f(5) : object "a" not found
> environment(new.f) <- e2
> new.f
function(u) u+a
<environment: 0x1feecec>
> new.f(5)
[1] 6
```

Manipulation of objects in environments is particularly useful in Chapter 3.

A.9 Time series objects

There are different ways to handle time series data in R. In the base R system, there is a class named `ts`, which is suitable for unidimensional or multidimensional time series where time instants are assumed to be equally spaced. Many other classes for storing data of time series type are provided by additional packages such as `its`, `tseries`, and `zoo`. They can handle irregularly spaced time series even with abstract indexing instead of time. In this book, we mainly work with the base `ts` objects, so we briefly describe their structure.

These objects assume the time variable is regularly spaced; hence, to initialize such objects, one needs to specify the starting date, the time increment δ or the frequency (i.e., the reciprocal of δ), or the ending date (in the latter case, the values of the time series are eventually recycled). In the next example, we generate a vector of pseudo random numbers[5] from the Gaussian distribution, cumulate them, and create a `ts` object with them. We use different values for the `frequency` parameter to show how dates are handled in the simplest way (more sophisticated options exist to treat dates).

```
> set.seed(123)
> z <- cumsum(rnorm(10))
> x <- ts(z, start=1980)
> x
Time Series:
Start = 1980
End = 1989
Frequency = 1
 [1] -0.5604756 -0.7906531  0.7680552  0.8385636  0.9678513
 [6]  2.6829163  3.1438325  1.8787713  1.1919184  0.7462564
> x <- ts(z, start=1980, frequency=4)
> x
           Qtr1       Qtr2       Qtr3       Qtr4
1980 -0.5604756 -0.7906531  0.7680552  0.8385636
1981  0.9678513  2.6829163  3.1438325  1.8787713
1982  1.1919184  0.7462564
> x <- ts(z, start=1980, frequency=12)
> x
            Jan        Feb        Mar        Apr
1980 -0.5604756 -0.7906531  0.7680552  0.8385636
            May        Jun        Jul        Aug
1980  0.9678513  2.6829163  3.1438325  1.8787713
            Sep        Oct
1980  1.1919184  0.7462564
```

Once a `ts` object is created, several methods and accessor functions are available in the base R system to handle, analyze, and modify their structure. We give a brief example below.

```
> time(x)
          Jan      Feb      Mar      Apr      May      Jun
1980 1980.000 1980.083 1980.167 1980.250 1980.333 1980.417
          Jul      Aug      Sep      Oct
1980 1980.500 1980.583 1980.667 1980.750
> deltat(x)
[1] 0.08333333
> frequency(x)
[1] 12
> start(x)
[1] 1980    1
> end(x)
[1] 1980   10
```

Finally, what matters to the subject of this book is the ability to resample data from a `ts` object. This is particularly relevant in estimation methods

[5] We discuss briefly R's pseudo random number generators in Section 1.2.

that imply a simulation step. The relevant function is `window`, of which we show some applications below.

```
> window(x, end=c(1980,6))
            Jan        Feb        Mar        Apr
1980 -0.5604756 -0.7906531  0.7680552  0.8385636
            May        Jun
1980  0.9678513  2.6829163
> window(x, freq=4)
           Qtr1       Qtr2       Qtr3       Qtr4
1980 -0.5604756  0.8385636  3.1438325  0.7462564
```

In Chapter 3, we mainly work by resampling the time series using the time step between observations. The following example is more specific to the subject of inference for discretely observed diffusion processes and is also discussed in Chapter 3.

```
> set.seed(132)
> z <- cumsum(rnorm(100))
> x <- ts(z, deltat=0.1)
> x
Time Series:
Start = c(1, 1)
End = c(10, 10)
Frequency = 10
  [1]  0.474111397 -0.080833369 -0.090790662  0.976202446
  [5]  0.263974326 -0.056693942 -1.294612588 -2.379483005
  [9] -1.510482696  0.805597503  1.120447528 -0.150929590
 [13]  0.411930315 -0.047739687  0.762566467  1.667115120

(...) # some output has been eliminated

 [93] 11.699791682 10.796491623  9.914504234 10.106141068
 [97] 10.614473960 11.270690115 10.763255231 10.252251090
> window(x, deltat=0.4)
Time Series:
Start = 1
End = 10.6
Frequency = 2.5
 [1]  0.4741114  0.2639743 -1.5104827  0.4119303  0.4583497
 [6]  0.3119781  0.2396707  1.4266107  3.4782968  0.6377071
[11]  3.4589132  2.2149872  4.2695622  5.9369322  4.6732305
[16]  4.4013621  8.4027980  7.7126426  7.5110659  8.1040908
[21]  7.9757358 10.6676656 10.3474873 11.6997917 10.6144740
```

A.10 R Scripts

Quite frequently a Monte Carlo study can take some time to execute. In this case, the obvious way to proceed is to leave the computer alone to work on some R simulation code. These programs are just ASCII files, usually but not necessarily with the extension `.R`, that contain several lines of code to be executed by R. During an interactive session, one might want to use the command `source("myScript.R")`, where `myScript.R` is an ASCII file in the current working directory, to execute the R code in the script. Under Unix-like shells or the MS-DOS command line, it is possible to use the commands

```
R CMD BATCH myScript.R
```

for Unix-like shells and

```
Rcmd BATCH myScript.R
```

for the MS-DOS command line. In this case, R will produce the corresponding output of the script in a file with the extension changed into `.Rout`.

A.11 Miscellanea

To save all the output of an interactive session, it is possible to use the command `sink` as follows.

```
> sink("output.txt") # opens the file
# you do your analysis. All the output goes into output.txt
> sink() # closes the file
```

In parallel, it is possible to save and reload the list of commands typed during an interactive session using `save.history` and `load.history`.

The current working directory can be changed or located using `setwd` and `getwd` commands. Under some systems and when this concept applies, the default starting working directory is the current working directory of the user (for example, in a Unix shell). The content of a directory can be seen using `dir`.

Finally, most of the GUIs have capabilities for installing and loading packages, and these are easily accessible from the menus of the GUI but system dependent, so we don't discuss them here. We just mention that several `*.packages` functions (`download`, `install`, etc.) exist and can be used directly from the R console.

Installed R packages can be loaded into R using `library(package)` (e.g., `library(sde)`) and the main (index) help page for each package can be accessed by

```
> library(help=package)
```

Appendix B: The sde Package

This appendix contains the documentation pages of the `sde` package available on CRAN (Comprehensive R Archive Network) (`http://CRAN.R-Project.org`). To install the `sde` package on your version of R, type the following line in the R console.

```
> install.packages("sde")
```

If you don't have enough privileges to install software on your machine or account, you will need the help of your system administrator. Once the package has been installed, you can actually use it by loading the code with

```
> library(sde)
```

A short list of help topics, corresponding to most of the commands in the package, is available by typing

```
> library(help=sde)
```

The examples presented in this book are contained in text files named `exY.ZZ.R`, where `Y` is the chapter in which example `ZZ` is contained. These files can be found in the directory called `book` inside the installed version of the `sde` package; i.e., they are accessible with

```
> setwd(file.path(.libPaths()[1],"sde","book"))
> dir()
 [1] "ch1.R"     "ch2.R"     "ch3.R"     "ch4.R"     "ex1.01.R" "ex1.02.R"
 [7] "ex1.03.R" "ex1.04.R" "ex1.05.R" "ex1.06.R" "ex1.07.R" "ex1.08.R"
[13] "ex1.09.R" "ex1.10.R" "ex1.11.R" "ex1.12.R" "ex1.13.R" "ex1.14.R"
[19] "ex1.15.R" "ex2.01.R" "ex2.02.R" "ex2.03.R" "ex2.04.R" "ex2.05.R"
[25] "ex2.06.R" "ex2.07.R" "ex2.08.R" "ex2.09.R" "ex2.10.R" "ex2.11.R"
[31] "ex2.12.R" "ex2.13.R" "ex2.14.R" "ex2.15.R" "ex2.16.R" "ex2.17.R"
[37] "ex2.18.R" "ex3.01.R" "ex3.02.R" "ex3.03.R" "ex3.04.R" "ex3.05.R"
[43] "ex3.06.R" "ex3.07.R" "ex3.08.R" "ex4.01.R" "ex4.02.R" "ex4.03.R"
[49] "ex4.04.R" "ex4.05.R" "ex4.06.R" "ex4.07.R"
```

and executed via, for example, the `source` command

```
> source("ex1.01R")
```

although we suggest opening each file in a text editor and copying and pasting the code line-by-line into the R console to get immediate feedback.

BM	*Brownian motion, Brownian bridge, and geometric Brownian motion simulators*

Description

Brownian motion, Brownian bridge, and geometric Brownian motion simulators.

Usage

```
BBridge(x=0, y=0, t0=0, T=1, N=100)
BM(x=0, t0=0, T=1, N=100)
GBM(x=1, r=0, sigma=1, T=1, N=100)
```

Arguments

`x`	initial value of the process at time `t0`.
`y`	terminal value of the process at time `T`.
`t0`	initial time.
`r`	the interest rate of the GBM.
`sigma`	the volatility of the GBM.
`T`	final time.
`N`	number of intervals in which to split `[t0,T]`.

Details

These functions return an invisible `ts` object containing a trajectory of the process calculated on a grid of `N+1` equidistant points between `t0` and `T`; i.e., `t[i] = t0 + (T-t0)*i/N, i in 0:N`. `t0=0` for the geometric Brownian motion.
The function `BBridge` returns a trajectory of the Brownian bridge starting at `x` at time `t0` and ending at `y` at time `T`; i.e.,

$$\{B(t), t_0 \le t \le T | B(t_0) = x, B(T) = y\}.$$

The function `BM` returns a trajectory of the translated Brownian motion $B(t), t \ge 0 | B(t_0) = x$; i.e., $x + B(t - t_0)$ for `t >= t0`. The standard Brownian motion is obtained choosing `x=0` and `t0=0` (the default values).
The function `GBM` returns a trajectory of the geometric Brownian motion starting at `x` at time `t0=0`; i.e., the process

$$S(t) = x \exp\{(r - \sigma^2/2)t + \sigma B(t)\}.$$

Value

`X`	an invisible `ts` object

Examples

```
plot(BM())
plot(BBridge())
plot(GBM())
```

---

cpoint *Volatility change-point estimator for diffusion processes*

Description

Volatility change-point estimator for diffusion processes based on least squares.

Usage

```
cpoint(x, mu, sigma)
```

Arguments

`x`	a `ts` object.
`mu`	a function of `x` describing the drift coefficient.
`sigma`	a function of `x` describing the diffusion coefficient.

Details

The function returns a list of elements containing the discrete `k0` and continuous `tau0` change-point instant, the estimated volatilities before (`theta1`) and after (`theta2`) the time change. The model is assumed to be of the form

$$\mathrm{d}X_t = b(X_t)\mathrm{d}t + \theta\sigma(X_t)\mathrm{d}W_t$$

where `theta` = `theta1` for `t<=tau0` and `theta` = `theta2` otherwise.
If the drift coefficient is unknown, the model

$$\mathrm{d}X_t = b(X_t)\mathrm{d}t + \theta\mathrm{d}W_t$$

is considered and `b` is estimated nonparametrically.

Value

`X`	a list

Examples

```
tau0 <- 0.6
k0 <- ceiling(1000*tau0)
set.seed(123)
X1 <- sde.sim(X0=1, N=2*k0, t0=0, T=tau0, model="CIR",
              theta=c(6,2,1))
X2 <- sde.sim(X0=X1[2*k0+1], N=2*(1000-k0), t0=tau0,
   T=1, model="CIR", theta=c(6,2,3))

Y <- ts(c(X1,X2[-1]), start=0, deltat=deltat(X1))
X <- window(Y,deltat=0.01)
DELTA <- deltat(X)
n <- length(X)

mu <- function(x) 6-2*x
sigma <- function(x) sqrt(x)

cp <- cpoint(X,mu,sigma)
cp
plot(X)
abline(v=tau0,lty=3)
abline(v=cp$tau0,col="red")

# nonparametric estimation
cpoint(X)
```

`DBridge` *Simulation of diffusion bridge*

Description

Simulation of diffusion bridge.

Usage

```
DBridge(x=0, y=0, t0=0, T=1, delta, drift, sigma, ...)
```

Arguments

`x`	initial value of the process at time `t0`.
`y`	terminal value of the process at time `T`.
`t0`	initial time.
`delta`	time step of the simulation.
`drift`	drift coefficient: an expression of two variables `t` and `x`.
`sigma`	diffusion coefficient: an expression of two variables `t` and `x`.
`T`	final time.
`...`	passed to the `sde.sim` function.

Details

The function returns a trajectory of the diffusion bridge starting at `x` at time `t0` and ending at `y` at time `T`.
The function uses the `sde.sim` function to simulate the paths internally. Refer to the `sde.sim` documentation for further information about the argument "`...`"

Value

`X`	an invisible `ts` object

Examples

```
d <- expression((3-x))
s <- expression(1.2*sqrt(x))
par(mar=c(3,3,1,1))
par(mfrow=c(2,1))
set.seed(123)
X <- DBridge(x=1.7,y=0.5, delta=0.01, drift=d, sigma=s)
plot(X)
X <- DBridge(x=1,y=5, delta=0.01, drift=d, sigma=s)
plot(X)
```

`dcElerian` *Approximated conditional law of a diffusion process by Elerian's method*

Description

Approximated conditional densities for $X(t)|X(t_0) = x_0$ of a diffusion process.

Usage

```
dcElerian(x, t, x0, t0, theta, d, s, sx, log=FALSE)
```

Arguments

`x`	vector of quantiles.
`t`	lag or time.
`x0`	the value of the process at time `t0`; see details.
`t0`	initial time.
`theta`	parameter of the process; see details.
`log`	logical; if TRUE, probabilities p are given as $\log(p)$.
`d`	drift coefficient as a function; see details.
`s`	diffusion coefficient as a function; see details.
`sx`	partial derivative w.r.t. `x` of the diffusion coefficient; see details.

Details

This function returns the value of the conditional density of $X(t)|X(t_0) = x_0$ at point `x`.
All the functions `d`, `s`, and `sx` must be functions of `t`, `x`, and `theta`.

Value

`x`	a numeric vector

`dcEuler` *Approximated conditional law of a diffusion process*

Description

Approximated conditional densities for $X(t)|X(t_0) = x_0$ of a diffusion process.

Usage

```
dcEuler(x, t, x0, t0, theta, d, s, log=FALSE)
```

Arguments

`x`	vector of quantiles.
`t`	lag or time.
`x0`	the value of the process at time `t0`; see details.
`t0`	initial time.
`theta`	parameter of the process; see details.
`log`	logical; if TRUE, probabilities p are given as $\log(p)$.
`d`	drift coefficient as a function; see details.
`s`	diffusion coefficient as a function; see details.

Details

This function returns the value of the conditional density of $X(t)|X(t_0) = x_0$ at point `x`.
The functions `d` and `s` must be functions of `t`, `x`, and `theta`.

Value

`x`	a numeric vector

`dcKessler` *Approximated conditional law of a diffusion process by Kessler's method*

Description

Approximated conditional densities for $X(t)|X(t_0) = x_0$ of a diffusion process.

Usage

```
dcKessler(x, t, x0, t0, theta, d, dx, dxx, s, sx, sxx,
          log=FALSE)
```

Arguments

`x`	vector of quantiles.
`t`	lag or time.
`x0`	the value of the process at time `t0`; see details.
`t0`	initial time.
`theta`	parameter of the process; see details.
`log`	logical; if TRUE, probabilities p are given as $\log(p)$.
`d`	drift coefficient as a function; see details.
`dx`	partial derivative w.r.t. `x` of the drift coefficient; see details.
`dxx`	second partial derivative wrt `x^2` of the drift coefficient; see details.
`s`	diffusion coefficient as a function; see details.
`sx`	partial derivative w.r.t. `x` of the diffusion coefficient; see details.
`sxx`	second partial derivative w.r.t. `x^2` of the diffusion coefficient; see details.

Details

This function returns the value of the conditional density of $X(t)|X(t_0) = x_0$ at point `x`.
All the functions `d`, `dx`, `dxx`, `dt`, `s`, `sx`, and `sxx` must be functions of `t`, `x`, and `theta`.

Value

`x`	a numeric vector

`dcOzaki` *Approximated conditional law of a diffusion process by Ozaki's method*

Description

Approximated conditional densities for $X(t)|X(t_0) = x_0$ of a diffusion process.

Usage

```
dcOzaki(x, t, x0, t0, theta, d, dx, s, log=FALSE)
```

Arguments

`x`	vector of quantiles.
`t`	lag or time.
`x0`	the value of the process at time `t0`; see details.
`t0`	initial time.
`theta`	parameter of the process; see details.
`log`	logical; if TRUE, probabilities p are given as $\log(p)$.
`d`	drift coefficient as a function; see details.
`dx`	partial derivative w.r.t. `x` of the drift coefficient; see details.
`s`	diffusion coefficient as a function; see details.

Details

This function returns the value of the conditional density of $X(t)|X(t_0) = x_0$ at point `x`.
All the functions `d`, `dx`, and `s` must be functions of `t`, `x`, and `theta`.

Value

`x`	a numeric vector

`dcShoji` *Approximated conditional law of a diffusion process by the Shoji-Ozaki method*

Description

Approximated conditional densities for $X(t)|X(t_0) = x_0$ of a diffusion process.

Usage

```
dcShoji(x, t, x0, t0, theta, d, dx, dxx, dt, s, log=FALSE)
```

Arguments

`x`	vector of quantiles.
`t`	lag or time.
`x0`	the value of the process at time `t0`; see details.
`t0`	initial time.
`theta`	parameter of the process; see details.
`log`	logical; if TRUE, probabilities p are given as $\log(p)$.
`d`	drift coefficient as a function; see details.
`dx`	partial derivative w.r.t. `x` of the drift coefficient; see details.
`dxx`	second partial derivative w.r.t. `x^2` of the drift coefficient; see details.
`dt`	partial derivative w.r.t. `t` of the drift coefficient; see details.
`s`	diffusion coefficient as a function; see details.

Details

This function returns the value of the conditional density of $X(t)|X(t_0) = x_0$ at point `x`.
All the functions `d`, `dx`, `dxx`, `dt`, and `s` must be functions of `t`, `x`, and `theta`.

Value

`x`	a numeric vector

`dcSim` *Pedersen's simulated transition density*

Description

Simulated transition density $X(t)|X(t_0) = x_0$X(t) — X(t0) = x0 of a diffusion process based on Pedersen's method.

Usage

```
dcSim(x0, x, t, d, s, theta, M=10000, N=10, log=FALSE)
```

Arguments

`x0`	the value of the process at time 0.
`x`	value in which to evaluate the conditional density.
`t`	lag or time.
`theta`	parameter of the process; see details.
`log`	logical; if TRUE, probabilities p are given as $\log(p)$.
`d`	drift coefficient as a function; see details.
`s`	diffusion coefficient as a function; see details.
`N`	number of subintervals; see details.
`M`	number of Monte Carlo simulations, which should be an even number; see details.

Details

This function returns the value of the conditional density of $X(t)|X(0) = x_0$ at point `x`.
The functions `d` and `s`, must be functions of `t`, `x`, and `theta`.

Value

`x`	a numeric vector

Examples

```
## Not run:
d1 <- function(t,x,theta) theta[1]*(theta[2]-x)
s1 <- function(t,x,theta) theta[3]*sqrt(x)

from <- 0.08
x <- seq(0,0.2, length=100)
sle10 <- NULL
sle2 <- NULL
sle5 <- NULL
true <- NULL
set.seed(123)
for(to in x){
 sle2 <- c(sle2, dcSim(from, to, 0.5, d1, s1,
    theta=c(2,0.02,0.15), M=50000,N=2))
 sle5 <- c(sle5, dcSim(from, to, 0.5, d1, s1,
    theta=c(2,0.02,0.15), M=50000,N=5))
 sle10 <- c(sle10, dcSim(from, to, 0.5, d1, s1,
    theta=c(2,0.02,0.15), M=50000,N=10))
 true <- c(true, dcCIR(to, 0.5, from, c(2*0.02,2,0.15)))
}
```

```
par(mar=c(5,5,1,1))
plot(x, true, type="l", ylab="conditional density")
lines(x, sle2, lty=4)
lines(x, sle5, lty=2)
lines(x, sle10, lty=3)
legend(0.15,20, legend=c("exact","N=2", "N=5", "N=10"),
   lty=c(1,2,4,3))
## End(Not run)
```

DWJ *Weekly closings of the Dow-Jones industrial average*

Description

This dataset contains the weekly closings of the Dow-Jones industrial average in the period July 1971–August 1974. These data were proposed to test change-point estimators. There are 162 data, and the main evidence found by several authors is that a change in the variance occurred around the third week of March 1973.

Usage

```
data(DWJ)
```

Examples

```
data(DWJ)
ret <- diff(DWJ)/DWJ[-length(DWJ)]

par(mfrow=c(2,1))
par(mar=c(3,3,2,1))
plot(DWJ,main="Dow-Jones closings",ylab="",type="p")
plot(ret,main="Dow-Jones returns",ylab="",type="p")

cp <- cpoint(ret)
cp
abline(v=cp$tau0,lty=3)

cp <- cpoint(window(ret,end=cp$tau0))
cp
abline(v=cp$tau0,lty=3)
```

EULERloglik *Euler approximation of the likelihood*

Description

Euler approximation of the likelihood of a process solution of a stochastic differential equation. These functions are useful to calculate approximated maximum likelihood estimators when the transition density of the process is not known.

Usage

```
EULERloglik(X, theta, d, s, log = TRUE)
```

Arguments

`X`	a ts object containing a sample path of an sde.
`theta`	vector of parameters.
`d,s`	drift and diffusion coefficients; see details.
`log`	logical; if TRUE, the log-likelihood is returned.

Details

The function `EULERloglik` returns the Euler approximation of the log-likelihood. The functions `s` and `d` are the drift and diffusion coefficients with arguments `(t,x,theta)`.

Value

`x`	a number

Examples

```
set.seed(123)
d <- expression(-1*x)
s <- expression(2)
sde.sim(drift=d, sigma=s) -> X

S <- function(t, x, theta) sqrt(theta[2])
B <- function(t, x, theta) -theta[1]*x

true.loglik <- function(theta){
 DELTA <- deltat(X)
 lik <- 0
 for(i in 2:length(X))
  lik <- lik + dnorm(X[i], mean=X[i-1]*exp(-theta[1]*DELTA),
  sd = sqrt((1-exp(-2*theta[1]*DELTA))*
            theta[2]/(2*theta[1])),TRUE)
 lik
```

```
}

xx <- seq(-3,3,length=100)
sapply(xx, function(x) true.loglik(c(x,4))) -> py
sapply(xx, function(x) EULERloglik(X,c(x,4),B,S)) -> pz

# true likelihood
plot(xx,py,type="l",xlab=expression(beta),ylab="log-likelihood")
lines(xx,pz, lty=2) # Euler
```

---

gmm *Generalized method of moments estimator*

Description

Implementation of the estimator of the generalized method of moments by Hansen.

Usage

```
gmm(X, u, dim, guess, lower, upper, maxiter=30, tol1=1e-3,
   tol2=1e-3)
```

Arguments

`X`	a ts object containing a sample path of an sde.
`u`	a function of `x`, `y`, and `theta` and `DELTA`; see details.
`dim`	dimension of parameter space; see details.
`guess`	initial value of the parameters; see details.
`lower`	lower bounds for the parameters; see details.
`upper`	upper bounds for the parameters; see details.
`tol1`	tolerance for parameters; see details.
`tol2`	tolerance for Q1; see details.
`maxiter`	maximum number of iterations at the second stage; see details.

Details

The function gmm minimizes at the first stage the function `Q(theta) = t(Gn(theta)) * Gn(theta)` with respect to `theta`, where `Gn(theta) = mean(u(X[i+1], X[i], theta))`. Then a matrix of weights `W` is obtained by inverting an estimate of the long-run covariance and the quadratic function `Q1(theta) = t(Gn(theta)) * W * Gn(theta)` with starting value `theta1` (the solution at the first stage). The second stage is iterated until the first of these conditions verifies: (1) that the number

of iterations reaches `maxiter`; (2) that the Euclidean distance between `theta1` and `theta2 < tol1`; (3) that `Q1 < tol2`.
The function `u` must be a function of `(u,y,theta,DELTA)` and should return a vector of the same length as the dimension of the parameter space. The sanity checks are left to the user.

Value

x	a list with parameter estimates, the value of Q1 at the minimum, and the Hessian

Examples

```
## Not run:
alpha <- 0.5
beta <- 0.2
sigma <- sqrt(0.05)
true <- c(alpha, beta, sigma)
names(true) <- c("alpha", "beta", "sigma")

x0 <- rsCIR(1,theta=true)
set.seed(123)
sde.sim(X0=x0,model="CIR",theta=true,N=500000,delta=0.001) -> X
X <- window(X, deltat=0.1)
DELTA = deltat(X)
n <- length(X)
X <- window(X, start=n*DELTA*0.5)
plot(X)

u <- function(x, y, theta, DELTA){
  c.mean <- theta[1]/theta[2] +
             (y-theta[1]/theta[2])*exp(-theta[2]*DELTA)
  c.var <- ((y*theta[3]^2 *
     (exp(-theta[2]*DELTA)-exp(-2*theta[2]*DELTA))/theta[2] +
          theta[1]*theta[3]^2*
          (1-exp(-2*theta[2]*DELTA))/(2*theta[2]^2)))
  cbind(x-c.mean,y*(x-c.mean), c.var-(x-c.mean)^2,
        y*(c.var-(x-c.mean)^2))
}

CIR.lik <- function(theta1,theta2,theta3) {
 n <- length(X)
 dt <- deltat(X)
 -sum(dcCIR(x=X[2:n], Dt=dt, x0=X[1:(n-1)],
   theta=c(theta1,theta2,theta3), log=TRUE))
}
```

```
fit <- mle(CIR.lik, start=list(theta1=.1,  theta2=.1,theta3=.3),
    method="L-BFGS-B",lower=c(0.001,0.001,0.001), upper=c(1,1,1))
# maximum likelihood estimates
coef(fit)

gmm(X,u, guess=as.numeric(coef(fit)), lower=c(0,0,0),
    upper=c(1,1,1))

true
## End(Not run)
```

`HPloglik` *Aït-Sahalia Hermite polynomial expansion approximation of the likelihood*

Description

Aït-Sahalia Hermite polynomial expansion and Euler approximation of the likelihood of a process solution of a stochastic differential equation. These functions are useful to calculate approximated maximum likelihood estimators when the transition density of the process is not known.

Usage

```
HPloglik(X, theta, M, F, s, log=TRUE)
```

Arguments

`X`	a ts object containing a sample path of an sde.
`theta`	vector of parameters.
`M`	list of derivatives; see details.
`F`	the transform function; see details.
`s`	drift and diffusion coefficient; see details.
`log`	logical; if TRUE, the log-likelihood is returned.

Details

The function `HPloglik` returns the Hermite polynomial approximation of the likelihood of a diffusion process transformed to have a unitary diffusion coefficient. The function `F` is the transform function, and `s` is the original diffusion coefficient. The list of functions `M` contains the transformed drift in `M[[1]]` and the subsequent six derivatives in `x` of `M[[1]]`. The functions `F`, `s`, and `M` have arguments `(t,x,theta)`.

Value

`x`	a number

Examples

```
set.seed(123)
d <- expression(-1*x)
s <- expression(2)
sde.sim(drift=d, sigma=s) -> X

M0 <- function(t, x, theta) -theta[1]*x
M1 <- function(t, x, theta) -theta[1]
M2 <- function(t, x, theta) 0
M3 <- function(t, x, theta) 0
M4 <- function(t, x, theta) 0
M5 <- function(t, x, theta) 0
M6 <- function(t, x, theta) 0
mu <- list(M0, M1, M2, M3, M4, M5, M6)

F <- function(t, x, theta) x/sqrt(theta[2])
S <- function(t, x, theta) sqrt(theta[2])

true.loglik <- function(theta) {
 DELTA <- deltat(X)
 lik <- 0
 for(i in 2:length(X))
  lik <- lik + dnorm(X[i], mean=X[i-1]*exp(-theta[1]*DELTA),
   sd = sqrt((1-exp(-2*theta[1]*DELTA))*theta[2]/
             (2*theta[1])),TRUE)
 lik
}

xx <- seq(-3,3,length=100)
sapply(xx, function(x) HPloglik(X,c(x,4),mu,F,S)) -> px
sapply(xx, function(x) true.loglik(c(x,4))) -> py

plot(xx,px,type="l",xlab=expression(beta),ylab="log-likelihood")
lines(xx,py, lty=3) # true
```

---

ksmooth	*Nonparametric invariant density, drift, and diffusion coefficient estimation*

Description

Implementation of simple Nadaraya-Watson nonparametric estimation of drift and diffusion coefficient, and plain kernel density estimation of the invariant density for a one-dimensional diffusion process.

Usage

```
ksdrift(x, bw, n = 512)
ksdiff(x, bw, n = 512)
ksdens(x, bw, n = 512)
```

Arguments

x	a ts object.
bw	bandwidth.
n	number of points in which to calculate the estimates.

Details

These functions return the nonparametric estimate of the drift or diffusion coefficients for data `x` using the Nadaraya-Watson estimator for diffusion processes.
`ksdens` returns the density estimates of the invariant density.
If not provided, the bandwidth `bw` is calculated using Scott's rule (i.e., `bw = len^(-1/5)*sd(x)`) where `len=length(x)` is the number of observed points of the diffusion path.

Value

val	an invisible list of x and y coordinates and an object of class density in the case of invariant density estimation

Examples

```
set.seed(123)
theta <- c(6,2,1)
X <- sde.sim(X0 = rsCIR(1, theta), model="CIR", theta=theta,
     N=1000,delta=0.1)

b <- function(x)
 theta[1]-theta[2]*x

sigma <- function(x)
 theta[3]*sqrt(x)

minX <- min(X)
maxX <- max(X)

par(mfrow=c(3,1))
curve(b,minX,maxX)
lines(ksdrift(X),lty=3)
```

```
curve(sigma,minX, maxX)
lines(ksdiff(X),lty=3)

f <-function(x) dsCIR(x, theta)
curve(f,minX,maxX)
lines(ksdens(X),lty=3)
```

---

linear.mart.ef *Linear martingale estimating function*

Description

Apply a linear martingale estimating function to find estimates of the parameters of a process solution of a stochastic differential equation.

Usage

```
linear.mart.ef(X, drift, sigma, a1, a2, guess, lower, upper,
      c.mean, c.var)
```

Arguments

`X`	a ts object containing a sample path of an sde.
`drift`	an expression for the drift coefficient; see details.
`sigma`	an expression for the diffusion coefficient; see details.
`a1, a2`	weights or instruments.
`c.mean`	expressions for the conditional mean.
`c.var`	expressions for the conditional variance.
`guess`	initial value of the parameters; see details.
`lower`	lower bounds for the parameters; see details.
`upper`	upper bounds for the parameters; see details.

Details

The function `linear.mart.ef` minimizes a linear martingale estimating function that is a particular case of the polynomial martingale estimating functions.

Value

`x`	a vector of estimates

Examples

```
set.seed(123)
d <- expression(-1 * x)
s <- expression(1)
x0 <- rnorm(1,sd=sqrt(1/2))
sde.sim(X0=x0,drift=d, sigma=s,N=1000,delta=0.1) -> X

d <- expression(-theta * x)

linear.mart.ef(X, d, s, a1=expression(-x), lower=0, upper=Inf,
  c.mean=expression(x*exp(-theta*0.1)),
  c.var=expression((1-exp(-2*theta*0.1))/(2*theta)))
```

---

rcBS *Black-Scholes-Merton or geometric Brownian motion process conditional law*

Description

Density, distribution function, quantile function, and random generation for the conditional law $X(t)|X(0) = x_0$ of the Black-Scholes-Merton process also known as the geometric Brownian motion process.

Usage

```
dcBS(x, Dt, x0, theta, log = FALSE)
pcBS(x, Dt, x0, theta, lower.tail = TRUE, log.p = FALSE)
qcBS(p, Dt, x0, theta, lower.tail = TRUE, log.p = FALSE)
rcBS(n=1, Dt, x0, theta)
```

Arguments

`x`	vector of quantiles.
`p`	vector of probabilities.
`Dt`	lag or time.
`x0`	the value of the process at time `t`; see details.
`theta`	parameter of the Black-Scholes-Merton process; see details.
`n`	number of random numbers to generate from the conditional distribution.
`log, log.p`	logical; if TRUE, probabilities p are given as $\log(p)$.
`lower.tail`	logical; if TRUE (default), probabilities are `P[X <= x]`; otherwise, `P[X > x]`.

Details

This function returns quantities related to the conditional law of the process solution of

$$dX_t = \theta_1 X_t dt + \theta_2 X_t dW_t.$$

Constraints: $\theta_3 > 0$.

Value

`x`	a numeric vector

Examples

```
rcBS(n=1, Dt=0.1, x0=1, theta=c(2,1))
```

---

rcCIR	*Conditional law of the Cox-Ingersoll-Ross process*

Description

Density, distribution function, quantile function and random generation for the conditional law $X(t + D_t)|X(t) = x_0$ of the Cox-Ingersoll-Ross process.

Usage

```
dcCIR(x, Dt, x0, theta, log = FALSE)
pcCIR(x, Dt, x0, theta, lower.tail = TRUE, log.p = FALSE)
qcCIR(p, Dt, x0, theta, lower.tail = TRUE, log.p = FALSE)
rcCIR(n=1, Dt, x0, theta)
```

Arguments

`x`	vector of quantiles.
`p`	vector of probabilities.
`Dt`	lag or time.
`x0`	the value of the process at time `t`; see details.
`theta`	parameter of the Ornstein-Uhlenbeck process; see details.
`n`	number of random numbers to generate from the conditional distribution.
`log, log.p`	logical; if TRUE, probabilities p are given as $\log(p)$.
`lower.tail`	logical; if TRUE (default), probabilities are `P[X <= x]`; otherwise `P[X > x]`.

Details

This function returns quantities related to the conditional law of the process solution of

$$\mathrm{d}X_t = (\theta_1 - \theta_2 X_t)\mathrm{d}t + \theta_3\sqrt{X_t}\mathrm{d}W_t.$$

Constraints: $2\theta_1 > \theta_3^2$, all θ positive.

Value

`x`	a numeric vector

Examples

```
rcCIR(n=1, Dt=0.1, x0=1, theta=c(6,2,2))
```

`rcOU` *Ornstein-Uhlenbeck or Vasicek process conditional law*

Description

Density, distribution function, quantile function, and random generation for the conditional law $X(t + D_t)|X(t) = x_0$ of the Ornstein-Uhlenbeck process, also known as the Vasicek process.

Usage

```
dcOU(x, Dt, x0, theta, log = FALSE)
pcOU(x, Dt, x0, theta, lower.tail = TRUE, log.p = FALSE)
qcOU(p, Dt, x0, theta, lower.tail = TRUE, log.p = FALSE)
rcOU(n=1, Dt, x0, theta)
```

Arguments

`x`	vector of quantiles.
`p`	vector of probabilities.
`Dt`	lag or time.
`x0`	the value of the process at time `t`; see details.
`theta`	parameter of the Ornstein-Uhlenbeck process; see details.
`n`	number of random numbers to generate from the conditional distribution.
`log, log.p`	logical; if TRUE, probabilities p are given as $\log(p)$.
`lower.tail`	logical; if TRUE (default), probabilities are `P[X <= x]`; otherwise `P[X > x]`.

Details

This function returns quantities related to the conditional law of the process solution of

$$dX_t = (\theta_1 - \theta_2 X_t)dt + \theta_3 dW_t.$$

Constraints: $\theta_2 > 0, \theta_3 > 0$.
Please note that the process is stationary only if $\theta_2 > 0$.

Value

`x`	a numeric vector

Examples

```
rcOU(n=1, Dt=0.1, x0=1, theta=c(0,2,1))
```

`rsCIR` *Cox-Ingersoll-Ross process stationary law*

Description

Density, distribution function, quantile function, and random generation of the stationary law for the Cox-Ingersoll-Ross process.

Usage

```
dsCIR(x, theta, log = FALSE)
psCIR(x, theta, lower.tail = TRUE, log.p = FALSE)
qsCIR(p, theta, lower.tail = TRUE, log.p = FALSE)
rsCIR(n=1, theta)
```

Arguments

`x`	vector of quantiles.
`p`	vector of probabilities.
`theta`	parameter of the Cox-Ingersoll-Ross process; see details.
`n`	number of random numbers to generate from the conditional distribution.
`log, log.p`	logical; if TRUE, probabilities p are given as $\log(p)$.
`lower.tail`	logical; if TRUE (default), probabilities are `P[X <= x]`; otherwise `P[X > x]`.

Details

This function returns quantities related to the stationary law of the process solution of

$$dX_t = (\theta_1 - \theta_2 X_t)dt + \theta_3\sqrt{X_t}dW_t.$$

Constraints: $2\theta_1 > \theta_3^2$, all θ positive.

Value

`x`	a numeric vector

Examples

```
rsCIR(n=1, theta=c(6,2,1))
```

`rsOU` *Ornstein-Uhlenbeck or Vasicek process stationary law*

Description

Density, distribution function, quantile function, and random generation for the stationary law of the Ornstein-Uhlenbeck process also known as the Vasicek process.

Usage

```
dsOU(x, theta, log = FALSE)
psOU(x, theta, lower.tail = TRUE, log.p = FALSE)
qsOU(p, theta, lower.tail = TRUE, log.p = FALSE)
rsOU(n=1, theta)
```

Arguments

`x`	vector of quantiles.
`p`	vector of probabilities.
`theta`	parameter of the Ornstein-Uhlenbeck process; see details.
`n`	number of random numbers to generate from the conditional distribution.
`log, log.p`	logical; if TRUE, probabilities p are given as $\log(p)$.
`lower.tail`	logical; if TRUE (default), probabilities are `P[X <= x]`; otherwise `P[X > x]`.

Details

This function returns quantities related to the stationary law of the process solution of

$$dX_t = (\theta_1 - \theta_2 X_t)dt + \theta_3 dW_t.$$

Contraints: $theta_2 > 0, \theta_3 > 0$.
Please note that the process is stationary only if $\theta_2 > 0$.

Value

`x`	a numeric vector

Examples

```
rsOU(n=1, theta=c(0,2,1))
```

`sde.sim` *Simulation of stochastic differential equation*

Description

Generic interface to different methods of simulation of solutions to stochastic differential equations.

Usage

```
sde.sim(t0 = 0, T = 1, X0 = 1, N = 100, delta, drift, sigma,
   drift.x, sigma.x, drift.xx, sigma.xx, drift.t,
   method = c("euler", "milstein", "KPS", "milstein2",
   "cdist","ozaki","shoji","EA"),
   alpha = 0.5, eta = 0.5, pred.corr = T, rcdist = NULL,
   theta = NULL, model = c("CIR", "VAS", "OU", "BS"),
   k1, k2, phi, max.psi = 1000, rh, A, M=1)
```

Arguments

`t0`	time origin.
`T`	horizon of simulation.
`X0`	initial value of the process.
`N`	number of simulation steps.
`M`	number of trajectories.
`delta`	time step of the simulation.
`drift`	drift coefficient: an expression of two variables `t` and `x`.
`sigma`	diffusion coefficient: an expression of two variables `t` and `x`.

`drift.x`	partial derivative of the drift coefficient w.r.t. `x`: a function of two variables `t` and `x`.
`sigma.x`	partial derivative of the diffusion coefficient w.r.t. `x`: a function of two variables `t` and `x`.
`drift.xx`	second partial derivative of the drift coefficient w.r.t. `x`: a function of two variables `t` and `x`.
`sigma.xx`	second partial derivative of the diffusion coefficient w.r.t. `x`: a function of two variables `t` and `x`.
`drift.t`	partial derivative of the drift coefficient w.r.t. `t`: a function of two variables `t` and `x`.
`method`	method of simulation; see details.
`alpha`	weight `alpha` of the predictor-corrector scheme.
`eta`	weight `eta` of the predictor-corrector scheme.
`pred.corr`	boolean: whether to apply the predictor-correct adjustment; see details.
`rcdist`	a function that is a random number generator from the conditional distribution of the process; see details.
`theta`	vector of parameters for `cdist`; see details.
`model`	model from which to simulate; see details.
`k1`	lower bound for `psi(x)`; see details.
`k2`	upper bound for `psi(x)`; see details.
`phi`	the function `psi(x) - k1`.
`max.psi`	upper value of the support of `psi` to search for its maximum.
`rh`	the rejection function; see details.
`A`	`A(x)` is the integral of the `drift` between `0` and `x`.

Details

The function returns a `ts` object of length `N+1`; i.e., `X0` and the new `N` simulated values if `M=1`. For `M>1`, an `mts` (multidimensional `ts` object) is returned, which means that `M` independent trajectories are simulated. If the initial value `X0` is not of the length `M`, the values are recycled in order to have an initial vector of the correct length. If `delta` is not specified, then `delta = (T-t0)/N`. If `delta` is specified, then `N` values of the solution of the sde are generated and the time horizon `T` is adjusted to be `N * delta`.
The function `psi` is `psi(x) = 0.5*drift(x)^2 + 0.5*drift.x(x)`.
If any of `drift.x`, `drift.xx`, `drift.t`, `sigma.x`, and `sigma.xx` are not specified, then numerical derivation is attempted when needed.
If `sigma` is not specified, it is assumed to be the constant function `1`.
The `method` of simulation can be one among: `euler`, `KPS`, `milstein`, `milstein2`, `cdist`, `EA`, `ozaki`, and `shoji`. No assumption on the coefficients or on `cdist` is checked: the user is responsible for using the right method for the process object of simulation.

The `model` is one among: `CIR`: Cox-Ingersoll-Ross, `VAS`: Vasicek, `OU` Ornstein-Uhlenbeck, `BS`: Black and Scholes. No assumption on the coefficient `theta` is checked: the user is responsible for using the right ones. If the `method` is `cdist`, then the process is simulated according to its known conditional distribution. The random generator `rcdist` must be a function of `n`, the number of random numbers; `dt`, the time lag; `x`, the value of the process at time `t - dt`; and the vector of parameters `theta`. For the exact algorithm method `EA`: if missing `k1` and `k2` as well as `A`, `rh` and `phi` are calculated numerically by the function.

Value

x	returns an invisible `ts` object

Examples

```
# Ornstein-Uhlenbeck process
set.seed(123)
d <- expression(-5 * x)
s <- expression(3.5)
sde.sim(X0=10,drift=d, sigma=s) -> X
plot(X,main="Ornstein-Uhlenbeck")

# Multiple trajectories of the O-U process
set.seed(123)
sde.sim(X0=10,drift=d, sigma=s, M=3) -> X
plot(X,main="Multiple trajectories of O-U")

# Cox-Ingersoll-Ross process
# dXt = (6-3*Xt)*dt + 2*sqrt(Xt)*dWt
set.seed(123)
d <- expression( 6-3*x )
s <- expression( 2*sqrt(x) )
sde.sim(X0=10,drift=d, sigma=s) -> X
plot(X,main="Cox-Ingersoll-Ross")

# Cox-Ingersoll-Ross using the conditional distribution "rcCIR"

set.seed(123)
sde.sim(X0=10, theta=c(6, 3, 2), rcdist=rcCIR,
        method="cdist") -> X
plot(X, main="Cox-Ingersoll-Ross")

set.seed(123)
sde.sim(X0=10, theta=c(6, 3, 2), model="CIR") -> X
plot(X, main="Cox-Ingersoll-Ross")
```

```
# Exact simulation
set.seed(123)
d <- expression(sin(x))
d.x <- expression(cos(x))
A <- function(x) 1-cos(x)
sde.sim(method="EA", delta=1/20, X0=0, N=500,
        drift=d, drift.x = d.x, A=A) -> X
plot(X, main="Periodic drift")
```

---

sdeAIC *Akaike's information criterion for diffusion processes*

Description

Implementation of the AIC statistics for diffusion processes.

Usage

```
sdeAIC(X, theta, b, s, b.x, s.x, s.xx, B, B.x, H, S, guess,
       ...)
```

Arguments

X	a ts object containing a sample path of an sde.
theta	a vector or estimates of the parameters.
b	drift coefficient of the model as a function of x and theta.
s	diffusion coefficient of the model as a function of x and theta.
b.x	partial derivative of b as a function of x and theta.
s.x	partial derivative of s as a function of x and theta.
s.xx	second-order partial derivative of s as a function of x and theta.
B	initial value of the parameters; see details.
B.x	partial derivative of B as a function of x and theta.
H	function of (x,y), the integral of B/s; optional.
S	function of (x,y), the integral of 1/s; optional.
guess	initial value for the parameters to be estimated; optional.
...	passed to the optim function; optional.

Details

The `sdeAIC` evaluates the AIC statistics for diffusion processes using Dacunha-Castelle and Florens-Zmirou approximations of the likelihood. The parameter `theta` is supposed to be the value of the true MLE estimator or the minimum contrast estimator of the parameters in the model. If missing or `NULL` and `guess` is specified, `theta` is estimated using the minimum contrast estimator derived from the locally Gaussian approximation of the density. If both `theta` and `guess` are missing, nothing can be calculated.
If missing, `B` is calculated as `b/s - 0.5*s.x` provided that `s.x` is not missing.
If missing, `B.x` is calculated as `b.x/s - b*s.x/(s^2)-0.5*s.xx`, provided that `b.x`, `s.x`, and `s.xx` are not missing.
If missing, both `H` and `S` are evaluated numerically.

Value

x the value of the AIC statistics

Examples

```
set.seed(123)
# true model generating data
dri <- expression(-(x-10))
dif <- expression(2*sqrt(x))
sde.sim(X0=10,drift=dri, sigma=dif,N=1000,delta=0.1) -> X

# we test the true model against two competing models
b <- function(x,theta) -theta[1]*(x-theta[2])
b.x <- function(x,theta)  -theta[1]+0*x

s <- function(x,theta) theta[3]*sqrt(x)
s.x <- function(x,theta) theta[3]/(2*sqrt(x))
s.xx <- function(x,theta) -theta[3]/(4*x^1.5)
# AIC for the true model
sdeAIC(X, NULL, b, s, b.x, s.x, s.xx, guess=c(1,1,1),
       lower=rep(1e-3,3), method="L-BFGS-B")

s <- function(x,theta) sqrt(theta[3]*+theta[4]*x)
s.x <- function(x,theta) theta[4]/(2*sqrt(theta[3]+theta[4]*x))
s.xx <- function(x,theta) -theta[4]^2/(4*(theta[3]+theta[4]*x)^1.5)
# AIC for competing model 1
sdeAIC(X, NULL, b, s, b.x, s.x, s.xx, guess=c(1,1,1,1),
       lower=rep(1e-3,4), method="L-BFGS-B")

s <- function(x,theta) (theta[3]+theta[4]*x)^theta[5]
```

```
s.x <- function(x,theta)
           theta[4]*theta[5]*(theta[3]+theta[4]*x)^(-1+theta[5])
s.xx <- function(x,theta) (theta[4]^2*theta[5]*(theta[5]-1)
                 *(theta[3]+theta[4]*x)^(-2+theta[5]))
# AIC for competing model 2
sdeAIC(X, NULL, b, s, b.x, s.x, s.xx, guess=c(1,1,1,1,1),
       lower=rep(1e-3,5), method="L-BFGS-B")
```

`SIMloglik` *Pedersen's approximation of the likelihood*

Description

Pedersen's approximation of the likelihood of a process solution of a stochastic differential equation. This function is useful to calculate approximated maximum likelihood estimators when the transition density of the process is not known. It is computationally intensive.

Usage

```
SIMloglik(X, theta, d, s,  M=10000, N=2, log=TRUE)
```

Arguments

`X`	a ts object containing a sample path of an sde.
`theta`	vector of parameters.
`d,s`	drift and diffusion coefficients; see details.
`log`	logical; if TRUE, the log-likelihood is returned.
`N`	number of subintervals; see details.
`M`	number of Monte Carlo simulations, which should be an even number; see details.

Details

The function `SIMloglik` returns the simulated log-likelihood obtained by Pedersen's method. The functions `s` and `d` are the drift and diffusion coefficients with arguments `(t,x,theta)`.

Value

`x`	a number

Examples

```
## Not run:
set.seed(123)
d <- expression(-1*x)
s <- expression(2)
sde.sim(drift=d, sigma=s,N=50,delta=0.01) -> X

S <- function(t, x, theta) sqrt(theta[2])
B <- function(t, x, theta) -theta[1]*x

true.loglik <- function(theta) {
 DELTA <- deltat(X)
 lik <- 0
 for(i in 2:length(X))
  lik <- lik + dnorm(X[i], mean=X[i-1]*exp(-theta[1]*DELTA),
   sd = sqrt((1-exp(-2*theta[1]*DELTA))*
              theta[2]/(2*theta[1])),TRUE)
 lik
}

xx <- seq(-10,10,length=20)
sapply(xx, function(x) true.loglik(c(x,4))) -> py
sapply(xx, function(x) EULERloglik(X,c(x,4),B,S)) -> pz
sapply(xx, function(x) SIMloglik(X,c(x,4),B,S,M=10000,N=5)) -> pw

plot(xx,py,type="l",xlab=expression(beta),
   ylab="log-likelihood",ylim=c(0,15)) # true
lines(xx,pz, lty=2) # Euler
lines(xx,pw, lty=3) # Simulated
## End(Not run)
```

`simple.ef` *Simple estimating functions of types I and II*

Description

Apply a simple estimating function to find estimates of the parameters of a process solution of a stochastic differential equation.

Usage

```
simple.ef(X, f, guess, lower, upper)
```

Arguments

X	a ts object containing a sample path of an sde.

`f`	a list of expressions of `x` and/or `y` and the parameters to be estimated; see details.
`guess`	initial value of the parameters; see details.
`lower`	lower bounds for the parameters; see details.
`upper`	upper bounds for the parameters; see details.

Details

The function `simple.ef` minimizes a simple estimating function of the form `sum_i f_i(x,y;theta) = 0` or `sum_i f_i(x;theta)` as a function of `theta`. The index `i` varies in `1:length(theta)`.
The list `f` is a list of expressions in `x` or `(x,y)`.

Value

`x`	a vector of estimates

Examples

```
set.seed(123);
# Kessler's estimator for O-H process
K.est <- function(x) {
  n.obs <- length(x)
  n.obs/(2*(sum(x^2)))
}

# Least squares estimators for the O-H process
LS.est <- function(x) {
  n <- length(x) -1
  k.sum <- sum(x[1:n]*x[2:(n+1)])
  dt <- deltat(x)
  ifelse(k.sum>0, -log(k.sum/sum(x[1:n]^2))/dt, NA)
}

d <- expression(-1 * x)
s <- expression(1)
x0 <- rnorm(1,sd=sqrt(1/2))
sde.sim(X0=x0,drift=d, sigma=s,N=2500,delta=0.1) -> X

# Kessler's estimator as estimating function
f <- list(expression(2*theta*x^2-1))
simple.ef(X, f, lower=0, upper=Inf)
K.est(X)

# Least Squares estimator as estimating function
f <- list(expression(x*(y-x*exp(-0.1*theta))))
simple.ef(X, f, lower=0, upper=Inf)
LS.est(X)
```

`simple.ef2`	*Simple estimating function based on the infinitesimal generator a the diffusion process*

Description

Apply a simple estimating function based on the infinitesimal generator of a diffusion to find estimates of the parameters of a process solution of that particular stochastic differential equation.

Usage

```
simple.ef2(X, drift, sigma, h, h.x, h.xx, guess, lower,
           upper)
```

Arguments

`X`	a `ts` object containing a sample path of an sde.
`drift`	an expression for the drift coefficient; see details.
`sigma`	an expression for the diffusion coefficient; see details.
`h`	an expression of `x` and the parameters to be estimated; see details.
`h.x`	an expression of `x` containing the first derivative of `h`; see details.
`h.xx`	an expression of `x` containing the second derivative of `h`; see details.
`guess`	initial value of the parameters; see details.
`lower`	lower bounds for the parameters; see details.
`upper`	upper bounds for the parameters; see details.

Details

The function `simple.ef2` minimizes the simple estimating function of the form `sum_i f_i(x;theta) = 0`, where `f` is the result of applying the infinitesimal generator of the diffusion to the function `h`. This involves the drift and diffusion coefficients plus the first two derivatives of `h`. If not provided by the user, the derivatives are calculated by the function.

Value

`x`	a vector of estimates

Examples

```
set.seed(123)
d <- expression(10 - x)
s <- expression(sqrt(x))
x0 <- 10
sde.sim(X0=x0,drift=d, sigma=s,N=1500,delta=0.1) -> X

# rather difficult problem unless a good initial guess is given
d <- expression(alpha + theta*x)
s <- expression(x^gamma)
h <- list(expression(x), expression(x^2), expression(x^2))
simple.ef2(X, d, s, h, lower=c(0,-Inf,0), upper=c(Inf,0,1))
```

References

1. Abramowitz, M., Stegun, I.A. (1964) *Handbook of Mathematical Functions*, Dover Publications, New York.
2. Ahn, D.H., Gao, B. (1999) A parametric nonlinear model of term structure dynamics, *Rev. Financial Stud.*, **12**, 721–762.
3. Aït-Sahalia, Y. (1996) Nonparametric pricing of interest rate derivative securities, *Econometrica*, **64**, 527–560.
4. Aït-Sahalia, Y. (1996) Testing continuous-time models of the spot interest rate, *Rev. Financial Stud.*, **9**(2), 385–426.
5. Aït-Sahalia, Y. (2002) Comment on "G. Durham, A. Gallant, Numerical techniques for maximum likelihood estimation of continuous-time diffusion processes, J. Bus. Econ. Stat., 20, 297–316". *J. Bus. Econ. Stat.*, **20**, 317–321.
6. Aït-Sahalia, Y. (2002) Maximum likelihood estimation of discretely sampled diffusions: a closed form approximation approach. *Econometrica*, **70**, 223–262.
7. Aït-Sahaia, Y, Lo, W. (1998) Nonparametric estimation of state price densities implicit in financial asset prices, *J. Finance*, **53**, 499–547.
8. Aït-Sahaia, Y. (1999) Transition densities for interest rate and other nonlinear diffusions, *J. Finance*, **54**, 1361–1395.
9. Akaike, H. (1973) Information theory and an extension of the maximum likelihood principle, *2nd International Symposium in Information Theory*, Petrov, B.N., Csaki, F., eds., Akademiai Kaido, Budapest, 267–281.
10. Akaike, H. (1974) A new look at the statistical model identification, *IEEE Trans. Autom. Control*, **AC-19**, 716–723.
11. Albert, J. (2007) *Bayesian Computation with R*, Use R Series, Springer, New York.
12. Alcock, J., Burrage, K. (2004) A genetic estimation algorithm for parameters of stochastic ordinary differential equations, *Comput. Stat. Data Anal.*, **47**, 255–275.
13. Arato, M. (1982) *Linear Stochastic Systems with Constant Coefficients: A Statistical Approach*, Lecture Notes in Control and Information Sciences, **45**, Springer, Berlin.
14. Arnold, L. (1974) *Stochastic Differential Equations: Theory and Applications*, John Wiley and Sons, New York.
15. Bai, J. (1994) Least squares estimation of a shift in linear processes, *J. Time Ser. Anal.*, **15**, 453–472.

16. Bai, J. (1997) Estimation of a change point in multiple regression models, *Rev. Econ. Stat.*, **79**, 551–563.
17. Bailey, N.T.J. (1957) *The Mathematical Theory of Epidemics*, Griffin, London.
18. Bandi, F., Phillips, P. (2003) Fully nonparametric estimation of scalar diffusion models, *Econometrica*, **71**, 241–283.
19. Banon, G. (1978) Nonparametric identification for diffusion processes, *SIAM J. Control Optim.*, **16**, 380–395.
20. Banks, H.T. (1975) *Modeling and Control in the Biological Sciences*, Lecture Notes in Biomathematics, **6**, Springer-Verlag, Berlin.
21. Barndorff-Nielsen, O.E. (1978) Hyperbolic distributions and distributions on hyperbolae, *Scand. J. Stat.*, **5**, 151–157.
22. Barndorff-Nielsen, O.E., Kent, J., Sørensen, M. (1982) Normal variance-mean mixtures and z-distributions, *Int. Stat. Rev.*, **50**, 145–159.
23. Bartholomew, D.J. (1973) *Stochastic Models for the Social Processes*, 2nd ed., Wiley, New York.
24. Basawa, I.V., Prakasa Rao, B.L.S. (1980) *Statistical Inference for Stochastic Processes*, Academic Press, London.
25. Beckers, S. (1980) The constant elasticity of variance model and its implications in option pricing, *J. Finance*, **35**(3), 661–673.
26. Bergstrom, A.R. (1990) *Continuous Time Econometric Modeling*, Oxford University Press, Oxford.
27. Beskos, A., Papaspiliopoulos, O., Roberts, G.O. (2004) Retrospective exact simulation of diffusion sample paths with applications, Working paper of Lancaster University. Available at `http://www.maths.lancs.ac.uk/papaspil/research.html`.
28. Beskos, A., Papaspiliopoulos, O., Roberts, G.O. (2007) A new factorization of diffusion measure with view towards simulation, Working paper of Lancaster University. Available at `http://www.maths.lancs.ac.uk/papaspil/research.html`.
29. Beskos, A. and Roberts, G.O. (2005) Exact simulation of diffusions, *Ann. Appl. Probab.*, **4**, 2422–2444.
30. Bibby, B., Jacobsen, M., Sørensen, M. (2007) Estimating functions for discretely sampled diffusion-type models, in *Handbook of Financial Econometrics*, Aït-Sahalia, Y., Hansen, L.P., eds, North-Holland, Amsterdam.
31. Bibby, B., Sørensen, M. (1995) Martingale estimating functions for discretely observed diffusion processes, *Bernoulli*, **1**, 17–39.
32. Bibby, B.M., Sørensen, M. (1997) A hyperbolic diffusion model for stock prices, *Finance Stochastics*, **1**, 25–41.
33. Bibby, B.M., Sørensen, M. (2001) Simplified estimating functions for diffusion models with a high-dimensional parameter, *Scand. J. Stat.*, **28**(1), 99–112.
34. Bibby, B.M., Skovgaard, I.M., Sørensen, M. (2005) Diffusion-type models with given marginals and autocorrelation function, *Bernoulli*, **1**, 191–220.
35. Billingsley, P. (1961) *Statistical Inference for Markov Processes*, University of Chicago Press, Chicago.
36. Black, F., Scholes, M.S. (1973) The pricing of options and corporate liabilities, *J. Polit. Econ.*, **81**, 637–654.
37. Bladt, M., Sørensen, M. (2007) Simple simulation of diffusion bridges with application to likelihood inference for diffusions, Working Paper, University of Copenhagen. Available at `http://www.math.ku.dk/~michael/diffusionbridgepreprint.pdf`.

38. Borodin, A.N., Salminen, P. (1996) *Handbook of Brownian Motion–Facts and Formulae*, Birkhäuser, Boston.
39. Bosq, D. (1998) *Nonparametric Statistics for Stochastic Processes*, 2nd ed., Springer-Verlag, New York.
40. Box, G.E.P., Jenkins, G.M. (1970) *Time Series Analysis: Forecasting and Control*, Holden-Day, San Francisco.
41. Brennan, M.J., Schwartz, E. (1979) A continuous-time approach to the pricing of bonds, *J. Banking Finance*, **3**, 133–155.
42. Brennan, M.J., Schwartz, E. (1980) Analyzing convertible securities, *J. Financial Quant. Anal.*, **15**(4), 907–929
43. Brennan, M.J., Schwartz, E.S. (1982) An equilibrium model of bond pricing and a test of market efficiency, *J. Financial Quant. Anal.*, **17**, 75–100.
44. Brugière, P. (1991) Estimation de la variance d'un processus de diffusion dans le cas multidimensionnel, *C. R. Acad. Sci.*, T. 312, Ser. I, **13**, 999–1005.
45. Brugière, P. (1993) Théorém de limite centrale pour un estimateur non paramétrique de la variance d'un processus de diffusion multidimensionnelle, *Ann. Inst. Henri Poincaré*, **29**(3), 357–389.
46. Canabarro, E. (1995) Where do one-factor interest models fail?, *J. Fixed Income*, **5**, 31–52.
47. Carmona, R. (2004) *Statistical Analysis of Financial Data in S-Plus*, Springer-Verlag, New York.
48. Christensen, B.J., Poulsen, R., Sørensen, M. (2001) Optimal inference for diffusion processes with applications to the short rate of interest, Working Paper No. 102, Centre for Analytical Finance, University of Aarhus.
49. Chan, K.C., Karolyi, G.A., Longstaff, F.A., Sanders, A.B. (1992) An empirical investigation of alternative models of the short-term interest rate, *J. Finance*, **47**, 1209–1227.
50. Chen, G., Choi, Y.K., Zhou, Y. (2005) Nonparametric estimation of structural change points in volatility models for time series, *J. Econ.*, **126**, 79–144.
51. Clement, E. (1997) Estimation of diffusion processes by simulated moment methods, *Scand. J. Stat.*, **24**, 353–369.
52. Cleur, E.M. (2001) Maximum likelihood estimates of a class of one-dimensional stochastic differential equation models from discrete data, *J. Time Ser. Anal.*, **22**(5), 505–515.
53. Cleur, E.M. (1999) One dimensional SDE models, low order numerical methods and simulation based estimation: a comparison of alternative estimators, *Comput. Econ.*, **13**, 477–497.
54. Cobb, L. (1978) Stochastic catastrophe models and multimodal distributions, *Behav. Sci.*, **23**, 360–374.
55. Cobb, L. (1981) Stochastic differential equations for the social sciences, in *Mathematical Frontiers of the Social and Policy Sciences*, Cobb, L., Thrall, M., eds, Westview Press, Boulder, co, 1–26.
56. Comte, F., Genon-Catalot, V., Rozenholc, Y. (2007) Penalized nonparametric mean square estimation of the coefficients of diffusion processes, *Bernoulli*, **13**(2), 514–543.
57. Cox, J.C. (1996), The constant elasticity of variance option pricing model, *J. Portfolio Manag.*, **3**, 15–17.
58. Cox, J.C., Ingersoll, J.E., Ross, S.A. (1980) An analysis of variable rate loan contracts, *J. Finance*, **35**(2), 389–403.

59. Cox, J.C., Ingersoll, J.E., Ross, S.A. (1985) A theory of the term structure of interest rates, *Econometrica*, **53**, 385–408.
60. Cramér, H. (1925) On some classes of series used in mathematical statistics, in *Proceedings of the Sixth Scandinavian Mathematical Congress*, 39–425.
61. Csörgö, M., Horváth, L. (1997) *Limit Theorems in Change-Point Analysis*, Wiley and Sons, New York.
62. Dacunha-Castelle, D., Florens-Zmirou, D. (1986) Estimation of the coefficients of a diffusion from discrete observations, *Stochastics*, **19**, 263–284.
63. Dalgaard, P. (2002) *Introductory Statistics with R*, Springer, New York.
64. De Gregorio, A., Iacus, S.M. (2007) Least squares volatility change point estimation for partially observed diffusion processes, Working Paper. Available at `http://arxiv.org/abs/0709.2967v1`.
65. Dellaportas, P., Friel, N., Roberts, G.O. (2004) Bayesian model selection for partially observed diffusion models, Preprint.
66. Delattre, S. (1997) Estimation du coefficient de diffusion pour un processus de diffusion en présance d'erreurs d'arrondi. Thesis, Université Paris-6.
67. Delattre, S., Jacod, J. (1997) A central limit theorem for normalized functions of the increments of a diffusion process, in the presence of round-off errors, *Bernoulli*, **3**(1), 1–28.
68. Devroye, L. (1986) Pseudorandom number generation by nonlinear method, *Int. Stat. Rev.*, **63**, 247–255.
69. Ditlevsen, P.D., Ditlevsen, S., Andersen, K.K. (2002) The fast climate fluctuations during the stadial and interstadial climate states, *Ann. Glaciol.*, **35**, 457–462.
70. Doss, H. (1977) Liens entre equations differentielles stochastiques et ordinaires, *Ann. Inst. Henri Poincaŕe*, **13**, 99–125.
71. Dothan, U.L. (1978), On the term structure of interest rates, *J. Financial Econ.*, **6**, 59–69.
72. Dyrting, S. (2004) Evaluating the noncentral chi-square distribution for the Cox-Ingersoll-Ross process, *Comput. Econ.*, **24**, 35–50.
73. Durett, R. (1996) *Stochastic Calculus: A Practical Introduction*, CRC Press, Boca Raton.
74. Durham, G., Gallant, A. (2002) Numerical techniques for maximum likelihood estimation of continuous-time diffusion processes, *J. Bus. Econ. Stat.*, **20**, 297–316.
75. Duffie, D., Glynn, P. (2004) Estimation of continuous-time Markov processes sampled at random time intervals, *Econometrica*, **72**(6), 1773–1808.
76. Elerian, O. (1998) A note on the existence of a closed form conditional density for the Milstein scheme, Working Paper, Nuffield College, Oxford University. Available at `http://www.nuff.ox.ac.uk/economics/papers/`.
77. Eberlein, E., Keller, U. (1995) Hyperbolic distributions in finance, *Bernoulli*, **1**, 281-299.
78. Elerian, O., Chib, S., Shepard, N. (2001) Likelihood inference for discretely observed nonlinear diffusions, *Econometrica*, **69**(6), 959–993.
79. Epanechnikov, V.A. (1969) Nonparametric estimates of multivariate probability density, *Theory Probab. Appl.*, **14**, 153–158.
80. Eraker, B. (2001) MCMC analysis of diffusion models with application to finance, *J. Bus. Econ. Stat.*, **19**, 177–191.
81. Fan, J. (2005) A selective overview of nonparametric methods in financial econometrics, *Stat. Sci.*, **20**(4), 317–337.

82. Feller, W. (1951a) Two singular diffusion problems, *Ann. of Math.*, **54**(1), 173–182.
83. Feller, W. (1951b) Diffusion processes in genetics, in *Proceedings of the 2nd Berkeley Symposium on Mathematical Statistics and Probability*, Neyman,J. ed., University of California Press, Berkeley, 227–246.
84. Feller, W. (1971) *An Introduction to Probability Theory and Its Applications*, Vol 2, 2nd ed., John Wiley & Sons, Inc., Princeton University, New York.
85. Florens, D. (1999) Estimation of the diffusion coefficient from the crossings, *Stat. Infer. Stochastic Process.*, **1**, 175–195.
86. Florens-Zmirou, D. (1989) Approximate discrete time schemes for statistics of diffusion processes, *Statistics*, **20**, 547–557.
87. Florens-Zmirou, D. (1993) On estimating the diffusion coefficient from discrete observations, *J. App. Prob.*, **30**, 790–804.
88. Fölmer, H., Schied, A. (2002) *Stochastic Finance: An Introduction in Discrete Time*, De Gruyter Studies in Mathematics, Berlin.
89. Forman, J.L., Sørensen, M. (2006) The Pearson diffusions: a class of statistically tractable diffusion processes, WP17 Nov.2006, Danish Center for Accounting and Finance. Available at `http://www.d-caf.dk/wp/dcaf-wp-17.pdf`.
90. Gallant, A.R., Long, J. (1997) Estimating stochastic diffusion equations efficiently by minimum chi-squared, *Biometrika*, **84**, 125–141.
91. Gallant, A.R., Tauchen, G. (1992) A nonparametric approach to nonlinear time series analysis: estimation and simulation, in *New Directions in Time Series Analysis, Part II*, Brillinger, D., Caines, P., Geweke, J., Parzen, E., Rosenblatt, M., Taqqu, M.S., eds., Springer-Verlag, New York, 71–92.
92. Gallant, A.R., Tauchen, G. (1996) Which moments to match, *Econ. Theory*, **12**, 657–681.
93. Gallant, A.R., Tauchen, G. (2001) EMM: a program for efficient method of moments estimations. Available at `http://www.econ.duke.edu/~get/emm.htm`.
94. Genon-Catalot, V., Laredo, C., Picard, D. (1992) Nonparametric estimation of the diffusion coefficient by wavelets methods, *Scand. J. Stat.*, **19**, 317–335.
95. Genon-Catalot, V., Jacod, J. (1993) On the estimation of the diffusion coefficient for multidimensional diffusion processes, *Ann. Inst. Henri Poincaré*, **29**, 119–151.
96. Genon-Catalot, V., Jacod, J. (1994) On the estimation of the diffusion coefficient for diffusion processes: random sampling, *Scand. J. Stat.*, **21**, 192–221.
97. Gloter, A., Jacod, J. (2001) Diffusions with measurement errors I+II, *ESAIM: Probab. Stat.*, **5**, 225–260.
98. Good, I.J. (1953) The population frequencies species and the estimation of population parameters, *Biometrika*, **40**, 237–260.
99. Gourieroux, C., Monfort, A., Renault, E. (1993) Indirect inference, *J. Appl. Econ.*, **8**, 85–118.
100. Gray, P. (2002) Bayesian estimation of financial models, *Accounting Finance*, **42**, 97–116.
101. Hall, A.R. (2005) *Generalized Method of Moments*, Advanced Texts in Econometrics, Oxford University Press, New York.
102. Hanes, D.P., Schall, J.D. (1988) Neural control of voluntary movement initiation, *Science*, **274**, 427–430.
103. Hansen, L.P. (1982) Large sample properties of generalized method of moments estimators, *Econometrica*, **50**(4), 1029–1054.

104. Hansen, L.P., Scheinkman, J.A. (1995) Back to the future: generating moment implications for continuous-time Markov processes, *Econometrica*, **63**, 767–804.
105. Hardle, W. (1990) *Applied Non-parametric Regression*, Cambridge University Press, Cambridge.
106. Heyde, C.C. (1997) *Quasi-likelihood and Its Applications: A General Approach to Optimal Parameter Estimation*, Springer-Verlag, New York.
107. Hille, E., Phillips, R.S. (1957) *Functional Analysis and Semigroups*, American Mathematical Society, Providence, RI.
108. Hoffmann, M. (1999) Adaptive estimation in diffusion processes, *Stochastic Process. Appl.*, **79**, 135–163.
109. Holden, A.V. (1976) *Models for Stochastic Activity of Neurones*, Springer-Verlag, New York.
110. Holland, C.J. (1976) On a formula in diffusion processes in population genetics, *Proc. Am. Math. Soc.*, **54**(1), 316–318.
111. Holmes, E.E. (2004) Beyond theory to application and evaluation: diffusion approximations for population viability analysis, *Ecol. Appl.*, **14**(4), 1272–1293.
112. Honore, P. (1997) Maximum likelihood estimation of non-linear continuous-time term-structure models, Working Paper 1997-7, Department of Finances, Aahrus School of Business. Available at `http://ssrn.com/abstract=7669`.
113. Hsu, D.A. (1977) Tests for variance shift at an unknown time point, *Appl. Stat.*, **26**(3), 279–284.
114. Hsu, D.A. (1979) Detecting shifts of parameter in gamma sequences with applications to stock price and air traffic flow analysis, *J. Am. Stat. Assoc.*, **74**(365), 31–40.
115. Hull, J. (2000) *Options, Futures and Other Derivatives*, Prentice-Hall, Englewood Cliffs, NJ.
116. Iacus, S.M. (2002) Statistical analysis of stochastic resonance with ergodic diffusion noise, *Stochastics Stochastics Reports*, **73**(3-4), 271–285.
117. Iacus, S.M., La Torre, D. (2006) IFSM representation of Brownian motion with applications to simulation, *Math Everywhere: Deterministic and Stochastic Modelling in Biomedicine, Economics and Industry, Dedicated to the 60th Birthday of Vincenzo Capasso*, Aletti, G., Burger, M., Micheletti, A., Morale, D., eds., Springer, New York.
118. Iacus, S.M, Uchida, M., Yoshida, N. (2006): Parametric estimation for partially hidden diffusion processes sampled at discrete times. Preprint. Available at `http://arxiv.org/abs/math.ST/0611781`.
119. Inclan, C., Tiao, G.C. (1994) Use of cumulative sums of squares for retrospective detection of change of variance, *J. Am. Stat. Assoc.*, **89**, 913–923.
120. Jacobsen, M. (2001a) Discretely observed diffusions: classes of estimating functions and small Δ-optimality, *Scand. J. Stat.*, **28**, 123–149.
121. Jacobsen, M. (2001b) Small Δ-optimal martingale estimating functions for discretely observed diffusions: A simulation study, Preprint No. 2, Department of Theoretical Statistics, University of Copenhagen.
122. Jacobsen, M. (2002) Optimality and small Δ-optimality of martingale estimating functions, *Bernoulli*, **8**, 643–668.
123. Jacod, J. (2000) Non-parametric kernel estimation of the coefficient of a diffusion, *Scand. J. Stat.*, **27**, 83–96.
124. Jacod, J., Shiryayev, A.N. (1987) *Limit Theorems for Stochastic Processes*, Springer-Verlag, New York.

125. Jäckel, P. (2002) *Monte Carlo Methods in Finance*, John Wiley and Sons, Trowbridge.
126. Jensen, B., Poulsen, R. (2002) Transition densities of diffusion processes: numerical comparison of approximation techniques, *J. Derivatives*, **9**, 18–32.
127. Jiang, G., Knight, J. (1997) A nonparametric approach to the estimation of diffusion processes, with an application to a short-term interest rate model, *Econ. Theory*, **13**, 615–645.
128. Jimenez, J.C., Biscay, R.J., Ozaki, T. (2006) Inference methods for discretely observed continuous-time stochastic volatility models: a commented overview, *Asia-Pacific Financial Markets*, **12**(2), 109–141.
129. Jørgensen, B. (1982) *Statistical Properties of the Generalized Inverse Gaussian Distribution*, Lecture Notes in Statistics 9, Springer-Verlag, New York.
130. Karatzas, I., Shrevre, S.E. (1988) *Brownian Motion and Stochastic Calculus*, Springer-Verlag, New York.
131. Karlin, S., Taylor, H.M. (1981) *A Second Course in Stochastic Processes*, Academic Press, New York.
132. Karlin, S., Tavare, S. (1983) A class of diffusion processes with killing arising in population genetics, *SIAM J. Appl. Math.*,**43**(1), 31–41.
133. Kelly, L., Platen, E., Sørensen, M. (2004) Estimation for discretely observed diffusions using transform functions, *J. Appl. Probab.*, **41A**, 99–118.
134. Kemma, A.G.Z., Vorst, A.C.F. (1990) A pricing method for options based on average asset values, *J. Banking Finance*, March, 113–129.
135. Kent, J. (1978) Time-reversible diffusions, *Adv. Appl. Probab.*, **10**, 819–835.
136. Kessler, M. (1996) *Estimation paramétrique des coefficients d'une diffusione ergodique à partir d'observations discrètes*, PhD thesis, Laboratoire de Probabilités, Université Paris VI.
137. Kessler, M. (1997) Estimation of an ergodic diffusion from discrete observations, *Scand. J. Stat.*, **24**, 211–229.
138. Kessler, M. (2000) Simple and explicit estimating functions for a discretely observed diffusion process, *Scand. J. Stat.*, **27**, 65–82.
139. Kessler, M., Paredes, S. (2002) Computational aspects related to martingale estimating equations for a discretely observed diffusion, *Scand. J. Stat.*, **29**(3), 425–440.
140. Kessler, M., Sørensen, M. (1999) Estimating equations based on eigenfunctions for a discretely observed diffusion process, *Bernoulli*, **5**, 299–314.
141. Kloden, P., Platen, E. (1999) *Numerical Solution of Stochastic Differential Equations*, Applied Mathematics, **23**, Third corrected printing, Springer, New York.
142. Kloden, P., Platen, E., Schurz, H. (2000) *Numerical Solution of SDE through Computer Experiments*, Springer, Berlin.
143. Kloden, P., Platen, E., Schurz, H., Sørensen, M. (1996) On effect of discretization on estimation of the drift parameters for diffusion processes, *J. Appl. Probab.*, **33**(4), 1061–1073.
144. Konishi, S., Kitagawa, G. (1996) Generalised information criteria in model selection, *Biometrika*, **83**, 875–890.
145. Konishi, S., Kitagawa, G. (2003) Asymptotic theory for information criteria in model selection–functional approach, C. R. Rao 80th birthday felicitation volume, Part IV, *J. Stat. Planning Infer.*, **114**, 45–61.
146. Küchler, U., Sørensen, M. (1997) *Exponential Families of Stochastic Processes*, Springer, New York.

147. Kushner, H.J. (1967) *Stochastic Stability and Control*, Academic Press, New York.
148. Kutoyants, Y. (1994) *Identification of Dynamical Systems with Small Noise*, Kluwer, Dordrecht.
149. Kutoyants, Y. (2004) *Statistical Inference for Ergodic Diffusion Processes*, Springer-Verlag, London.
150. Krylov, N.V., Zvonkin, A.K. (1981) On strong solutions of stochastic differential equations, *Sel. Math. Sov.*, **1**, 19–61.
151. Le Breton, A. (1976) On continuous and discrete sampling for parameter estimation in diffusion type processes, *Math. Programming Stud.*, **5**, 124–144.
152. Lee, S., Nishiyama, Y., Yoshida, N. (2006) Test for parameter change in diffusion processes by cusum statistics based on one-step estimators, *Ann. Inst. Stat. Mat.*, **58**, 211–222.
153. L'Ecuyer, P., Lemieux, C. (2002) Recent advances in randomized quasi-Monte Carlo methods, in *Modeling Uncertainty: An Examination of Stochastic Theory, Methods, and Applications*, Drodr, M., L'Ecuyer, P., Szidarovszki, F., eds, Kluwer Academic Publishers, Boston, 419–474,
154. Lamberton, D., Lapayre, B. (1991) *Une introductions au calcul Stochastique Applique à la Finance*, Collection Mathématiques et Applications, Ellipse.
155. Lange, K. (2002) *Mathematical and Statistical Methods for Genetic Analysis*, Springer-Verlag, New York.
156. Lapayre, B., Pardoux, É., Sentis, R. (2003) *Introduction to Monte-Carlo Methods for Transportation and Diffusion Equations*, Oxford University Press, New York (for the English version).
157. Lepage, T., Law, S., Tupper, P., Bryant, D. (2006) Continuous and tractable models for the variation of evolutionary rates, Preprint. Available at `http://arxiv.org/abs/math.PR/0506145`.
158. Liechty, J., Roberts, G. (2001) Markov chain Monte Carlo methods for switching diffusion models, *Biometrika*, **88**(2), 299–315.
159. Lipster, R.S., Shiryayev, A.N. (1997) *Statistics of Random Processes*, Volume 1, Springer-Verlag, New York.
160. Lo, A. (1988) Maximum likelihood estimation of generalized Itô processes with discretely sampled data, *Econ. Theory*, **4**, 231–247.
161. Masuda, H. (2006) Private communication.
162. Merton, R.C. (1973) Theory of rational option pricing, *Bell J. Econ. Manage. Sci.*, **4** (1), 141–183.
163. Mikosch, T. (1998) Elementary stochastic calculus with finance in view, World Scientific, Singapore.
164. Milstein, G.N. (1978) A method of second-order accuracy integration of stochastic differential equations, *Theory Probab. Appl.*, **23**, 396–401.
165. Negri, I. (1998) Stationary distribution function estimation for ergodic diffusion process, *Stat. Infer. Stochastic Process.*, **1**, 61–84.
166. Newey, W. K., West, K.D. (1994) Automatic lag selection in covariance matrix estimation, *Rev. Econ. Stud.*, **61**, 631–653.
167. Newton, N.J. (1994) Variance reduction for simulated diffusions, *SIAM J. Appl. Math.*, **54**(6), 1780–1805.
168. Nowman, K.B. (1997) Gaussian estimation of single-factor continuous time models of the term structure of interest rates, *J. Finance*, **52**(4), 1695–1706.
169. Nowman, K.B., Saltoğlu, B. (2003) Continuous time and nonparametric modelling of U.S. interest rate models, *I. Rev. Financial Anal.*, **12**, 25–34.

170. Øksendal, B. (1998) *Stochastic Differential Equations. An Introduction with Applications*, 5th ed., Springer-Verlag, Berlin.
171. Oliver, F.W.J. (1972) *Handbook of Mathematical Functions with Formulas, Graphs, and Mathematical Tables*, John Wiley & Sons, New York.
172. Osborne, M.F.M. (1959) Brownian motion in the stock market, *Oper. Res.*, **7**, 145–173.
173. Ozaki, T. (1985) Non-linear time series models and dynamical systems, in Handbook of Statistics, Volume 5, Hannan, E.J. ed., North-Holland, Amsterdam, 25–83.
174. Ozaki, T. (1992) A bridge between nonlinear time series models and nonlinear stochastic dynamical systems: a local linearization approach, *Stat. Sinica*, **2**, 25–83.
175. Ozaki, T. (1993) A local linearization approach to nonlinear filtering, *Int. J. Control*, **57**, 75–96.
176. Papanicolaou, G. (1995) Diffusions in random media, in *Surveys in Applied Mathematics*, Keller, J.B., McLaughin, D., Papanicolaou , G., eds, Plenum Press, New York, 205–255.
177. Pearson, K. (1895) Contributions to the mathematical theory of evolution II. Skew variation in homogeneous material, *Philos. Trans. R. Soc. London A*, **186**, 343–414.
178. Pearson, N.D., Sun, T.-S. (1994) Exploiting the conditional density in estimating the term structure: an application to the Cox, Ingersoll, and Ross model, *J. Finance*, **49**, 1279–1304.
179. Pedersen, A.R. (1994) Quasi-likelihood inference for discretely observed diffusion processes, Research Report No.295, Dept. Theor. Statist., University of Aarhus.
180. Pedersen, A.R. (1995) A new approach to maximum likelihood estimation for stochastic differential equations based on discrete observations. *Scand. J. Stat.*, **22**, 55–71.
181. Pedersen, A.R. (2000) Estimating the nitrous oxide emission rate from the soil surface by means of a diffusion model, *Scand. J. Stat.*, **27**, 385–403.
182. Penev, S., Raykov, T. (2000) A Wiener germ approximation of the noncentral chi square distribution and of its quantiles, *Comput. Stat.*, **15**(2), 219–228.
183. Perron, P. (1991) Testing consistency with varying sampling frequency, *Econ. Theory*, **7**, 341–368.
184. Phillips, P.C.B. (1973) The problem of identification in finite parameter continuous-time processes, *J. Econ.*, **1**, 351–362.
185. Phillips, P.C.B., Park, J. (1998) Nonstationary density estimation and kernel autoregression, Cowles Foundation Discussion Paper, No. 1181, Yale University.
186. Polson, N., Roberts, G. (1994) Bayes factors for discrete observations from diffusion processes, *Biometrika*, **81**(1), 11–26.
187. Poston, T., Stewart, I.N. (1978) *Catastrophe Theory and Its Applications*, Pittman, London.
188. Poulsen, P. (1998) Approximate maximum likelihood estimation of discretely observed diffusion processes, Working Paper no. 29, Centre for Analytical finance, Aarhus.
189. Prakasa Rao, B.L.S. (1972) Maximum likelihood estimation for Markov processes, *Ann. Inst. Stat, Math.*, **24**, 333–345.

190. Prakasa Rao, B.L.S. (1983) Asymptotic theory for nonlinear least squares estimator for diffusion processes, *Math. Oper. Stat. Ser. Stat.*, **14**, 195–209.
191. Prakasa Rao, B.L.S. (1990) Nonparametric density estimation for stochastic processes from sampled data, *Publ. Instit. Stat. Univ. Paris*, **35**, 51–83.
192. Prakasa Rao, B.L.S. (1999) *Statistical Inference for Diffusion Type Processes*, Oxford University Press, New York.
193. Revouz, D., Yor, M. (1991) *Continuous Martingales and Brownian Motion*, Springer-Verlag, Berlin.
194. Ricciardi, L.M. (1977) *Diffusion Processes and Related Topics in Biology*, Lecture Notes in Biomathematics, Springer, New York.
195. Ripley, B.D. (1987) *Stochastic Simulation*, Wiley, New York.
196. Roberts, G., Stramer, O. (2001) On inference for partial observations from partial observed nonlinear diffusion model using Metropolis-Hastings algorithm, *Biometrika*, **88**(3), 603–621.
197. Rogers, L.C.G., Williams, D. (1987) *Diffusions, Markov Processes, and Martingales, volume 2: Itô Calculus*, Wiley, New York.
198. Rümelin, W. (1983) Numerical treatment of stochastic differential equations, *SIAM J. Numerical Anal.*, **57**, 75–96.
199. Saltoğlu, B. (2003) Comparing forecasting ability of parametric and nonparametric methods: an application with Canadian monthly interest rates, *Appl. Financial Econ.*, **13**(3), 169–176.
200. Santa-Clara, P. (1995) Simulated likelihood estimation of diffusion with application to the short term interest rate, Anderson Graduate School of Management, University of California at Los Angeles.
201. Schroder, M. (1989) Computing the constant elasticity of variance option pricing formula, *J. Finance*, **44**(1), 211–219.
202. Schuecker, P., Böhringer, H., Arzner, K., Reiprich, T.H. (2001) Cosmic mass functions from Gaussian stochastic diffusion processes, *Astron. Astrophys.*, **370**, 715–728.
203. Scott, D.W.(1992) *Multivariate Density Estimation: Theory, Practice and Visualization*, John Wiley, New York.
204. Shoji, L. (1995) Estimation and inference for continuous time stochastic models, PhD thesis, Institute of Statistical Mathematics, Tokyo.
205. Shoji, L. (1998) A comparative study of maximum likelihood estimators for nonlinear dynamical system models, *Int. J. Control*, **71**(3), 391–404.
206. Shoji, L., Ozaki, T. (1997) Comparative study of estimation methods for continuous time stochastic processes, *J.Time Ser. Anal.*, **18**, 485–506.
207. Shoji, L., Ozaki, T. (1998) Estimation for nonlinear stochastic differential equations by a local linearization method, *Stochastic Anal. Appl.*, **16**, 733–752.
208. Shorack, G., Wellner, J. (1986) *Empirical Processes with Applications to Statistics*, Wiley Series in Probability and Statistics, John Wiley & Sons, New York.
209. Shreve, S.E. (2004) *Stochastic Calculus for Finance II: Continuous-Time Models*, Springer, New York.
210. Silverman, B.W. (1986), *Density Estimation for Statistics and Data Analysis*, Chapman and Hall, London.
211. Sørensen, H. (2001) Discretely observed diffusions: approximation of the continuous-time score function, *Scand. J. Stat.*, **28** (1), 113–121.
212. Sørensen, H. (2002) Estimation of diffusion parameters for discretely observed diffusion processes, *Bernoulli*, **8**(4), 491–508.

213. Sørensen, H. (2004) Parametric inference for diffusion processes observed at discrete points in time: a survey, *Int. Stat. Rev.*, **72**(3), 337–354.
214. Sørensen, M. (1997) Estimating functions for discretely observed diffusions: a review, in *Selected Proceedings of the Symposium on Estimating Functions*, Basawa, I.V., Godabme, V.P., Taylor, R.L., eds, Volume 32, IMS Lecture Notes, Institute of Mathematical Statistics, Hayward, CA, 305–325.
215. Sørensen, M. (1999) On asymptotics of estimating functions, *Braz. J. Probab. Stat.*, **13**, 111–136.
216. Sørensen, M. (2007) Exponential family inference for diffusion models, Research Report N.383, Department of Mathematical Sciences, University of Copenhagen. Available at `http://www.math.ku.dk/~michael/glm.pdf`
217. Sørensen, M., Uchida, M. (2003) Small-diffusion asymptotics for discretely sampled stochastic differential equations, *Bernoulli*, **9**, 1051–1069.
218. Stanton, R. (1997) A nonparametric model of term structure dynamics and the market price of interest rate risk, *J. Finance*, **52**, 1973–2002.
219. Stroock, D.W., Varadhan, S.R.S. (1979) *Multidimensional Diffusion Processes*, Springer-Verlag, New York.
220. Takeuchi, K. (1976) Distribution of information statistics and criteria for adequacy of models, *Math. Sci.*, **153**, 12–18 (in Japanese).
221. Tuerlink, F., Maris, E., Ratcliff, R., De Boeck, P. (2001) A comparison of four methods for simulating the diffusion process, *Behav. Res. Methods Instrum. Comput.*, **33**(4), 443–456.
222. Uchida, M. (2003) Estimation for dynamical systems with small noise from discrete observations, *J. Jpn. Stat. Soc*, **33**, 157–167.
223. Uchida, M., Yoshida, N. (2005) AIC for ergodic diffusion processes from discrete observations, preprint MHF 2005-12, March 2005, Faculty of Mathematics, Kyushu University, Fukuoka, Japan.
224. Uhlenbeck, G.E., Ornstein, L.S. (1930) On the theory of Brownian motion, *Phys. Rev.*, **36**, 823–841.
225. Vasicek, O. (1977) An equilibrium characterization of the term structure, *J. Financial Econ.*, **5**, 177–188.
226. Venables, W. N., Ripley, B. D. (2000), *S Programming*, Springer, New York.
227. von Neumann, J. (1951) Various techniques used in connection with random digits, in *The Monte Carlo Method*, Householder, A.S., et al., eds, National Bureau of Standards Applied Mathematics Series, **12**, 36–38.
228. Wichern, D.W., Miller, R.B., Hsu, D.A. (1976) Changes of variance in first-order autoregressive time series models with an application, *Appl. Stat.*, **25**(3), 248–256.
229. Wichura, M.J. (1988) Algorithm AS 241: the percentage points of the normal distribution, *Appl. Stat.*, **37**, 477–484.
230. Wong, E. (1964) The construction of a class of stationary Markov processes, in *Stochastic Processes in Mathematical Physics and Engineering*, Bellman, R., ed., American Mathematical Society, Providence, RI, 264–276.
231. Yoshida, N. (1992) Estimation for diffusion processes from discrete observation, *J. Multivar. Anal.*, **41**(2), 220–242.
232. Zeeman, E.C. (1977) *Catastrophe Theory: Selected Papers*, Addison-Wesley, Reading, MA.
233. Zhou, H. (2001) Finite sample properties of EMM, GMM, QMLE, and MLE for a square-root interest rate diffusion model, *J. Comput. Finance*, **5**, 89–122.

Index

Springer Series in Statistics

(continued from page ii)

Ibrahim/Chen/Sinha: Bayesian Survival Analysis
Jiang: Linear and Generalized Linear Mixed Models and Their Applications
Jolliffe: Principal Component Analysis, 2nd edition
Konishi/Kitagawa: Information Criteria and Statistical Modeling
Knottnerus: Sample Survey Theory: Some Pythagorean Perspectives
Kosorok: Introduction to Empirical Processes and Semiparametric Inference
Küchler/Sørensen: Exponential Families of Stochastic Processes
Kutoyants: Statistical Inference for Ergodic Diffusion Processes
Lahiri: Resampling Methods for Dependent Data
Lavallée: Indirect Sampling
Le Cam: Asymptotic Methods in Statistical Decision Theory
Le Cam/Yang: Asymptotics in Statistics: Some Basic Concepts, 2nd edition
Le/Zidek: Statistical Analysis of Environmental Space-Time Processes
Liese/Miescke: Statistical Decision Theory: Estimation, Testing, Selection
Liu: Monte Carlo Strategies in Scientific Computing
Manski: Partial Identification of Probability Distributions
Mielke/Berry: Permutation Methods: A Distance Function Approach, 2nd edition
Molenberghs/Verbeke: Models for Discrete Longitudinal Data
Mukerjee/Wu: A Modern Theory of Factorial Designs
Nelsen: An Introduction to Copulas, 2nd edition
Pan/Fang: Growth Curve Models and Statistical Diagnostics
Politis/Romano/Wolf: Subsampling
Ramsay/Silverman: Applied Functional Data Analysis: Methods and Case Studies
Ramsay/Silverman: Functional Data Analysis, 2nd edition
Reinsel: Elements of Multivariate Time Series Analysis, 2nd edition
Rosenbaum: Observational Studies, 2nd edition
Rosenblatt: Gaussian and Non-Gaussian Linear Time Series and Random Fields
Särndal/Swensson/Wretman: Model Assisted Survey Sampling
Santner/Williams/Notz: The Design and Analysis of Computer Experiments
Schervish: Theory of Statistics
Shaked/Shanthikumar: Stochastic Orders
Shao/Tu: The Jackknife and Bootstrap
Simonoff: Smoothing Methods in Statistics
Song: Correlated Data Analysis: Modeling, Analytics, and Applications
Sprott: Statistical Inference in Science
Stein: Interpolation of Spatial Data: Some Theory for Kriging
Taniguchi/Kakizawa: Asymptotic Theory for Statistical Inference for Time Series
Tanner: Tools for Statistical Inference: Methods for the Exploration of Posterior Distributions and Likelihood Functions, 3rd edition
Tillé: Sampling Algorithms
Tsaitis: Semiparametric Theory and Missing Data
van der Laan/Robins: Unified Methods for Censored Longitudinal Data and Causality
van der Vaart/Wellner: Weak Convergence and Empirical Processes: With Applications to Statistics
Verbeke/Molenberghs: Linear Mixed Models for Longitudinal Data
Weerahandi: Exact Statistical Methods for Data Analysis

Printed in the United States of America